Structural Mechanics Fundamentals

Structural Mechanics Fundamentals

Alberto Carpinteri

CRC Press
Taylor & Francis Group
Boca Raton London New York

CRC Press is an imprint of the
Taylor & Francis Group, an **informa** business

A SPON PRESS BOOK

CRC Press
Taylor & Francis Group
6000 Broken Sound Parkway NW, Suite 300
Boca Raton, FL 33487-2742

First issued in paperback 2019

© 2014 by Alberto Carpinteri
CRC Press is an imprint of Taylor & Francis Group, an Informa business

No claim to original U.S. Government works

ISBN-13: 978-0-415-58032-8 (pbk)

Library of Congress Cataloging-in-Publication Data

Carpinteri, A.
   Structural mechanics fundamentals / author, Alberto Carpinteri.
      pages cm
   Includes bibliographical references and index.
   ISBN 978-0-415-58032-8 (pbk.)
   1. Structural engineering. I. Title.

   TA633.C354 2013
   624.1'7--dc23   2012037625

Visit the Taylor & Francis Web site at
http://www.taylorandfrancis.com

and the CRC Press Web site at
http://www.crcpress.com

To my family

Contents

Preface

This book intends to provide a complete and uniform treatment of the most fundamental and traditional topics in structural mechanics. It represents the second edition of a substantial part (12 chapters over 20) of my previous book *Structural Mechanics: A Unified Approach*, published in 1997 by E & FN SPON, an imprint of Chapman & Hall.

After introducing the basic topics of the geometry of areas and of kinematics and statics of rigid body systems, the mechanics of linear elastic solids (beams, plates and 3-D solids) is presented, adopting a matrix formulation which is particularly useful for numerical applications. The analysis of strain and stress around a material point is carried out considering the tensorial character of these physical quantities. The linear elastic constitutive law is then introduced, with the related Clapeyron's and Betti's theorems. The kinematic, static and constitutive equations, once composed within the elastic problem, provide an operator equation which has as its unknown the generalized displacement vector. Moreover, constant reference is made to duality, that is to the strict correspondence between statics and kinematics that emerges as soon as the corresponding operators are rendered explicit, and it is at once seen how each of these is the adjoint of the other. The implication of the principle of virtual work by the static–kinematic duality is emphasized, as well as the inverse implication. Once introduced the Saint Venant problem with all the six elementary loading characteristics, the theory of beam systems (statically determinate or indeterminate) is presented, with the solution of numerous examples and the plotting of the corresponding diagrams of axial force, shearing force and bending moment obtained both analytically and graphically. For the examination of indeterminate beam systems, both the methods of forces and energy are applied.

This book is the fruit of many years of teaching in Italian universities, formerly at the University of Bologna and currently at the Politecnico di Torino, where I have been professor of structural mechanics since 1986. It has been written to be used as a text for graduate or undergraduate students of either architecture or engineering, as well as to serve as a useful reference for research workers and practising engineers. It has been my endeavour to update and modernize a basic, and in some respects dated, discipline by merging classical topics with ones that have taken shape in more recent times.

Finally, I wish to express my most sincere gratitude to all those colleagues, collaborators and students, who, having attended my lectures or having read the original manuscript, have, with their suggestions and comments, contributed to the text as it appears in its definitive form. I further wish to thank my master's student, Francesco Armenti, for helping me with the proof corrections and Dr. Amedeo Manuello for his precious advice in realizing the front cover.

Alberto Carpinteri
Torino, Italy

Acknowledgements

I would like to thank the following Colleagues for their teaching activity according to the contents of the present volume, and for their attentive revision of some chapters of it: Giulio Ventura, Giuseppe Lacidogna, Stefano Invernizzi, Pietro Cornetti, Marco Paggi, Amedeo Manuello, Mauro Corrado, Alberto Sapora; as well as the following Ph.D. Students: Gianfranco Piana, Sandro Cammarano, Federico Accornero.

Author

Alberto Carpinteri received his doctoral degrees in nuclear engineering cum laude (1976) and mathematics cum laude (1981) from the University of Bologna (Italy). After two years at the Consiglio Nazionale delle Ricerche, he was appointed assistant professor at the University of Bologna in 1980.

Carpinteri moved to the Politecnico di Torino in 1986 as professor and became the chair of solid and structural mechanics and the director of the Fracture Mechanics Laboratory. During this period, he held different positions of responsibility, including head of the Department of Structural Engineering (1989–1995) and founding member and director of the Post-graduate School of Structural Engineering (1990–).

Prof. Carpinteri was a visiting scientist at Lehigh University, Bethlehem, Pennsylvania (1982–1983), and was appointed a fellow of several academies and professional institutions, including the European Academy of Sciences (2009–), the International Academy of Engineering (2010–), the Turin Academy of Sciences (2005–) and the American Society of Civil Engineers (1996–).

Prof. Carpinteri was the president of various scientific associations and research institutions, as follows: the International Congress on Fracture, ICF (2009–2013); the European Structural Integrity Society, ESIS (2002–2006); the International Association of Fracture Mechanics for Concrete and Concrete Structures, IA-FraMCoS (2004–2007); the Italian Group of Fracture, IGF (1998–2005); and the National Research Institute of Metrology, INRIM (2011–2013). He was appointed a member of the Congress Committee of the International Union of Theoretical and Applied Mechanics, IUTAM (2004–2012); a member of the executive board of the Society for Experimental Mechanics, SEM (2012–2014); a member of the editorial board of 13 international journals; and the editor in chief of the journal *Meccanica* (Springer, IF = 1.568). He is also the author or editor of over 750 publications, of which more than 300 are papers in refereed international journals (ISI h-Index = 31, more than 3800 citations) and 43 are books.

Prof. Carpinteri has received numerous honours and awards, as follows: the Robert L'Hermite Medal from RILEM (1982), the Griffith Medal from ESIS (2008), the Swedlow Memorial Lecture Award from ASTM (2011) and the Inaugural Paul Paris Gold Medal from ICF (2013), among others.

Chapter 1

Introduction

1.1 PRELIMINARY REMARKS

Structural mechanics is the science that studies the **structural response** of solid bodies subjected to external loading. The structural response takes the form of **strains and internal stresses**.

The variation of shape generally involves relative and absolute displacements of the points of the body. The simplest case that can be envisaged is that of a string, one end of which is held firm while a tensile load is applied to the other end. The percentage lengthening or stretching of the string naturally implies a displacement, albeit small, of the end where the force is exerted. Likewise, a membrane, stretched by a system of balanced forces, will dilate in two dimensions, and its points will undergo relative and absolute displacements. Also three-dimensional bodies, when subjected to stress by a system of balanced forces, undergo, point by point and direction by direction, a dilation or a contraction, as well as an angular distortion. Similarly, beams and horizontal plates bend, imposing a certain curvature, respectively, to their axes and to their middle planes, and differentiated deflections to their points.

As regards internal stresses, these can be considered as exchanged between the single (even infinitesimal) parts which make up the body. In the case of the string, the tension is transmitted continuously from the end on which the force is applied right up to the point of constraint. Each elementary segment is thus subject to two equal and opposite forces exerted by the contiguous segments. Likewise, each elementary part of a membrane will be subject to four mutually perpendicular forces, two equal and opposite pairs. In three-dimensional bodies, each elementary part is subject to normal and tangential forces. The former generate dilations and contractions, whilst the latter produce angular distortions. Finally, each element of beam or plate that is bent is subject to self-balanced pairs of moments.

In addition to the shape and properties of the body, it is the external loading applied and the constraints imposed that determine the structural response. The constraints react to the external loads, exerting on the body additional loads called **constraint reactions**. These reactions are *a priori* unknown. In the case where the constraints are not redundant from the kinematic point of view, the calculation of the constraint reactions can be made considering the body as being perfectly rigid and applying only the cardinal equations of statics. In the alternative case where the constraints are redundant, the calculation of the constraint reactions requires, in addition to **equations of equilibrium**, the so-called **equations of congruence**. These equations are obtained by eliminating the redundant constraints, replacing them with the constraint reactions exerted by them and imposing the abeyance of the constraints that have been eliminated. The procedure presupposes that the strains and displacements, produced both by the external loading and by the reactions of the constraints that have been eliminated, are known. A simple example may suffice to illustrate these concepts.

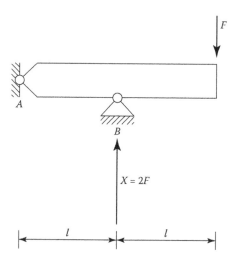

Figure 1.1

Let us consider a bar hinged at point A and supported at point B, subjected to the end force F (Figure 1.1). The reaction X produced by the support B is obtained by imposing equilibrium with regard to rotation about hinge A:

$$F(2l) = Xl \Rightarrow X = 2F \tag{1.1}$$

The equation of equilibrium with regard to vertical translation provides, on the other hand, the reaction of hinge A. The problem is thus **statically determinate** or **isostatic**.

Let us now consider the same bar hinged, not only at A but also at two points B_1 and B_2, distant $\frac{2}{3}l$ and $\frac{4}{3}l$, respectively, from point A (Figure 1.2a). The condition of equilibrium with regard to rotation gives us an equation with two unknowns:

$$F(2l) + X_1 \frac{2}{3}l = X_2 \frac{4}{3}l \tag{1.2}$$

Thus, the pairs of reactions X_1 and X_2 which ensure rotational equilibrium are infinite, but only one of these also ensures congruence, i.e. abeyance of the conditions of constraint. The vertical displacement in both B_1 and B_2 must in fact be zero.

To determine the constraint reactions, we thus proceed to eliminate one of the two hinges B_1 or B_2, for example, B_1, and we find out how much point B_1 rises owing to the external force F (Figure 1.2b) and how much it drops owing to the unknown reaction X_1 (Figure 1.2c). The condition of congruence consists of putting the total displacement of B_1 equal to zero:

$$\upsilon(F) = \upsilon(X_1) \tag{1.3}$$

The equation of equilibrium (1.2) and the equation of congruence (1.3) together solve the problem, which is said to be **statically indeterminate** or **hyperstatic**.

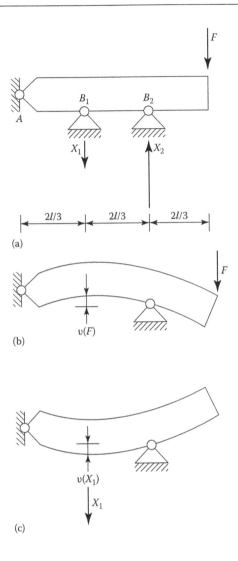

Figure 1.2

1.2 CLASSIFICATION OF STRUCTURAL ELEMENTS

As has already been mentioned in the preliminary remarks, the structural elements which combine to make up the load-bearing structures of civil and industrial constructions, as well as any naturally occurring structure such as rock masses, plants or skeletons, can fit into one of three distinct categories:

1. One-dimensional elements (e.g. ropes, struts, beams, arches)
2. Two-dimensional elements (e.g. membranes, plates, slabs, vaults, shells)
3. Three-dimensional elements (stubby solids)

In the case of one-dimensional elements, for example, beams (Figure 1.3), one of the three dimensions, the length, is much larger than the other two which compose the cross section. Hence, it is possible to neglect the latter two dimensions and consider the entire element as concentrated along the line forming its centroidal axis. In our calculations, features which

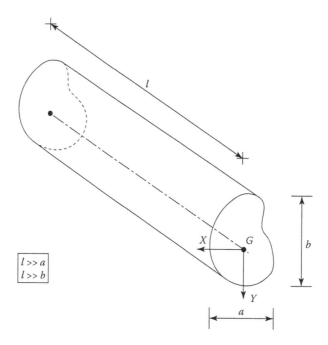

Figure 1.3

represent the geometry of the cross section and, consequently, the three dimensionality of the element will thus be used. Ropes are elements devoid of flexural and compressive stiffness and are able only to bear states of tensile stress. Bars, however, present a high axial stiffness, both in compression (struts) and in tension (tie rods), whilst their flexural stiffness is poor. Beams and, more generally, arches (or curvilinear beams) also present a high degree of flexural stiffness, provided that materials having particularly high tensile strength are used. In the case of stone materials and concrete, which present very low tensile strength, straight beams are reinforced to stand up to bending stresses, whilst arches are traditionally shaped so that only internal compressive stresses are produced.

When, in the cross section of a beam, one dimension is clearly smaller than the others (Figure 1.4), the beam is said to be **thin walled.** Beams of this sort can be easily produced by rolling or welding metal plate and prove to be extremely efficient from the point of view of the ratio of flexural strength to the amount of material employed.

In the case of two-dimensional elements, for example, flat plates (Figure 1.5a) or plates with double curvature (Figure 1.5b), one of the three dimensions, the thickness, is much smaller than the other two, which compose the middle surface. It is thus possible to neglect the thickness and to consider the entire element as being concentrated in its middle surface. Membranes are elements devoid of flexural and compressive stiffness and are able to withstand only states of biaxial traction. Also plates that are of a small thickness present a low flexural stiffness and are able to bear loads only in their middle plane. Thick plates (also referred to as slabs), instead, also withstand bending stresses, provided that materials having particularly high tensile strength are used. In the case of stone materials and concrete, flat plates are, on the other hand, ribbed and reinforced, while vaults and domes are traditionally shaped so that only internal compressive stresses are produced (for instance, in arched dams).

Finally, in the case of so-called **stubby solids,** the three dimensions are all comparable to one another and hence the analysis of the state of strain and internal stress must be three dimensional, without any particular simplifications or approximations.

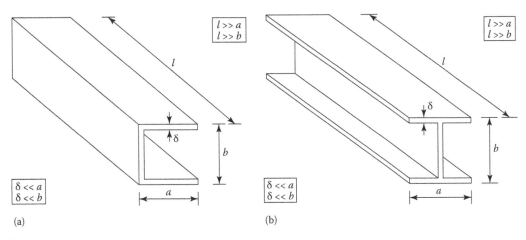

$l \gg a$
$l \gg b$

l

δ

b

$\delta \ll a$
$\delta \ll b$

a

(a)

$l \gg a$
$l \gg b$

l

δ

b

$\delta \ll a$
$\delta \ll b$

a

(b)

Figure 1.4

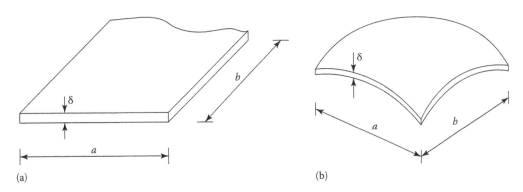

δ

b

a

(a)

δ

a

b

(b)

Figure 1.5

1.3 STRUCTURAL TYPES

The single structural elements, introduced in the previous section, are combined to form load-bearing structures. Usually, for buildings of a civil type, one-dimensional and two-dimensional elements are connected together. The characteristics of the individual elements and the way in which they are connected one to another and to the ground together define the structural type, which can be extremely varied, according to the purposes for which the building is designed.

In many cases, the two-dimensional elements do not have a load-bearing function (e.g. the walls of buildings in reinforced concrete), and hence it is necessary to highlight graphically and calculate only the so-called **framework**, made up exclusively of one-dimensional elements. This framework, according to the type of constraint which links together the various beams, will then be said to be **trussed** or **framed.** In the former case, the calculation is made by inserting hinges which connect the beams together, whereas in the latter case the beams are considered as built into one another. In real situations, however, beams are never connected by frictionless hinges or with perfectly rigid joints. Figures 1.6 through 1.11 show some examples of load-bearing frameworks: a timber-beam bridge, a truss in reinforced concrete, an arch centre, a plane steel frame, a grid and a three-dimensional frame.

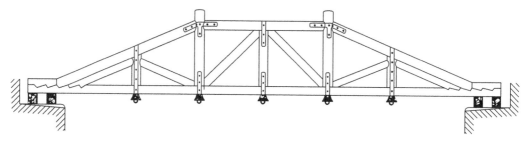

Figure 1.6

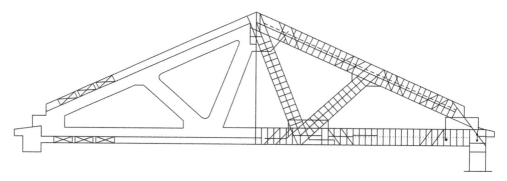

Figure 1.7

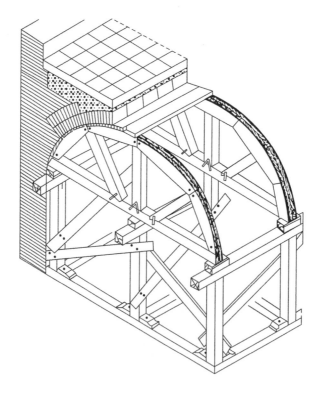

Figure 1.8

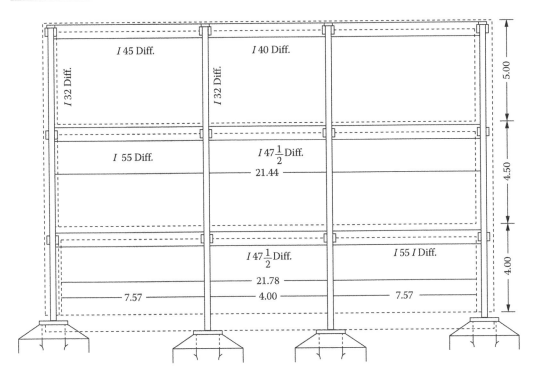

Figure 1.9

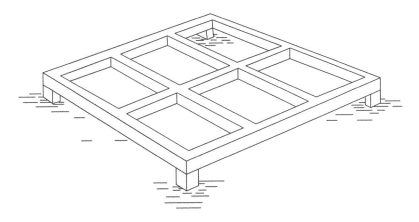

Figure 1.10

Also in the case of **bridges**, it is usually possible to identify a load-bearing structure consisting of one-dimensional elements. The road surface of an **arch bridge** is supported by a parabolic beam which is subject to compression and, if well designed, is devoid of dangerous internal flexural stresses. The road surface can be built to rest above the arch by means of struts (Figure 1.12) or can be suspended beneath the arch by means of tie rods (Figure 1.13). Inverting the static scheme and using a primary load-bearing element subject exclusively to tensile stress, we arrive at the structure of **suspension bridges** (Figure 1.14). In these, the road surface hangs from a parabolic cable by means of tie rods. The cable is, of course, able to withstand only tensile stresses, which are, however, transmitted onto two compressed piers.

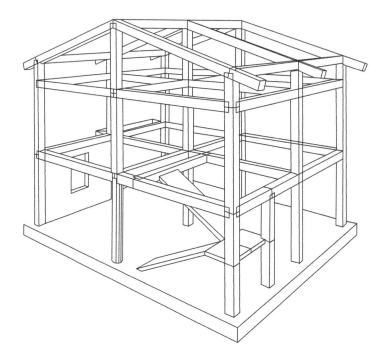

Figure 1.11

(a)

Figure 1.12

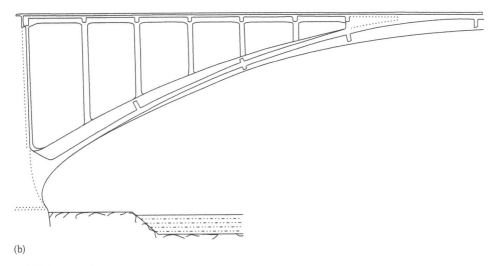

(b)

Figure 1.12 (continued)

As regards two-dimensional structural elements, it is advantageous to exploit the same static principles already met with in the case of bridges. To avoid, for example, dangerous internal stresses of a flexural nature, the usual approach is to use **vaults** or **domes** having double curvature, which present parabolic sections in both of the principal directions (Figure 1.15a). A variant is provided by the so-called **cross vault** (Figure 1.15b), consisting of two mutually intersecting cylindrical vaults. Membranes, on the other hand, can assume the form of hyperbolic paraboloids, with saddle points and curvatures of opposite sign (Figure 1.15c). In the so-called **prestressed membranes**, both those cables with the concavity facing upwards and those with the concavity facing downwards are subject to tensile stress.

(a)

Figure 1.13

(continued)

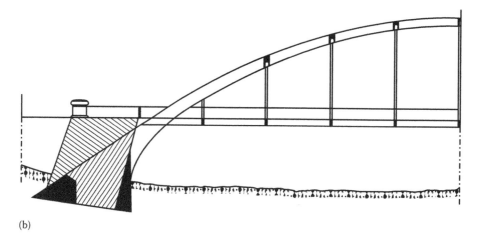

(b)

Figure 1.13 (continued)

(a)

Figure 1.14

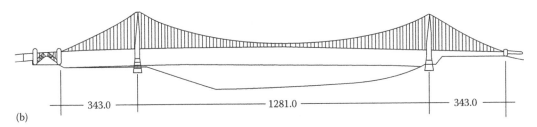

343.0   1281.0   343.0

(b)

Figure 1.14 (continued)

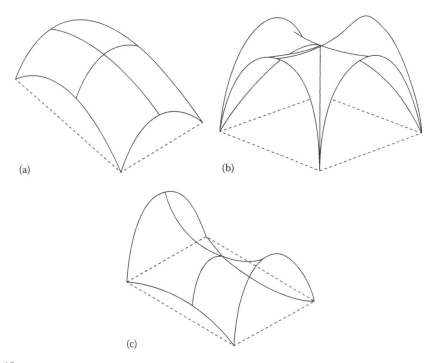

(a)   (b)

(c)

Figure 1.15

1.4 EXTERNAL LOADING AND CONSTRAINT REACTIONS

The strains and internal stresses of a structure obviously depend on the external loads applied to it. These can be of varying nature according to the structure under consideration. In the civil engineering field, the loads are usually represented by the **weight load**, both of the structural elements themselves (**permanent loads**) and of persons, vehicles or objects (**live loads**).

Figure 1.16 represents two load diagrams, used in the early decades of the last century, of horse-drawn carts and carriages. The forces are considered as concentrated and, of course, proceeding over the road surface. Figure 1.17 illustrates the load diagram of a roller, and Figure 1.18 that of a hoisting device. Figure 1.19 compares the permanent load diagrams of two beams, one with constant cross section and the other with linearly variable cross section.

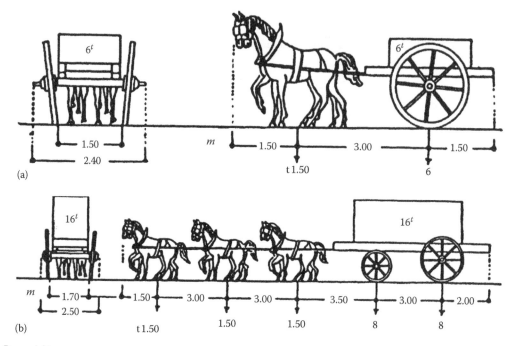

(a)

(b)

Figure 1.16

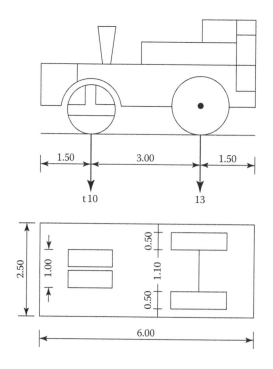

Figure 1.17

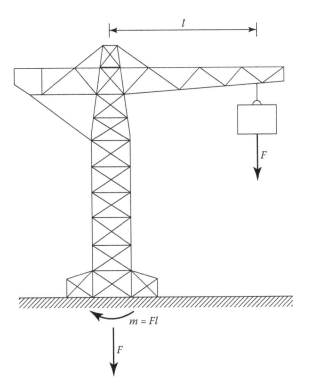

Figure 1.18

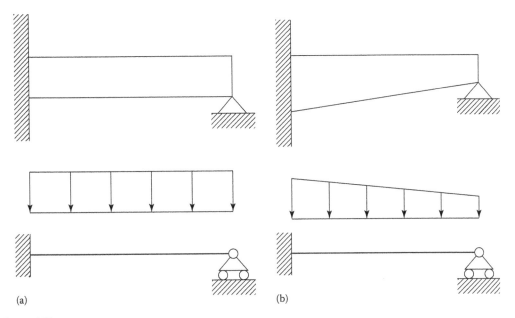

(a)   (b)

Figure 1.19

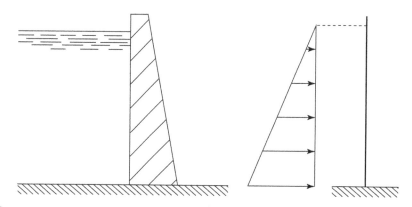

Figure 1.20

Other loads of a mechanical nature are **hydraulic loads** and **pneumatic loads.** Figure 1.20 shows how the thrust of water against a dam can be represented with a triangular distributed load. Then there are **inertial forces,** which act on rotating mechanical components, such as the blades of a turbine, or on the floors of a storeyed building, following ground vibration caused by an earthquake (Figure 1.21). A similar system of horizontal forces can represent the action of the wind on the same building.

In addition to external loading, the structural elements undergo the action of the other structural elements connected to them, including the action of the foundation. These kinds of action are more correctly termed **constraint reactions,** those exchanged between elements being **internal** and those exchanged with the foundation being **external.** The nature of the constraint reaction depends on the conformation and mode of operation of the constraint which connects the two parts.

Figure 1.22 gives examples of some types of **beam support** to the foundation. In the case of Figure 1.22a, we have a pillar in reinforced concrete; in Figure 1.22b, we have joints that are used in bridges, and in Figure 1.22c, a roller support. In all cases, the constraint reaction

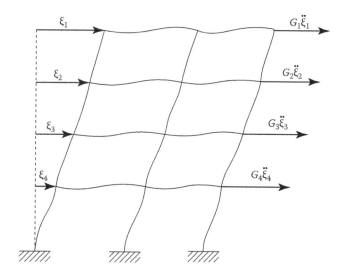

Figure 1.21

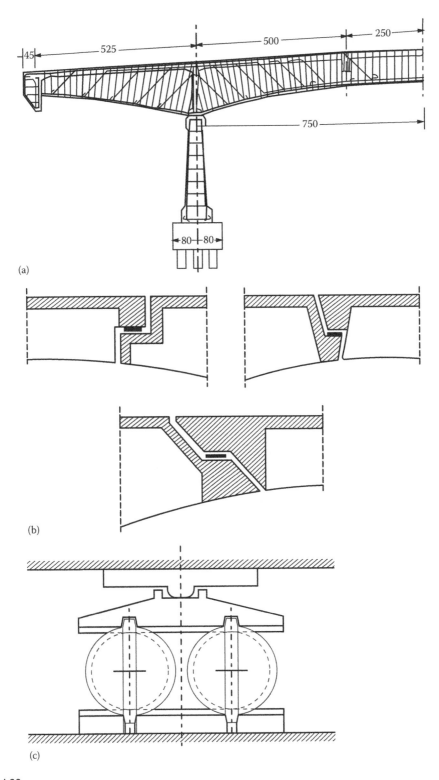

(a)

(b)

(c)

Figure 1.22

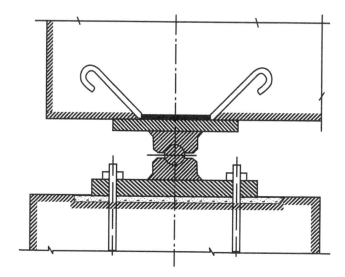

Figure 1.23

exchanged between the foundation and the structural part is constituted by a vertical force, no constraint being exerted horizontally, except for friction.

Figure 1.23 shows the detailed scheme of a hinge connecting a part in reinforced concrete to the foundation. The hinge allows only relative rotations between the two connected parts and hence reacts with a force that passes through its centre. In the case illustrated, there will thus be the possibility of a horizontal reaction, as well as a vertical one.

Figure 1.24 illustrates the joint between two timber beams, built with joining plates and riveting. Similar joints are made for steel girders by means of bolting or welding. This constraint is naturally more severe than a simple hinge, and yet in practice it proves to be much less rigid than a perfectly fixed joint. In the designing of trusses, it is customary to model the joint with a hinge, neglecting the exchange of moment between the two parts. The effect of making such an assumption is, in fact, that of guaranteeing a greater margin of safety.

1.5 STRUCTURAL COLLAPSE

If the loading exerted on a structure exceeds a certain limit, the consequence is the complete collapse or, at any rate, the failure of the structure itself. The loss of stability can occur in different ways depending on the shape and dimensions of the structural elements, as well as on the material of which these are made. In some cases, the constraints and joints can fail, with the result that rigid mechanisms are created, with consequent large displacements, toppling over, etc. In other cases, the structural elements themselves can give way; the mechanisms of structural collapse can be divided schematically into three distinct categories:

1. Buckling
2. Yielding
3. Brittle fracturing

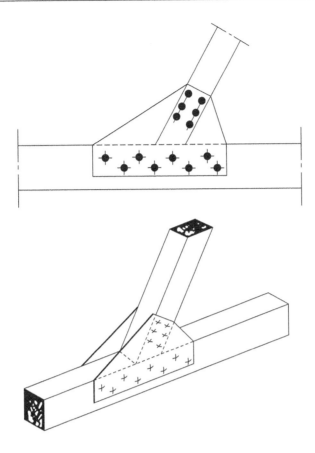

Figure 1.24

In real situations, however, many cases of structural collapse occur in such a way as to involve two of these mechanisms, if not all three.

Buckling, or instability of elastic equilibrium, is the type of structural collapse which involves slender structural elements, subject prevalently to compression, such as struts of trusses, columns of frameworks, piers and arches of bridges, valve stems, crankshafts, ceiling shells, submarine hulls, etc. This kind of collapse often occurs even before the material of which the element is made has broken or yielded.

Unlike buckling, **yielding,** or plastic deformation, involves also the material itself and occurs in a localized manner in one or more points of the structure. When, with the increase in load, plastic deformation has taken place in a sufficient number of points, the structure can give way altogether since it has become hypostatic, i.e. it has become a mechanism. This type of generalized structural collapse usually involves structures built of rather ductile material, such as metal frames and plates, which are mainly prone to bending.

Finally, **brittle fracturing** is of a localized origin, as is plastic deformation, but spreads throughout the structure and hence constitutes a structural collapse of a generalized nature. This type of collapse affects prevalently one- and two-dimensional structural elements of considerable thickness (bridges, dams, ships, large ceilings and vessels, etc.), large three-dimensional elements (rock masses, the Earth's crust, etc.), brittle materials (high-strength steel and concrete, rocks, ceramics, glass, etc.) and tensile conditions.

As, with the decrease in their degree of slenderness, certain structures, subject prevalently to compression and bending, very gradually pass from a collapse due to buckling to one due to plastic deformation, likewise, as we move down the size scale, other structures, prone to tension and bending, gradually pass from a collapse due to brittle fracturing to one due to plastic deformation.

1.6 NUMERICAL MODELS

With the development of electronics technology and the production of computers of ever-increasing power and capacity, structural analysis has undergone, in the last four or five decades, a remarkable metamorphosis. Calculations which were carried out manually by individual engineers, with at most the help of the traditional graphical methods, can now be performed using computer software.

Up to a few years ago, since the calculation of strains and internal stresses of complex structures could not be handled in such a way as to obtain an exact result, such calculations were carried out using a procedure of approximation. These approximations, at times, were somewhat crude and, in certain cases, far from being altogether realistic. Today numerical models allow us to consider enormous numbers of points, or nodes, with their corresponding displacements and corresponding strains and internal stresses. The so-called **finite-element method** is both a discretization method, since it considers a finite number, albeit a very large one, of structural nodes, and an interpolation method, since it allows us to estimate the static and kinematic quantities even outside the nodes.

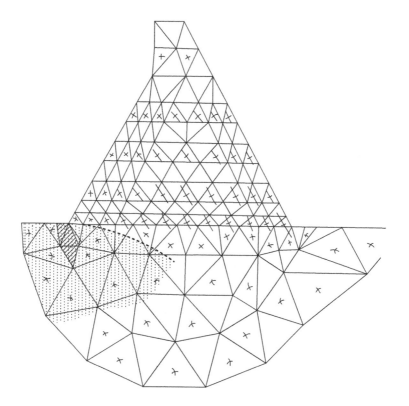

Figure 1.25

The enormous amount of information to be handled is organized and ordered in a matrix form by the computer. In this way, the language itself of structural analysis has taken on a different appearance, undoubtedly more synthetic and homogeneous. This means that, for every type of structural element, it is possible to write static, kinematic and constitutive equations having the same form. Once discretized, these provide a matrix of global stiffness which presents a dimension equal to the number of degrees of freedom considered. This matrix, multiplied by the vector of the nodal displacements, which constitutes the primary unknown of the problem, provides the vector of the external forces applied to the nodes; this represents the known term of the problem. Once this matrix equation has been resolved, taking into account any boundary conditions, it is then possible to arrive at the nodal strains and nodal internal stresses.

As an illustration of these mathematical techniques, a number of **finite-element meshes** are presented. They correspond to a buttress dam (Figure 1.25), a rock mass with a tunnel system (Figure 1.26), an eye hook (Figure 1.27), two mechanical components having supporting functions (Figure 1.28), a concrete vessel for a nuclear reactor (Figure 1.29) and an arch dam (Figure 1.30).

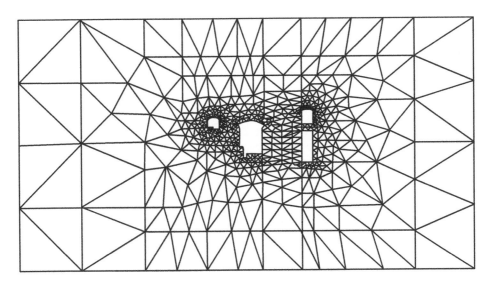

Figure 1.26

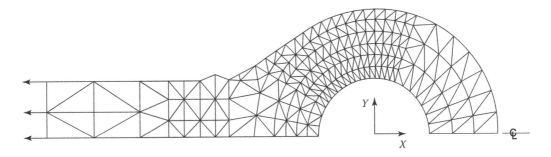

Figure 1.27

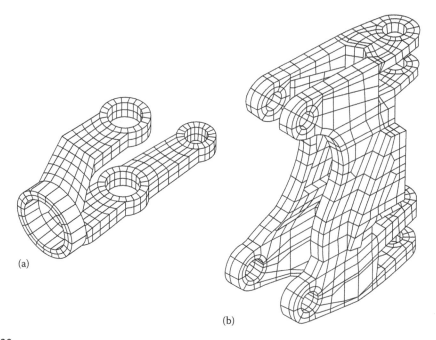

(a)

(b)

Figure 1.28

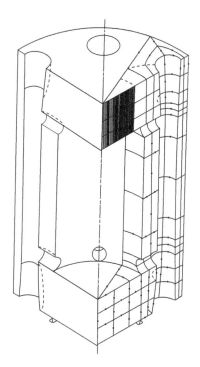

Figure 1.29

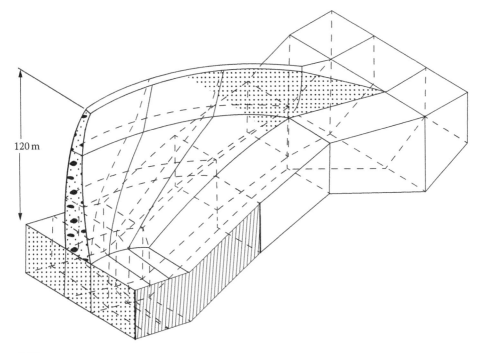

120 m

Figure 1.30

Chapter 2

Geometry of areas

2.1 INTRODUCTION

When analysing beam resistance, it is necessary to consider the geometrical features of the corresponding right sections. These features, as will emerge more clearly hereafter, amount to a scalar quantity, the **area**; a vector quantity, the position of the **centroid**; and a tensor quantity, consisting of the **central directions** and the **central moments of inertia**.

The laws of transformation, by translation and rotation of the reference system, both of the vector of static moments and of the tensor of moments of inertia, will be considered. It will thus become possible also to calculate composite sections, consisting of the combination of a number of elementary parts, and the graphical interpretation (due to Mohr) of this calculation will be given.

Particular attention will be paid to the cases of sections presenting symmetry, whether axial or polar, and of thin-walled beam sections, which have already been mentioned in the introductory chapter and for which a simplified calculation is possible. A number of examples will close the chapter.

2.2 LAWS OF TRANSFORMATION OF THE POSITION VECTOR

The coordinates x, y of a point of the plane in the XY reference system are linked to the coordinates $\bar{x}$, $\bar{y}$ of the same point in the **translated reference system** $\overline{XY}$ (Figure 2.1) by the following relations:

$$\bar{x} = x - x_0 \tag{2.1a}$$

$$\bar{y} = y - y_0 \tag{2.1b}$$

where x_0, y_0 are the coordinates of the origin $\bar{O}$ of the translated system, with respect to the original XY axes.

The laws of transformation (2.1) can be reproposed in a vector form as follows:

$$\{\bar{r}\} = \{r\} - \{r_0\} \tag{2.2}$$

where $\{r\}$ indicates the position vector $[x, y]^T$ of the generic point in the original reference system, with $\{\bar{r}\}$ being the position vector $[\bar{x}, \bar{y}]^T$ in the translated system and with $\{r_0\}$ being the position vector $[x_0, y_0]^T$ of the origin $\bar{O}$ of the translated system in the original reference system.

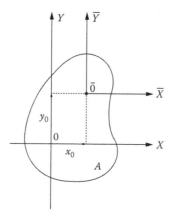

Figure 2.1

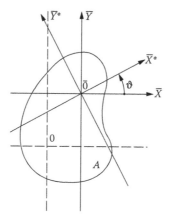

Figure 2.2

The coordinates $\bar{x}$, $\bar{y}$ of a point of the plane $\overline{XY}$ are linked to the coordinates $\bar{x}^*$, $\bar{y}^*$ of the same point in the rotated reference system $\overline{X}^*\overline{Y}^*$ (Figure 2.2) *via* the following relations:

$$\bar{x}^* = \bar{x}\cos\vartheta + \bar{y}\sin\vartheta \tag{2.3a}$$

$$\bar{y}^* = -\bar{x}\sin\vartheta + \bar{y}\cos\vartheta \tag{2.3b}$$

where ϑ indicates the angle of rotation of the second reference system with respect to the first (positive if the rotation is counterclockwise).

These transformation laws can be reproposed in a matrix form as follows:

$$\{\bar{r}^*\} = [N]\{\bar{r}\} \tag{2.4}$$

where

$$[N] = \begin{bmatrix} \cos\vartheta & \sin\vartheta \\ -\sin\vartheta & \cos\vartheta \end{bmatrix} \tag{2.5}$$

is the orthogonal matrix of rotation.

2.3 LAWS OF TRANSFORMATION OF THE STATIC MOMENT VECTOR

Consider the area A in the XY reference system (Figure 2.1). The definition of **static moment vector**, relative to the area A and calculated in the XY reference system, is given by the following two-component vector:

$$\{S\} = \begin{bmatrix} S_y \\ S_x \end{bmatrix} = \begin{bmatrix} \int_A x\,\mathrm{d}A \\ \int_A y\,\mathrm{d}A \end{bmatrix} = \int_A \{r\}\mathrm{d}A \tag{2.6}$$

The static moment vector, again referred to the area A, calculated in the translated $\overline{XY}$ system, can be expressed in the following way:

$$\{\bar{S}\} = \begin{bmatrix} S_{\bar{y}} \\ S_{\bar{x}} \end{bmatrix} = \begin{bmatrix} \int_A \bar{x}\,\mathrm{d}A \\ \int_A \bar{y}\,\mathrm{d}A \end{bmatrix} = \int_A \{\bar{r}\}\mathrm{d}A \tag{2.7}$$

Applying the transformation law (2.2), Equation 2.7 becomes

$$\{\bar{S}\} = \int_A \{r\}\mathrm{d}A - \{r_0\}\int_A \mathrm{d}A \tag{2.8}$$

since $\{r_0\}$ is a constant vector. Recalling definition (2.6), we obtain finally the static moment vector transformation law for translations of the reference system:

$$\{\bar{S}\} = \{S\} - A\{r_0\} \tag{2.9}$$

Vector relation (2.9) is equivalent to the following two scalar relations:

$$S_{\bar{y}} = S_y - Ax_0 \tag{2.10a}$$

$$S_{\bar{x}} = S_x - Ay_0 \tag{2.10b}$$

The reference system, translated with respect to the original one, for which both static moments vanish, is determined by the following position vector:

$$x_G = \frac{S_y}{A} \tag{2.11a}$$

$$y_G = \frac{S_x}{A} \tag{2.11b}$$

The origin G of this particular reference system is termed the **centroid** of area A and is a characteristic point of the area itself, in the sense that it is altogether independent of the choice of the original XY system.

Now consider the reference system $\bar{X}^*\bar{Y}^*$, rotated with respect to the $\overline{XY}$ system (Figure 2.2). The static moment vector, relative to area A and calculated in the rotated system $\bar{X}^*\bar{Y}^*$, may be expressed using the law (2.4):

$$\{\bar{S}^*\} = \int_A \{\bar{r}^*\}dA = [N]\int_A \{\bar{r}\}dA \tag{2.12}$$

where $[N]$ is the constant matrix (2.5). Finally, recalling definition (2.7), the static moment vector transformation law for rotations of the reference system is obtained as follows:

$$\{\bar{S}^*\} = [N]\{\bar{S}\} \tag{2.13}$$

The matrix relation (2.13) is equivalent to the following two scalar relations:

$$S_{\bar{y}^*} = S_{\bar{y}} \cos \vartheta + S_{\bar{x}} \sin \vartheta \tag{2.14a}$$

$$S_{\bar{x}^*} = -S_{\bar{y}} \sin \vartheta + S_{\bar{x}} \cos \vartheta \tag{2.14b}$$

From Equations 2.14, two important conclusions may be drawn.

1. The static moments are zero with respect to any pair of centroidal orthogonal axes.
2. If the origin $\bar{O}$ of the reference system does not coincide with the centroid G of area A, there exists no angle of rotation ϑ of the reference system for which the static moments both vanish. In fact, from Equations 2.14, we obtain

$$S_{\bar{y}^*} = 0 \quad \text{for } \vartheta = \arctan\left(-\frac{S_{\bar{y}}}{S_{\bar{x}}}\right) \tag{2.15a}$$

$$S_{\bar{x}^*} = 0 \quad \text{for } \vartheta = \arctan\left(+\frac{S_{\bar{x}}}{S_{\bar{y}}}\right) \tag{2.15b}$$

The conditions (2.15) are not, however, compatible.

If we consider a reference system $\bar{X}^*\bar{Y}^*$ obtained by translating and then rotating the original XY system (Figures 2.1 and 2.2), it is possible to formulate the general static moment vector transformation law for rototranslations of the reference system, combining the foregoing partial laws (2.9) and (2.13):

$$\{\bar{S}^*\} = [N](\{S\} - A\{r_0\}) \tag{2.16}$$

The inverse rototranslation formula may be obtained from the previous one by premultiplying both members by $[N]^T = [N]^{-1}$

$$\{S\} = [N]^T\{\bar{S}^*\} + A\{r_0\} \tag{2.17}$$

2.4 LAWS OF TRANSFORMATION OF THE MOMENT OF INERTIA TENSOR

Consider the following matrix product (referred to as the **dyadic product**):

$$\{r\}\{r\}^{\mathrm{T}} = \begin{bmatrix} x \\ y \end{bmatrix}[x \; y] = \begin{bmatrix} x^2 & xy \\ yx & y^2 \end{bmatrix} \tag{2.18}$$

The definition of the **moment of inertia tensor**, relative to area A and calculated in the XY reference system, is given by the following symmetric (2×2) tensor:

$$[I] = \begin{bmatrix} I_{yy} & I_{xy} \\ I_{yx} & I_{xx} \end{bmatrix} = \begin{bmatrix} \int\limits_{A} x^2\, \mathrm{d}A & \int\limits_{A} xy\, \mathrm{d}A \\ \int\limits_{A} yx\, \mathrm{d}A & \int\limits_{A} y^2\, \mathrm{d}A \end{bmatrix} \tag{2.19}$$

Taking into account relation (2.18), definition (2.19) can be expressed in the following compact form:

$$[I] = \int\limits_{A} \{r\}\{r\}^{\mathrm{T}}\, \mathrm{d}A \tag{2.20}$$

The moment of inertia tensor, relative again to area A and calculated in the translated reference system $\overline{XY}$ (Figure 2.1), can be expressed as follows:

$$[\overline{I}] = \int\limits_{A} \{\overline{r}\}\{\overline{r}\}^{\mathrm{T}}\, \mathrm{d}A \tag{2.21}$$

And, thus, applying the position vector transformation law for the translations of the reference system (Equation 2.2), we obtain

$$[\overline{I}] = \int\limits_{A} (\{r\} - \{r_0\})(\{r\} - \{r_0\})^{\mathrm{T}}\, \mathrm{d}A \tag{2.22}$$

Since the transpose of the sum of two matrices is equal to the sum of the transposes, we have

$$[\overline{I}] = \int\limits_{A} (\{r\} - \{r_0\})(\{r\}^{\mathrm{T}} - \{r_0\}^{\mathrm{T}})\, \mathrm{d}A$$

$$= \int\limits_{A} \{r\}\{r\}^{\mathrm{T}}\, \mathrm{d}A - \int\limits_{A} \{r\}\mathrm{d}A\, \{r_0\}^{\mathrm{T}} - \{r_0\}\int\limits_{A} \{r\}^{\mathrm{T}}\mathrm{d}A + \{r_0\}\{r_0\}^{\mathrm{T}}\int\limits_{A} \mathrm{d}A \tag{2.23}$$

Finally, recalling definitions (2.6) and (2.20), we obtain the law of transformation of the moment of inertia tensor for translations of the reference system

$$[\overline{I}] = [I] + A\{r_0\}\{r_0\}^{\mathrm{T}} - \{r_0\}\{S\}^{\mathrm{T}} - \{S\}\{r_0\}^{\mathrm{T}} \tag{2.24}$$

The matrix relation (2.24) can be rendered explicit as follows:

$$I_{\bar{x}\bar{x}} = I_{xx} + Ay_0^2 - 2y_0S_x \tag{2.25a}$$

$$I_{\bar{y}\bar{y}} = I_{yy} + Ax_0^2 - 2x_0S_y \tag{2.25b}$$

$$I_{\bar{x}\bar{y}} = I_{\bar{y}\bar{x}} = I_{xy} + Ax_0y_0 - x_0S_x - y_0S_y \tag{2.25c}$$

The earlier relations simplify in the case where the origin of the primitive XY reference system coincides with the centroid G of area A. In this case, we have

$$S_x = S_y = 0 \tag{2.26}$$

and Equations 2.25 assume the form of well-known **Huygens' laws**:

$$I_{\bar{x}\bar{x}} = I_{x_G x_G} + Ay_0^2 \tag{2.27a}$$

$$I_{\bar{y}\bar{y}} = I_{y_G y_G} + Ax_0^2 \tag{2.27b}$$

$$I_{\bar{x}\bar{y}} = I_{x_G y_G} + Ax_0y_0 \tag{2.27c}$$

As regards relations (2.27a) and (2.27b), it may be noted how the centroidal moment of inertia is the minimum of all those corresponding to an infinite number of parallel straight lines.

Now consider the moment of inertia tensor, relative to area A and calculated in the rotated reference system $\bar{X}^*\bar{Y}^*$ (Figure 2.2):

$$[\bar{I}^*] = \int_A \{\bar{r}^*\}\{\bar{r}^*\}^{\mathrm{T}} \, dA \tag{2.28}$$

Using the law (2.4) of transformation of the position vector for rotations of the reference system, we have

$$[\bar{I}^*] = \int_A ([N]\{\bar{r}\})([N]\{\bar{r}\})^{\mathrm{T}} \, dA \tag{2.29}$$

Now applying the law by which the transpose of the product of two matrices is equal to the inverse product of the transposes, we have

$$[\bar{I}^*] = \int_A ([N]\{\bar{r}\})(\{\bar{r}\}^{\mathrm{T}}[N^{\mathrm{T}}]) \, dA \tag{2.30}$$

Exploiting the associative law and carrying the constant matrices $[N]$ and $[N]^T$ outside the integral sign, Equation 2.30 becomes

$$[\bar{I}^*] = [N] \int_A \{\bar{r}\}\{\bar{r}\}^T \, \mathrm{d}A [N]^T \tag{2.31}$$

Finally, recalling definition (2.21), we obtain the law of transformation of the moment of inertia tensor for rotations of the reference system

$$[\bar{I}^*] = [N][\bar{I}][N]^T \tag{2.32}$$

Matrix relation (2.32) can be rendered explicit as follows:

$$I_{\bar{x}^*\bar{x}^*} = I_{\bar{x}\bar{x}} \cos^2 \vartheta + I_{\bar{y}\bar{y}} \sin^2 \vartheta - 2 I_{\bar{x}\bar{y}} \sin \vartheta \cos \vartheta \tag{2.33a}$$

$$I_{\bar{y}^*\bar{y}^*} = I_{\bar{x}\bar{x}} \sin^2 \vartheta + I_{\bar{y}\bar{y}} \cos^2 \vartheta + 2 I_{\bar{x}\bar{y}} \sin \vartheta \cos \vartheta \tag{2.33b}$$

$$I_{\bar{x}^*\bar{y}^*} = I_{\bar{y}^*\bar{x}^*} = I_{\bar{x}\bar{y}} \cos 2\vartheta + \frac{1}{2}(I_{\bar{x}\bar{x}} - I_{\bar{y}\bar{y}}) \sin 2\vartheta \tag{2.33c}$$

Two important conclusions can be derived from Equations 2.33:

1. The sum of the two moments of inertia I_{xx} and I_{yy} remains constant as the angle of rotation ϑ varies. We have in fact

$$I_{\bar{x}^*\bar{x}^*} + I_{\bar{y}^*\bar{y}^*} = I_{\bar{x}\bar{x}} + I_{\bar{y}\bar{y}} \tag{2.34}$$

This sum is the first scalar invariant of the moment of inertia tensor and can be interpreted as the **polar moment of inertia** of area A with respect to the origin of the reference system:

$$I_p = \int_A r^2 \, \mathrm{d}A \tag{2.35}$$

2. Equating to zero the expression of the product of inertia $I_{\bar{x}^*\bar{y}^*}$, it is possible to obtain the angle of rotation ϑ_0 which renders the moment of inertia tensor diagonal:

$$I_{\bar{x}^*\bar{y}^*} = I_{\bar{y}^*\bar{x}^*} = 0 \quad \text{for}$$

$$\vartheta_0 = \frac{1}{2} \arctan\left(\frac{2 I_{\bar{x}\bar{y}}}{I_{\bar{y}\bar{y}} - I_{\bar{x}\bar{x}}}\right), \quad -\frac{\pi}{4} < \vartheta_0 < \frac{\pi}{4} \tag{2.36}$$

Substituting Equation 2.36 in (2.33a and 2.33b), the so-called **principal moments of inertia** are determined. The two orthogonal directions defined by the angle ϑ_0 are referred to as the **principal directions of inertia**. It can be demonstrated how the principal moments of inertia are, in one case, the minimum and, in the other, the maximum of all the moments of inertia $I_{\bar{x}^*\bar{x}^*}$ and $I_{\bar{y}^*\bar{y}^*}$, which we have as the angle of rotation ϑ varies. When the axes, in addition to being principal, are also centroidal, they are referred to as **central**, as are the corresponding moments of inertia.

The general law of transformation of the moment of inertia tensor for rototranslations of the reference system (Figures 2.1 and 2.2) is obtained by combining the partial laws (2.24) and (2.32):

$$[\bar{I}^*] = [N]([I] + A\{r_0\}\{r_0\}^T - \{r_0\}\{S\}^T - \{S\}\{r_0\}^T)[N]^T \tag{2.37}$$

The inverse rototranslation formula may be obtained from (2.37) by premultiplying both sides of the equation by $[N]^T$ and postmultiplying them by $[N]$ and inserting Equation 2.17:

$$[I] = [N]^T[\bar{I}^*][N] + [N]^T\{\bar{S}^*\}\{r_0\}^T + \{r_0\}\{\bar{S}^*\}^T[N] + A\{r_0\}\{r_0\}^T \tag{2.38}$$

2.5 PRINCIPAL AXES AND MOMENTS OF INERTIA

Using well-known trigonometric formulas, relation (2.33a) becomes

$$I_{\bar{x}^*\bar{x}^*} = I_{\bar{x}\bar{x}}\frac{1+\cos2\vartheta}{2} + I_{\bar{y}\bar{y}}\frac{1-\cos2\vartheta}{2} - I_{\bar{x}\bar{y}}\sin2\vartheta \tag{2.39}$$

Via Equation 2.36, we obtain

$$I_{\bar{x}^*\bar{x}^*}(\vartheta_0) = \frac{I_{\bar{x}\bar{x}}+I_{\bar{y}\bar{y}}}{2} + \frac{I_{\bar{x}\bar{x}}-I_{\bar{y}\bar{y}}}{2}\cos2\vartheta_0 + \frac{I_{\bar{x}\bar{x}}-I_{\bar{y}\bar{y}}}{2}\tan2\vartheta_0\sin2\vartheta_0 \tag{2.40}$$

and hence

$$I_{\bar{x}^*\bar{x}^*}(\vartheta_0) = \frac{I_{\bar{x}\bar{x}}+I_{\bar{y}\bar{y}}}{2} + \frac{I_{\bar{x}\bar{x}}-I_{\bar{y}\bar{y}}}{2}\frac{1}{\cos2\vartheta_0} \tag{2.41}$$

Since we know from trigonometry that

$$\frac{1}{\cos2\vartheta_0} = (1+\tan^2 2\vartheta_0)^{\frac{1}{2}} \tag{2.42}$$

it is possible to apply Equation 2.36 once more:

$$\frac{1}{\cos2\vartheta_0} = \left(1 + \frac{4I^2_{\bar{x}\bar{y}}}{(I_{\bar{y}\bar{y}}-I_{\bar{x}\bar{x}})^2}\right)^{\frac{1}{2}}$$

$$= \begin{cases} \dfrac{1}{I_{\bar{x}\bar{x}}-I_{\bar{y}\bar{y}}}\left((I_{\bar{x}\bar{x}}-I_{\bar{y}\bar{y}})^2 + 4I^2_{\bar{x}\bar{y}}\right)^{\frac{1}{2}} & \text{when } I_{\bar{x}\bar{x}} > I_{\bar{y}\bar{y}} \\[3mm] \dfrac{1}{I_{\bar{y}\bar{y}}-I_{\bar{x}\bar{x}}}\left((I_{\bar{x}\bar{x}}-I_{\bar{y}\bar{y}})^2 + 4I^2_{\bar{x}\bar{y}}\right)^{\frac{1}{2}} & \text{when } I_{\bar{x}\bar{x}} < I_{\bar{y}\bar{y}} \end{cases} \tag{2.43}$$

Then, indicating $I_{\bar{x}^*\bar{x}^*}(\vartheta_0)$ with the simpler notation I_ξ, we have

$$
I_\xi = \begin{cases} \dfrac{I_{\bar{x}\bar{x}} + I_{\bar{y}\bar{y}}}{2} + \dfrac{1}{2}\left((I_{\bar{x}\bar{x}} - I_{\bar{y}\bar{y}})^2 + 4I^2_{\bar{x}\bar{y}}\right)^{\frac{1}{2}} & \text{when } I_{\bar{x}\bar{x}} > I_{\bar{y}\bar{y}} \\[3mm] \dfrac{I_{\bar{x}\bar{x}} + I_{\bar{y}\bar{y}}}{2} - \dfrac{1}{2}\left((I_{\bar{x}\bar{x}} - I_{\bar{y}\bar{y}})^2 + 4I^2_{\bar{x}\bar{y}}\right)^{\frac{1}{2}} & \text{when } I_{\bar{x}\bar{x}} < I_{\bar{y}\bar{y}} \end{cases}
\tag{2.44}
$$

Likewise, indicating $I_{\bar{y}^*\bar{y}^*}(\vartheta_0)$ with I_η, we have

$$
I_\eta = \begin{cases} \dfrac{I_{\bar{x}\bar{x}} + I_{\bar{y}\bar{y}}}{2} - \dfrac{1}{2}\left((I_{\bar{x}\bar{x}} - I_{\bar{y}\bar{y}})^2 + 4I^2_{\bar{x}\bar{y}}\right)^{\frac{1}{2}} & \text{when } I_{\bar{x}\bar{x}} > I_{\bar{y}\bar{y}} \\[3mm] \dfrac{I_{\bar{x}\bar{x}} + I_{\bar{y}\bar{y}}}{2} + \dfrac{1}{2}\left((I_{\bar{x}\bar{x}} - I_{\bar{y}\bar{y}})^2 + 4I^2_{\bar{x}\bar{y}}\right)^{\frac{1}{2}} & \text{when } I_{\bar{x}\bar{x}} < I_{\bar{y}\bar{y}} \end{cases}
\tag{2.45}
$$

We can thus conclude that, when the $\overline{XY}$ axes, by rotation, become the principal axes, the order relation is conserved:

$$
I_{\bar{x}\bar{x}} > I_{\bar{y}\bar{y}} \Rightarrow I_\xi > I_\eta
\tag{2.46a}
$$

$$
I_{\bar{x}\bar{x}} < I_{\bar{y}\bar{y}} \Rightarrow I_\xi < I_\eta
\tag{2.46b}
$$

When

$$
I_{\bar{x}\bar{x}} = I_{\bar{y}\bar{y}}, \quad I_{\bar{x}\bar{y}} \neq 0
\tag{2.47}
$$

relation (2.36) is not defined and thus it makes no difference whether the $\overline{XY}$ reference system is rotated by $\pi/4$ clockwise or counterclockwise ($\vartheta_0 = \pm\pi/4$) in order to obtain the principal directions.

Moreover, when

$$
I_{\bar{x}\bar{x}} = I_{\bar{y}\bar{y}}, \quad I_{\bar{x}\bar{y}} = 0
\tag{2.48}
$$

all the rotated reference systems $\overline{X}^*\overline{Y}^*$ are principal systems, for any angle of rotation ϑ_0. The areas that satisfy conditions (2.48) are said to be **gyroscopic**. As will be seen in the next section, it is possible to give a synthetic graphical interpretation of cases (2.47) and (2.48).

2.6 MOHR'S CIRCLE

With the aim of introducing the graphical method of **Mohr's circle**, let us consider the inverse problem of the one previously solved: Given an area A, and its principal axes of inertia $\xi\eta$ and the corresponding principal moments known, with respect to a point O of the plane (Figure 2.3), we intend to express the moments of inertia with respect to a reference system rotated by an angle ϑ, counterclockwise with respect to the principal reference system.

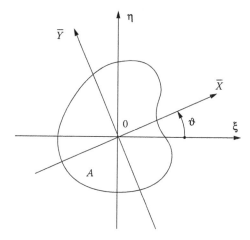

Figure 2.3

Applying Equations 2.33, and since $I_{\xi\eta} = 0$, we have

$$I_{\overline{xx}} = I_\xi \cos^2 \vartheta + I_\eta \sin^2 \vartheta \tag{2.49a}$$

$$I_{\overline{yy}} = I_\xi \sin^2 \vartheta + I_\eta \cos^2 \vartheta \tag{2.49b}$$

$$I_{\overline{xy}} = \frac{I_\xi - I_\eta}{2} \sin 2\vartheta \tag{2.49c}$$

The trigonometry formulas used previously give

$$I_{\overline{xx}} = \frac{I_\xi + I_\eta}{2} + \frac{I_\xi - I_\eta}{2} \cos 2\vartheta \tag{2.50a}$$

$$I_{\overline{yy}} = \frac{I_\xi + I_\eta}{2} - \frac{I_\xi - I_\eta}{2} \cos 2\vartheta \tag{2.50b}$$

$$I_{\overline{xy}} = \frac{I_\xi - I_\eta}{2} \sin 2\vartheta \tag{2.50c}$$

Relations (2.50a and 2.50c) constitute the parametric equations of a circumference having as its centre

$$C\left(\frac{I_\xi + I_\eta}{2}, 0\right) \tag{2.51a}$$

and as its radius

$$R = \frac{I_\xi - I_\eta}{2} \tag{2.51b}$$

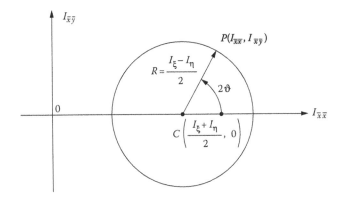

Figure 2.4

in Mohr's plane (Figure 2.4). The earlier circumference represents all the pairs $(I_{\overline{xx}}, I_{\overline{xy}})$ which succeed one another as the angle ϑ (Figure 2.3) varies. Note that, since $I_{\overline{xx}}$ is in any case positive, we have in fact a Mohr's half-plane.

Let us now reconsider the direct problem: Given the moments of inertia with respect to the two generic orthogonal axes $\overline{X}\overline{Y}$ (Figure 2.3), determine the principal axes and moments of inertia. This determination has already been made analytically in Section 2.5. We shall now proceed to repropose it graphically using Mohr's circle (Figure 2.5):

1. The first operation to be carried out is to identify the two notable points P and P' on Mohr's plane:

$$P(I_{\overline{xx}}, I_{\overline{yy}}), \quad P'(I_{\overline{yy}}, -I_{\overline{xy}}) \tag{2.52}$$

2. The intersection C of the segment PP' with the axis $I_{\overline{xx}}$ identifies the centre of Mohr's circle, while the segments CP and CP' represent two radii of that circle.
3. Draw through the point P the line parallel to the axis $I_{\overline{xx}}$ and through P' the line parallel to the axis $I_{\overline{xy}}$. These two lines meet in point P^*, called the **pole**, again belonging to Mohr's circle.

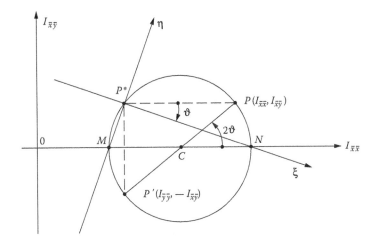

Figure 2.5

4. The lines joining pole P^* with points M and N of the $I_{\overline{xx}}$ axis, which are the intersections of the circumference with the axis, give the directions of the two principal axes of inertia. Naturally, points M and N each have the value of one of the two principal moments of inertia as abscissa. In particular, in Figure 2.5, the abscissa of M is I_η, while the abscissa of N is I_ξ, since we have assumed $I_{\overline{xx}} > I_{\overline{yy}}$. Pole P^* can obviously also fall in one of the three remaining quadrants corresponding to Mohr's circle.

The graphical construction described earlier and shown in Figure 2.5 is justified by noting that the circumferential angle $\widehat{PP^*N}$ is half of the corresponding central angle $\widehat{PCN} = 2\vartheta$ and that thus its amplitude is equal to the angle ϑ.

2.7 AREAS PRESENTING SYMMETRY

An area is said to present **oblique axial symmetry** (Figure 2.6a) when there exists a straight line s which cuts the area into two parts, and a direction s' conjugate with this straight line, such that, if we consider a generic point P, belonging to the area and the line PC, parallel to the direction s' and we draw on that line the segment $\overline{CP'} = \overline{PC}$ on the opposite side of P with respect to s, the point P' still belongs to the area. When the angle α between the directions of the lines s and s' is equal to $90°$, then we have **right axial symmetry** (Figure 2.6b).

It is easy to verify that the centroid of a section having axial symmetry lies on the corresponding axis of symmetry. The centroid relative to the pair of symmetrical elementary areas located in P and P' coincides in fact with point C (Figure 2.6). Applying the so-called **distributive law** of the centroid, it is possible to think of concentrating the whole area on the axis of symmetry s, and thus the global centroid is sure to lie on the same line s.

In the case of an area presenting right symmetry (Figure 2.6b), the axis of symmetry is also a central axis of inertia. In fact, it is centroidal and, with respect to it and to any orthogonal axis, the product of inertia $I_{ss'}$ vanishes by symmetry.

When there are two or more axes of symmetry (oblique or right), since the centroid must belong to each axis, it coincides with their intersection (Figure 2.7). In the case of double right symmetry (Figure 2.7a), the axes of symmetry are also central axes of inertia.

An area is said to present **polar symmetry** (Figure 2.8) when there exists a point C such that, if we consider a generic point P belonging to the area and the line PC joining the two

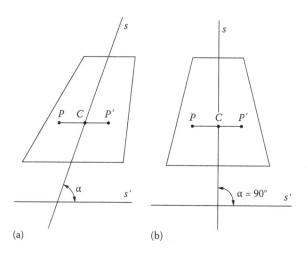

(a)　　　　　　　　　　　　(b)

Figure 2.6

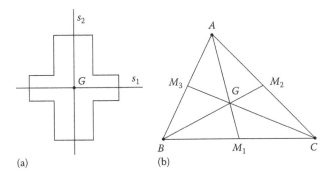

Figure 2.7

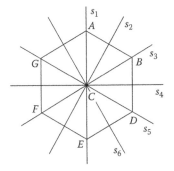

Figure 2.8

points, and we draw on this line the segment $\overline{CP'} = \overline{PC}$ on the side opposite to P with respect to C, the point P' still belongs to the area.

It is immediately verifiable that the centroid of a section having polar symmetry coincides with its geometrical centre C. The centroid corresponding to the pair of symmetrical elementary areas in P and P' coincides, in fact, with point C (Figure 2.8). Applying the distributive law of the centroid, it is possible to think of concentrating the whole area in point C, and thus the global centroid must certainly coincide with the same point C.

It is interesting to note how an n-tuple right symmetry area, with n being an even number ($2 \leq n < \infty$), is also a polar symmetry area (Figure 2.9), whereas a polar symmetry area is not necessarily also an n-tuple right symmetry area.

Areas having n-tuple right symmetry, with n being an odd number ($3 \leq n < \infty$), do not, however, present polar symmetry, even though they are gyroscopic areas, as also are those with n even.

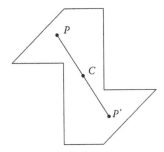

Figure 2.9

2.8 ELEMENTARY AREAS

If, on an XY plane, we assign n disjoint areas, A_1, A_2,..., A_n, the distributive law of static moments and, respectively, that of the moments of inertia are defined as follows (Figure 2.10):

$$S\left(\bigcup_{i=1}^{n} A_i\right) = \sum_{i=1}^{n} S(A_i) \tag{2.53a}$$

$$I\left(\bigcup_{i=1}^{n} A_i\right) = \sum_{i=1}^{n} I(A_i) \tag{2.53b}$$

where S and I indicate generically a static moment and a moment of inertia, respectively, calculated with respect to the coordinate axes (S_x, S_y, I_{xx}, I_{yy}, I_{xy}).

In determining the static and inertial characteristics of composite areas, it is necessary to exploit the aforementioned laws. These derive from the integral nature of the definitions which have previously been given of first- and second-order moments. The first law expresses the fact that the static moment of a composite area (i.e. of the disjoint union of more than one elementary area) is equal to the sum of the static moments of the single areas. The second law refers to the moments of inertia and is altogether analogous.

Since it is therefore possible to reduce the calculation of composite areas to the calculation of simpler areas, the importance of calculating once and for all the static and inertial features of elementary areas emerges clearly. In the sequel, we shall examine the rectangle, the right triangle and the annulus sector.

Consider the **rectangle** having base b and height h (Figure 2.11). From the definition of centroid, we immediately obtain the static moments in the XY reference system:

$$S_x = Ay_G = \frac{1}{2}bh^2 \tag{2.54a}$$

$$S_y = Ax_G = \frac{1}{2}hb^2 \tag{2.54b}$$

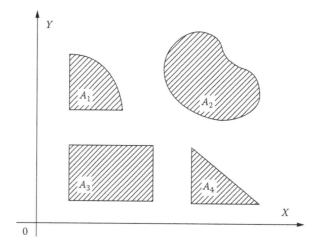

Figure 2.10

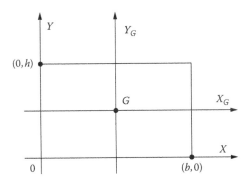

Figure 2.11

For the central moments of inertia, we have

$$I_{x_G x_G} = \int\limits_{-b/2}^{+b/2} \int\limits_{-h/2}^{+h/2} y^2 \mathrm{d}x\,\mathrm{d}y = \frac{bh^3}{12} \tag{2.55a}$$

and likewise

$$I_{y_G y_G} = \frac{hb^3}{12} \tag{2.55b}$$

It is then possible to obtain the inertia tensor in the XY reference system by applying Huygens' laws (2.27):

$$I_{xx} = I_{x_G x_G} + Ay_0^2 = \frac{bh^3}{12} + \frac{bh^3}{4} = \frac{bh^3}{3} \tag{2.56a}$$

$$I_{yy} = I_{y_G y_G} + Ax_0^2 = \frac{hb^3}{12} + \frac{hb^3}{4} = \frac{hb^3}{3} \tag{2.56b}$$

$$I_{xy} = I_{x_G y_G} + Ax_0 y_0 = 0 + bh\left(-\frac{b}{2}\right)\left(-\frac{h}{2}\right) = \frac{b^2 h^2}{4} \tag{2.56c}$$

Consider the **right triangle** having base b and height h (Figure 2.12). As is well known, its centroid coincides with the point of intersection of the three medians, which are at the same time axes of oblique symmetry. The moment of inertia with respect to the axis X_C of the triangle MOP is equal to the moment of inertia with respect to the axis X_C of rectangle $NOPQ$. The latter, in fact, is obtained from the former by suppressing triangle MNC and adding triangle CQP. These two triangles are equal and arranged symmetrically with respect to the axis X_C. Hence,

$$I_{x_C x_C} = \frac{1}{3} b \left(\frac{h}{2}\right)^3 = \frac{bh^3}{24} \tag{2.57}$$

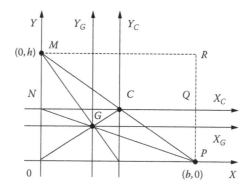

Figure 2.12

Applying the inverse of Huygens' law, we obtain

$$I_{x_G x_G} = \frac{bh^3}{24} - \left(\frac{bh}{2}\right)\left(\frac{h}{6}\right)^2 = \frac{bh^3}{36} \tag{2.58}$$

Finally, applying Huygens' law, we have

$$I_{xx} = \frac{bh^3}{36} + \left(\frac{bh}{2}\right)\left(\frac{h}{3}\right)^2 = \frac{bh^3}{12} \tag{2.59}$$

Likewise

$$I_{yy} = \frac{hb^3}{12} \tag{2.60}$$

As regards the product of inertia I_{xy}, its integral definition can be applied:

$$I_{xy} = \int\limits_0^b \int\limits_0^{h(b-x)/b} xy \, dx \, dy = \frac{b^2 h^2}{24} \tag{2.61}$$

The product of inertia $I_{x_G y_G}$ may be obtained from I_{xy}, *via* the inverse application of Huygens' law:

$$I_{x_G y_G} = \frac{b^2 h^2}{24} - \left(\frac{bh}{2}\right)\left(-\frac{b}{3}\right)\left(-\frac{h}{3}\right) = -\frac{b^2 h^2}{72} \tag{2.62}$$

Consider the **annulus sector** of internal radius R_1, external radius R_2 and angular amplitude φ (Figure 2.13). The static moment of the sector with respect to the X axis is

$$S_x = \int\limits_0^\varphi \int\limits_{R_1}^{R_2} (r \sin\varphi) r \, dr d\varphi = \frac{1}{3}(1 - \cos\varphi)\left(R_2^3 - R_1^3\right) \tag{2.63a}$$

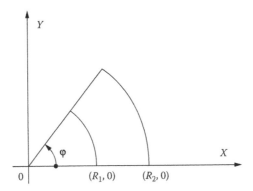

Figure 2.13

Likewise, the static moment with respect to the Y axis is

$$S_y = \int\limits_0^\varphi \int\limits_{R_1}^{R_2} (r\cos\varphi) r\, dr\, d\varphi = \frac{1}{3}\sin\varphi\left(R_2^3 - R_1^3\right) \tag{2.63b}$$

Also in the calculation of the moments of inertia, it is possible to apply the definition:

$$I_{xx} = \int\limits_0^\varphi \int\limits_{R_1}^{R_2} (r\sin\varphi)^2 r\, dr\, d\varphi = \frac{1}{8}(\varphi - \sin\varphi\cos\varphi)\left(R_2^4 - R_1^4\right) \tag{2.64a}$$

$$I_{yy} = \int\limits_0^\varphi \int\limits_{R_1}^{R_2} (r\cos\varphi)^2 r\, dr\, d\varphi = \frac{1}{8}(\varphi + \sin\varphi\cos\varphi)\left(R_2^4 - R_1^4\right) \tag{2.64b}$$

$$I_{xy} = \int\limits_0^\varphi \int\limits_{R_1}^{R_2} (r\sin\varphi)(r\cos\varphi) r\, dr\, d\varphi = \frac{1}{16}(1 - \cos 2\varphi)\left(R_2^4 - R_1^4\right) \tag{2.64c}$$

In the particular case of a circle of radius R, we have

$$R_1 = 0, \quad R_2 = R, \quad \varphi = 2\pi$$

and thus

$$S_x = S_y = 0 \tag{2.65a}$$

$$I_{xx} = I_{yy} = \frac{\pi R^4}{4}, \quad I_{xy} = 0 \tag{2.65b}$$

The static moments and the product of inertia are zero by symmetry. Another way to obtain the moments of inertia I_{xx} and I_{yy} of the circle is that of calculating the polar moment as given by Equation 2.35:

$$I_p = I_{xx} + I_{yy} = \int_0^{2\pi} \int_0^R (r^2) r \, dr \, d\varphi = \frac{\pi R^4}{2} \tag{2.66}$$

Since $I_{xx} = I_{yy}$, once more we obtain Equation 2.65b.

2.9 THIN-WALLED SECTIONS

A section is said to be **thin walled** when one of its dimensions (the thickness δ) is clearly smaller than the others (Figure 2.14). In these cases, the section is represented and calculated as if its whole area were concentrated in its midline m. This approximate calculation approaches the exact result, the smaller the thickness is, compared to the other dimensions of the section.

Consider a **rectilinear segment** of length l and thickness δ (Figure 2.15). The moment of inertia $I_{x_G x_G} = I_{xx} = l\delta^3/12$ can be neglected with regard to all the other quantities. This moment is, in fact, an infinitesimal of a higher order, as it is proportional to the infinitesimal quantity δ raised to the third power, whilst the other quantities are proportional to the quantity δ raised to the first power.

When Huygens' law is applied, for the calculation of the moments of inertia of a rectilinear segment with respect to the translated axes p and n (Figure 2.15), it is important to distinguish the two cases. In fact, for the calculation of I_{pp}, it is possible to consider the entire area concentrated in the centroid and to neglect the local moment

$$I_{pp} = \delta l d_p^2 \tag{2.67a}$$

while for the calculation of I_{nn}, in addition to the contribution of translation, it is necessary to include also the local contribution, which is not a negligible quantity:

$$I_{nn} = \frac{\delta l^3}{12} + \delta l d_n^2 \tag{2.67b}$$

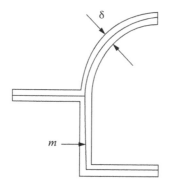

Figure 2.14

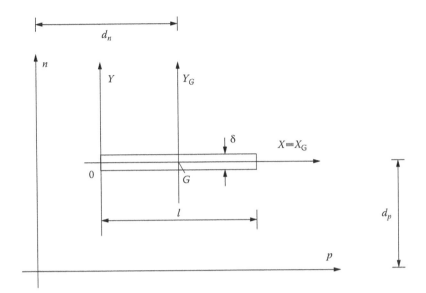

Figure 2.15

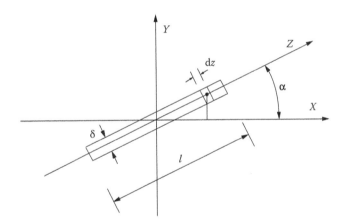

Figure 2.16

If we imagine inclining the rectilinear segment by an angle α with respect to the X axis (Figure 2.16), the moment of inertia with respect to the X axis then becomes

$$I_{xx} = \int_A y^2 dA = \int_{-l/2}^{+l/2} (z \sin \alpha)^2 \delta dz \qquad (2.68)$$

where Z is the longitudinal axis of the segment. Evaluating the integral, we obtain

$$I_{xx} = \frac{\delta l^3}{12} \sin^2 \alpha \qquad (2.69a)$$

and, likewise,

$$I_{yy} = \frac{\delta l^3}{12} \cos^2 \alpha \tag{2.69b}$$

$$I_{xy} = \frac{\delta l^3}{12} \sin \alpha \cos \alpha \tag{2.69c}$$

It may be noted how, for $\alpha = 0, \pi/2$, we obtain once more the results already found.

Consider an **arc of circumference** of radius R, angular amplitude φ and thickness δ (Figure 2.17). To define the static and inertial characteristics, it is possible to reconsider the formulas of Section 2.8 and to particularize them for

$$R_1 \simeq R_2 \simeq R; \quad R_2 - R_1 = \delta \ll R \tag{2.70}$$

For example, Equation 2.63a is transformed as follows:

$$S_x = \frac{1}{3}(1 - \cos \varphi)(R_2 - R_1)\left(R_2^2 + R_2 R_1 + R_1^2\right) \simeq (1 - \cos \varphi)\delta R^2 \tag{2.71a}$$

And, likewise, Equation 2.63b becomes

$$S_y = \sin \varphi \, \delta R^2 \tag{2.71b}$$

Then, as regards the moments of inertia, from Equations 2.64 and 2.70, we obtain

$$I_{xx} = \frac{1}{2}(\varphi - \sin \varphi \cos \varphi)\delta R^3 \tag{2.72a}$$

$$I_{yy} = \frac{1}{2}(\varphi + \sin \varphi \cos \varphi)\delta R^3 \tag{2.72b}$$

$$I_{xy} = \frac{1}{4}(1 - \cos 2\varphi)\delta R^3 \tag{2.72c}$$

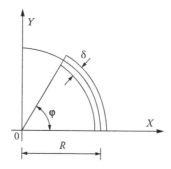

Figure 2.17

2.10 EXAMPLES OF CALCULATION

Five examples of calculation are given in the following for five sections, without any particular comments. In each case, the areas, the static moments, the coordinates of the centroid, the moments of inertia in the centroidal system, the central directions and moments of inertia are calculated and listed in order and for each elementary part of the section as well as for the entire section.

Example 1 (Figure 2.18a) concerns an L-section made up of two rectangular plates and of a triangular angle iron (Figure 2.18b). This section does not present particular symmetries

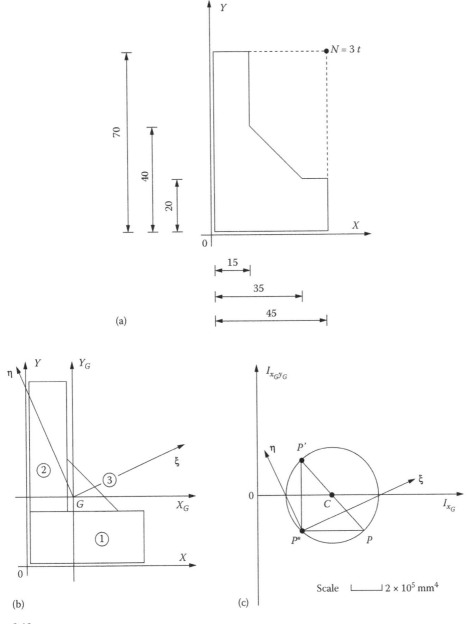

(a)

(b)

(c)

Scale $\llcorner\quad\lrcorner 2 \times 10^5$ mm^4

Figure 2.18

and, thus, for the determination of the global centroid, requires the calculation of the areas and the static moments. The latter are obtained by multiplying the partial areas by the coordinates of the partial centroids. For the calculation of the moments of inertia in the centroidal system $X_G Y_G$, use has, instead, been made of Huygens' formulas (2.27), which add the local moment of inertia to the moment of translation. Angle ϑ_0 of counterclockwise rotation, which provides the central directions, may be obtained analytically from Equation 2.36, just as the central moments may be deduced from Equations 2.44 and 2.45. Figure 2.18c gives the graphical construction of Mohr's circle, with the definition of the aforesaid quantities.

Example 2 (Figure 2.19a) concerns an H-section which may be obtained ideally by removing a square and a semicircle from the rectangle circumscribing the section (Figure 2.19b). The axis of symmetry Y is also centroidal and principal, and thus central. The ordinate y_G of the global centroid remains unknown, and hence only the static moments with respect to

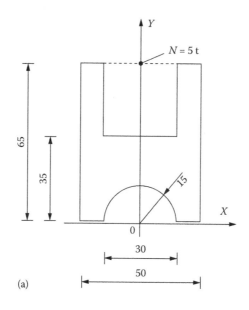

(a)

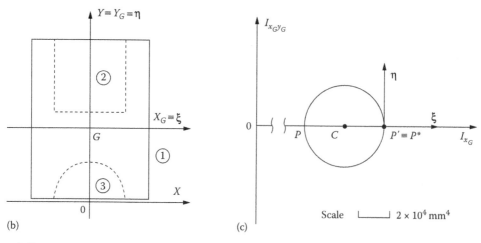

(b) (c)

Figure 2.19

the X axis are to be calculated. Whereas the moments $I_{x_G}^{(1)}$ and $I_{x_G}^{(2)}$, relative to the rectangle and to the square, are obtained by applying Equation 2.27a, for the moment $I_{x_G}^{(3)}$ of the semi-circle, Equation 2.25a has been resorted to. The partial products of inertia are all zero by symmetry. In this case, Mohr's graphical construction is of little significance, on account of the axial symmetry (Figure 2.19c).

Example 3 (Figure 2.20a) concerns a closed thin-walled section made up of three plates inclined by 60° with respect to one another (Figure 2.20b). To calculate the moments of inertia, Huygens' laws (2.27) have been used, whilst the local moments of the inclined segments have been evaluated using Equations 2.69. Figure 2.20c presents Mohr's graphical construction.

Example 4 (Figure 2.21a) concerns a closed thin-walled section made up of three plane plates and one cylindrical plate (Figure 2.21b). The static moment S_x, relative to the circular segment, has been calculated according to Equation 2.71b, while the moment of inertia $I_{x_G}^{(4)}$ has been evaluated, applying the law of transformation by translation (2.25a) and, locally, Equation 2.72b. The moment of inertia $I_{y_G}^{(4)}$ has, instead, been evaluated by simply applying Equation 2.72a. Mohr's circle for this case is represented in Figure 2.21c.

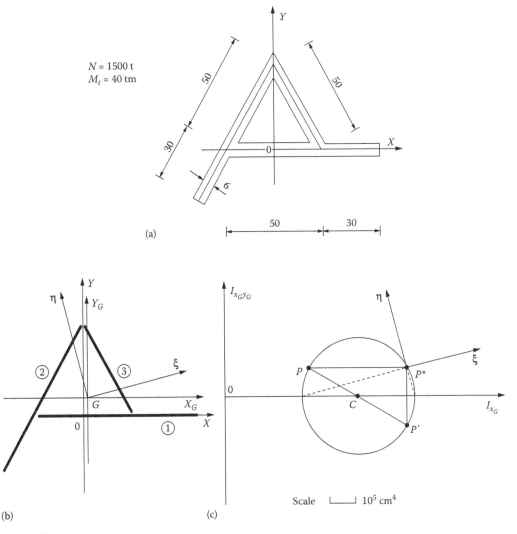

(a)

(b)   (c)

Figure 2.20

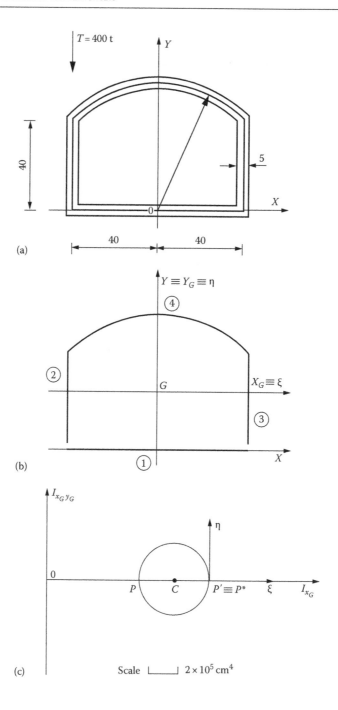

Figure 2.21

Finally, Example 5 (Figure 2.22a) regards a thin-walled section having polar symmetry, made up of three plane plates (Figure 2.22b). In this case, calculation of the position of the centroid serves no purpose, as the polar symmetry causes it to coincide with the geometrical centre of the area. Mohr's circle (Figure 2.22c) provides confirmation of the central directions and moments determined analytically.

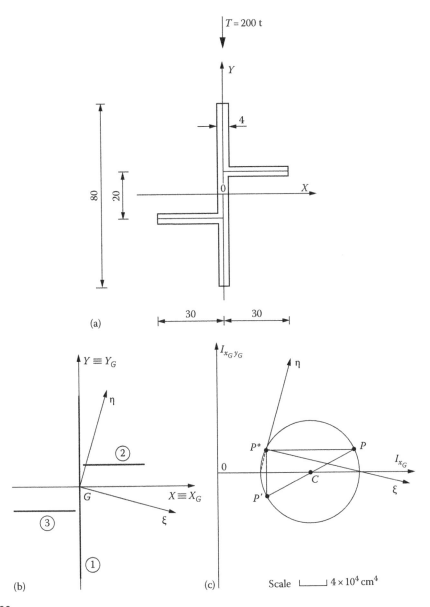

Figure 2.22

Example 1

$$\begin{cases} A^{(1)} = 45 \times 20 \, \text{mm}^2 = 900 \, \text{mm}^2 \\ A^{(2)} = 50 \times 15 \, \text{mm}^2 = 750 \, \text{mm}^2 \\ A^{(3)} = \dfrac{1}{2} \times 20 \times 20 \, \text{mm}^2 = 200 \, \text{mm}^2 \end{cases}$$

$$A = \sum_{i=1}^{3} A^{(i)} = 1,850 \, \text{mm}^2$$

$$\begin{cases} S_x^{(1)} = A^{(1)}y_G^{(1)} = 900 \times 10 = 9{,}000\,\text{mm}^3 \\ S_x^{(2)} = A^{(2)}y_G^{(2)} = 750 \times 45 = 33{,}750\,\text{mm}^3 \\ S_x^{(3)} = A^{(3)}y_G^{(3)} = 200 \times 26.67 = 5{,}333\,\text{mm}^3 \end{cases}$$

$$S_x = \sum_{i=1}^{3} S_x^{(i)} = 48{,}083\,\text{mm}^3$$

$$\begin{cases} S_y^{(1)} = A^{(1)}x_G^{(1)} = 900 \times 22.5 = 20{,}250\,\text{mm}^3 \\ S_y^{(2)} = A^{(2)}x_G^{(2)} = 750 \times 7.5 = 5{,}625\,\text{mm}^3 \\ S_y^{(3)} = A^{(3)}x_G^{(3)} = 200 \times 21.67 = 4{,}333\,\text{mm}^3 \end{cases}$$

$$S_y = \sum_{i=1}^{3} S_y^{(i)} = 30{,}208\,\text{mm}^3$$

$$\begin{cases} x_G = \dfrac{S_y}{A} = 16.33\,\text{mm} \\ y_G = \dfrac{S_x}{A} = 25.99\,\text{mm} \end{cases}$$

$$\begin{cases} I_{x_G}^{(1)} = \dfrac{45 \times 20^3}{12} + 900 \times 15.99^2 = 260{,}112\,\text{mm}^4 \\ I_{x_G}^{(2)} = \dfrac{15 \times 50^3}{12} + 750 \times (45 - 25.99)^2 = 427{,}285\,\text{mm}^4 \\ I_{x_G}^{(3)} = \dfrac{20 \times 20^3}{36} + 200 \times (26.67 - 25.99)^2 = 4{,}536\,\text{mm}^4 \end{cases}$$

$$I_{x_G} = \sum_{i=1}^{3} I_{x_G}^{(i)} = 691{,}933\,\text{mm}^4$$

$$\begin{cases} I_{y_G}^{(1)} = \dfrac{20 \times 45^3}{12} + 900 \times (22.5 - 16.33)^2 = 186{,}100\,\text{mm}^4 \\ I_{y_G}^{(2)} = \dfrac{50 \times 15^3}{12} + 750 \times (16.33 - 7.5)^2 = 72{,}583\,\text{mm}^4 \\ I_{y_G}^{(3)} = \dfrac{20 \times 20^3}{36} + 200 \times (21.67 - 16.33)^2 = 10{,}133\,\text{mm}^4 \end{cases}$$

$$I_{y_G} = \sum_{i=1}^{3} I_{y_G}^{(i)} = 268{,}816\,\text{mm}^4$$

$$\begin{cases} I_{x_G y_G}^{(1)} = -900 \times (22.5 - 16.33)(25.99 - 10) = -88,792 \, \text{mm}^4 \\ I_{x_G y_G}^{(2)} = -750 \times (16.33 - 7.5)(45 - 25.99) = -125,941 \, \text{mm}^4 \\ I_{x_G y_G}^{(3)} = -\dfrac{20^2 \times 20^2}{72} + 200 \times (21.67 - 16.33)(26.67 - 25.99) = -1,500 \, \text{mm}^4 \end{cases}$$

$$I_{x_G y_G} = \sum_{i=1}^{3} I_{x_G y_G}^{(i)} = -216,186 \, \text{mm}^4$$

$$\vartheta_0 = \frac{1}{2} \arctan \frac{2 \times (-216,186)}{268,816 - 691,933} = 22.81°$$

$$I_\xi = \frac{691,933 + 268,816}{2} + \frac{1}{2} \left((691,933 - 268,816)^2 + 4 \times (216,186)^2 \right)^{\frac{1}{2}}$$

$$= 480,374 + 302,478 = 782,852 \, \text{mm}^4$$

$$I_\eta = 480,374 - 302,478 = 177,896 \, \text{mm}^4$$

Example 2

$$\begin{cases} A^{(1)} = 50 \times 65 = 3,250 \, \text{mm}^2 \\ A^{(2)} = -30 \times 30 \, \text{mm}^2 = -900 \, \text{mm}^2 \\ A^{(1)} = -\dfrac{1}{2} \times 3.1415 \times 15^2 = -353.42 \, \text{mm}^2 \end{cases}$$

$$A = \sum_{i=1}^{3} A^{(i)} = 1,996.58 \, \text{mm}^2$$

$$\begin{cases} S_x^{(1)} = A^{(1)} y_G^{(1)} = 3,250 \times 32.5 = 105,625 \, \text{mm}^3 \\ S_x^{(2)} = A^{(2)} y_G^{(2)} = -900 \times 50 = -45,000 \, \text{mm}^3 \\ S_x^{(3)} = -\dfrac{2}{3}(15)^3 = -2,250 \, \text{mm}^3 \end{cases}$$

$$\begin{cases} S_x = \sum_{i=1}^{3} S_x^{(i)} = 58,375 \, \text{mm}^3 \\ S_y = 0 \end{cases}$$

$$\begin{cases} x_G = 0 \\ y_G = \dfrac{S_x}{A} = 29.24 \, \text{mm} \end{cases}$$

$$\begin{cases} I_{x_G}^{(1)} = \dfrac{50 \times 65^3}{12} + 3{,}250 \times (32.5 - 29.24)^2 = 1{,}178{,}810.5 \, \text{mm}^4 \\[3mm] I_{x_G}^{(2)} = -\dfrac{30 \times 30^3}{12} - 900 \times (50 - 29.24)^2 = -455{,}379.84 \, \text{mm}^4 \\[3mm] I_{x_G}^{(3)} = -\dfrac{\pi}{8} \times 15^4 - 353.42 \times (29.24)^2 + 2 \times 29.24 \times \dfrac{2}{3} \times 15^3 = -190{,}465.99 \, \text{mm}^4 \end{cases}$$

$$I_{x_G} = \sum_{i=1}^{3} I_{x_G}^{(i)} = 532{,}956 \, \text{mm}^4$$

$$\begin{cases} I_{y_G}^{(1)} = \dfrac{65 \times 50^3}{12} = 677{,}083.33 \, \text{mm}^4 \\[3mm] I_{y_G}^{(2)} = -\dfrac{30 \times 30^3}{12} = -67{,}500 \, \text{mm}^4 \\[3mm] I_{y_G}^{(3)} = -\dfrac{\pi}{8} \times 15^4 = -19{,}879.80 \, \text{mm}^4 \end{cases}$$

$$I_{y_G} = \sum_{i=1}^{3} I_{y_G}^{(i)} = 589{,}702 \, \text{mm}^4$$

$$\begin{cases} I_\xi = I_{x_G} = 532{,}956 \, \text{mm}^4 \\[2mm] I_\eta = I_{y_G} = 589{,}702 \, \text{mm}^4 \end{cases}$$

Example 3

$$\begin{cases} A^{(1)} = 80 \times 6 = 480 \, \text{cm}^2 \\[2mm] A^{(2)} = 80 \times 6 = 480 \, \text{cm}^2 \\[2mm] A^{(3)} = 50 \times 6 = 300 \, \text{cm}^2 \end{cases}$$

$$A = \sum_{i=1}^{3} A^{(i)} = 1{,}260 \, \text{cm}^2$$

$$\begin{cases} S_x^{(1)} = 0 \\[2mm] S_x^{(2)} = A^{(2)} y_G^{(2)} = 480 \times 8.66 = 4{,}156.8 \, \text{cm}^3 \\[2mm] S_x^{(3)} = A^{(3)} y_G^{(3)} = 300 \times 21.65 = 6{,}495 \, \text{cm}^3 \end{cases}$$

$$S_x = \sum_{i=1}^{3} S_x^{(i)} = 10{,}652 \, \text{cm}^3$$

$$\begin{cases} S_y^{(1)} = A^{(1)} y_G^{(1)} = 480 \times 15 = 7,200 \, cm^3 \\ S_y^{(2)} = A^{(2)} y_G^{(2)} = -480 \times 20 = -9,600 \, cm^3 \\ S_y^{(3)} = A^{(3)} x_G^{(3)} = 300 \times 12.5 = 3,750 \, cm^3 \end{cases}$$

$$S_y = \sum_{i=1}^{3} S_y^{(i)} = 1,350 \, cm^3$$

$$\begin{cases} x_G = \dfrac{S_y}{A} = 1.07 \, cm \\ y_G = \dfrac{S_x}{A} = 8.45 \, cm \end{cases}$$

$$\begin{cases} I_{x_G}^{(1)} = 480 \times (8.45)^2 = 34,273.2 \, cm^4 \\ I_{x_G}^{(2)} = \dfrac{6 \times 80^3}{12} (0.866)^2 + 480 \times (8.66 - 8.45)^2 = 192,009.9 \, cm^4 \\ I_{x_G}^{(3)} = \dfrac{6 \times 50^3}{12} (0.866)^2 + 300 \times (21.65 - 8.45)^2 = 99,144.25 \, cm^4 \end{cases}$$

$$I_{x_G} = \sum_{i=1}^{3} I_{x_G}^{(i)} = 325,446 \, cm^4$$

$$\begin{cases} I_{y_G}^{(1)} = \dfrac{6 \times 80^3}{12} + 480 \times (15 - 1.07)^2 = 349,000 \, cm^4 \\ I_{y_G}^{(2)} = \dfrac{6 \times 80^3}{12} (0.5)^2 + 480 \times (20 + 1.07)^2 = 277,000 \, cm^4 \\ I_{y_G}^{(3)} = \dfrac{6 \times 50^3}{12} (0.5)^2 + 300 \times (12.5 - 1.07)^2 = 55,000 \, cm^4 \end{cases}$$

$$I_{y_G} = \sum_{i=1}^{3} I_{y_G}^{(i)} = 681,000 \, cm^4$$

$$\begin{cases} I_{x_G y_G}^{(1)} = -480 \times (15 - 1.07) \times 8.45 = -56,500 \, cm^4 \\ I_{x_G y_G}^{(2)} = \dfrac{6 \times 80^3}{12} \times (0.866) \times (0.5) - 480 \times (8.66 - 8.45) \times (20 + 1.07) = 109,000 \, cm^4 \\ I_{x_G y_G}^{(3)} = \dfrac{6 \times 50^3}{12} \times (0.866) \times (-0.5) + 300 \times (12.5 - 1.07) \times (21.65 - 8.45) = 18,000 \, cm^4 \end{cases}$$

$$I_{x_G y_G} = \sum_{i=1}^{3} I_{x_G y_G}^{(i)} = 70,500\,\text{cm}^4$$

$$\vartheta_0 = \frac{1}{2}\arctan\frac{2\times 70,500}{681,000 - 325,000} = 10.8°$$

$$I_\xi = \frac{325,000 + 681,000}{2} - \frac{1}{2}\big((681,000 - 325,000)^2 + 4\times(70,500)^2\big)^{\frac{1}{2}}$$

$$= 503,000 - 191,000 = 312,000\,\text{cm}^4$$

$$I_\eta = 503,000 + 191,000 = 694,000\,\text{cm}^4$$

Example 4

$$\begin{cases} A^{(1)} = 80\times 5 = 400\,\text{cm}^2 \\ A^{(2)} = A^{(3)} = 40\times 5 = 200\,\text{cm}^2 \\ A^{(4)} = \dfrac{\pi}{2}\times(56.57)\times 5 = 444\,\text{cm}^2 \end{cases}$$

$$A = \sum_{i=1}^{4} A^{(i)} = 1,244\,\text{cm}^2$$

$$\begin{cases} S_x^{(1)} = 0 \\ S_x^{(2)} = S_x^{(3)} = 200\times 20 = 4,000\,\text{cm}^3 \\ S_x^{(4)} = 2\times 5\times(56.57)^2\times(0.707) = 22,600\,\text{cm}^3 \end{cases}$$

$$S_x = \sum_{i=1}^{4} S_x^{(i)} = 30,600\,\text{cm}^3$$

$$S_y = 0$$

$$\begin{cases} x_G = 0 \\ y_G = \dfrac{S_x}{A} = 24.61\,\text{cm} \end{cases}$$

$$\begin{cases} I_{x_G}^{(1)} = 400\times(24.61)^2 = 242,000\,\text{cm}^4 \\ I_{x_G}^{(2)} = I_{x_G}^{(3)} = \dfrac{5\times 40^3}{12} + 200\times(24.61 - 20)^2 = 31,000\,\text{cm}^4 \\ I_{x_G}^{(4)} = \left(\dfrac{\pi}{4} + \dfrac{1}{2}\right)\times 5\times(56.57)^3 + 444 + (24.61)^2 - 2\times(24.61)\times(22,600) = 320,000\,\text{cm}^4 \end{cases}$$

$$I_{x_G} = \sum_{i=1}^{4} I_{x_G}^{(i)} = 624,000\,\mathrm{cm}^4$$

$$\begin{cases} I_{y_G}^{(1)} = \dfrac{5 \times 80^3}{12} = 213,000\,\mathrm{cm}^4 \\[2mm] I_{y_G}^{(2)} = I_{y_G}^{(3)} = 200 \times 40^2 = 320,000\,\mathrm{cm}^4 \\[2mm] I_{y_G}^{(4)} = \left(\dfrac{\pi}{4} - \dfrac{1}{2}\right) \times 5 \times (56.57)^3 = 258,000\,\mathrm{cm}^4 \end{cases}$$

$$I_{y_G} = \sum_{i=1}^{4} I_{y_G}^{(i)} = 1,111,000\,\mathrm{cm}^4$$

$$\begin{cases} I_\xi = I_{x_G} = 624,000\,\mathrm{cm}^4 \\[2mm] I_\eta = I_{y_G} = 1,111,000\,\mathrm{cm}^4 \end{cases}$$

Example 5

$$\begin{cases} I_{x_G}^{(1)} = \dfrac{4 \times 80^3}{12} = 170,667\,\mathrm{cm}^4 \\[2mm] I_{x_G}^{(2)} = I_{x_G}^{(3)} = 120 \times 10^2 = 12,000\,\mathrm{cm}^4 \end{cases}$$

$$I_{x_G} = \sum_{i=1}^{3} I_{x_G}^{(i)} = 194,667\,\mathrm{cm}^4$$

$$\begin{cases} I_{x_G}^{(1)} = 0 \\[2mm] I_{x_G}^{(2)} = I_{x_G}^{(3)} = \dfrac{4 \times 30^3}{3} = 36,000\,\mathrm{cm}^4 \end{cases}$$

$$I_{y_G} = \sum_{i=1}^{3} I_{y_G}^{(i)} = 72,000\,\mathrm{cm}^4$$

$$\begin{cases} I_{x_G y_G}^{(1)} = 0 \\[2mm] I_{x_G y_G}^{(2)} = I_{x_G y_G}^{(3)} = 120 \times 15 \times 10 = 18,000\,\mathrm{cm}^4 \end{cases}$$

$$I_\xi = \frac{194,667 + 72,000}{2} + \frac{1}{2}\left((194,667 - 72,000)^2 + 4 \times (36,000)^2\right)^{\frac{1}{2}}$$

$$= 133,000 + 71,000 = 204,000\,\mathrm{cm}^4$$

$$I_\eta = 133,000 - 71,000 = 62,000\,\mathrm{cm}^4$$

$$\vartheta_0 = \frac{1}{2}\arctan\frac{2 \times 36,000}{72,000 - 194,667} = -15.2°$$

Chapter 3

Kinematics and statics of rigid body systems

3.1 INTRODUCTION

The kinematics and statics of rigid systems are intimately connected. The movements prevented by mutual and external constraints are in fact strictly related to the reactive forces exerted by the constraints themselves. More particularly, it will be noted, and subsequently rigorously demonstrated, how the static matrix is the transpose of the kinematic one and *vice versa*, a property that will re-present itself also in other chapters devoted to elastic body systems. In every case, this property will show itself to be a consequence of the principle of virtual work, each of the two theorems implying the other.

After defining plane constraints from the twin viewpoints of kinematics and statics and investigating the concept of duality from the algebraic standpoint in depth, the same concept will be reproposed from the graphical point of view, with the presentation of various examples of statically indeterminate, statically determinate and hypostatic constraint. Particular attention will be paid to the condition of ill-disposed constraint (for which the system has a rigid deformed configuration) in the framework of the hypothesis of linearized constraints. In this case, the solution of the equilibrium equations proves to be impossible.

3.2 DEGREES OF FREEDOM OF A MECHANICAL SYSTEM

The **degrees of freedom** of a mechanical system represent the number of generalized coordinates that are necessary and sufficient to describe its configuration. A system with g degrees of freedom can thus be arranged according to ∞^g different configurations.

Consider, for instance, the case of a material point forced to move, in three-dimensional space, on a surface of equation $f(x, y, z) = 0$ (Figure 3.1). The degrees of freedom, which originally are three, are reduced to two by the constraint f, which binds the coordinates of the point. In the same way, it may be stated that, in the case where the point is forced to follow a skew curve of Equations $f_1(x, y, z) = 0, f_2(x, y, z) = 0$, the degrees of freedom are further reduced to one, the curve being a geometric variety of one dimension only.

Imagine then connecting a material point A with a fixed system of reference (e.g. the foundation) by means of a rigid rod OA (Figure 3.2) and connecting, by means of another rod AB, the point A to a second point B. If we assume that both connecting rods are not extensible, the **rigidity constraints** that these impose on the two points may be represented by the equations of two circumferences: the first, with radius l_1, centred in the origin O, and the second, with radius l_2, with the centre travelling along the first circumference:

$$x_A^2 + y_A^2 = l_1^2 \tag{3.1a}$$

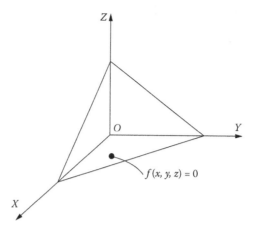

Figure 3.1

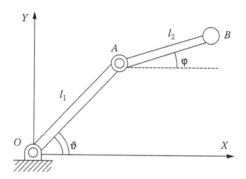

Figure 3.2

$$(x_A - x_B)^2 + (y_A - y_B)^2 = l_2^2 \tag{3.1b}$$

The number of residual degrees of freedom (two) will be given by the difference between the degrees of freedom of the two unconstrained material points (four) and the degrees of constraint (two).

In the case of three material points connected together by three connecting rods (rigidity constraints) as in Figure 3.3, we have originally six degrees of freedom. In the plane, in fact, the position of each point can be identified by two coordinates. On the other hand, the three rigidity constraints

$$(x_1 - x_2)^2 + (y_1 - y_2)^2 = l_{12}^2 \tag{3.2a}$$

$$(x_2 - x_3)^2 + (y_2 - y_3)^2 = l_{23}^2 \tag{3.2b}$$

$$(x_3 - x_1)^2 + (y_3 - y_1)^2 = l_{13}^2 \tag{3.2c}$$

reduce the degrees of freedom of the system to three ($g = 6 - 3 = 3$). Note that the system made up of three points and three connecting rods is rigid and its position in the plane can

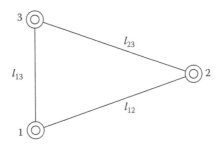

Figure 3.3

be defined once three relevant items are known (e.g. the coordinates of the centroid and the angle of orientation).

By adding a fourth point to the earlier system and connecting it to two of the previous points by means of a pair of connecting rods, we again obtain a rigid system with three degrees of freedom in the plane ($g = 8 - 5 = 3$). In fact, two new degrees of freedom are introduced, but at the same time, these are eliminated with the two connecting rods (Figure 3.4). The same happens if a fifth point is added, and so on.

Whereas in the plane, five is the minimum number of connecting rods required to connect four material points rigidly (Figure 3.4), in space this number rises to six (Figure 3.5), so as to form a tetrahedron. In fact, the original $4 \times 3 = 12$ degrees of freedom are reduced to 6, which is the number of degrees of freedom of a rigid body in three-dimensional space. The generalized coordinates of a body can therefore be considered the Cartesian coordinates of the centroid plus the three Euler angles. Often, in the pages that follow, the rigidity constraint will be, to use the term generally adopted, **linearized**. This is to say that only infinitesimal

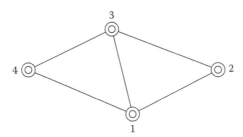

Figure 3.4

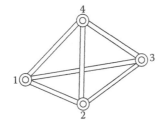

Figure 3.5

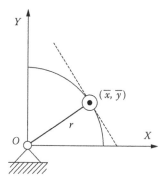

Figure 3.6

displacements about the initial configuration will be considered, and it will thus be possible to equate the circular trajectories with the rectilinear tangential ones. The simplest case is that of a material point connected to the foundation by a connecting rod. Obviously, such an elementary system has one degree of freedom and can be defined as **hypostatic**. The trajectory imposed on the point is the circumference of centre O and radius r (Figure 3.6), although, circumscribing the kinematic analysis about any initial position of coordinates $(\bar{x}, \bar{y})$, it is possible to assume as local trajectory an infinitesimal segment of the tangent

$$x\bar{x} + y\bar{y} = r^2 \tag{3.3}$$

The elementary displacements are hence considered to be vectors perpendicular to the radius vector. This statement derives from the kinematic theory of the rigid body in three-dimensional space.

As is well known from rational mechanics, the relation that links the elementary displacements of two generic points P and O of a rigid body undergoing a rototranslational motion is the following (Figure 3.7):

$$\{ds_P\} = \{ds_O\} + \{d\varphi\} \wedge \{P - O\} \tag{3.4}$$

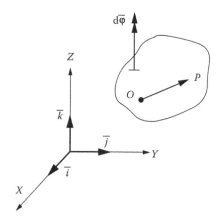

Figure 3.7

where $\{d\varphi\}$, termed **rotation vector,** is the vector that has as its axis of application, that of instantaneous rotation; as its sense, the feet–head sense of an observer who sees the body rotate counterclockwise; and as its magnitude, the value of the infinitesimal angle of rotation.

With $\bar{i}, \bar{j}, \bar{k}$ as the unit vectors of the reference axes X, Y and Z, relation (3.4) can take on the following form:

$$(u_P - u_O)\bar{i} + (v_P - v_O)\bar{j} + (w_P - w_O)\bar{k} \tag{3.5}$$

$$= \det \begin{bmatrix} \bar{i} & \bar{j} & \bar{k} \\ \varphi_x & \varphi_y & \varphi_z \\ (x_P - x_O) & (y_P - y_O) & (z_P - z_O) \end{bmatrix}$$

where the determinant of the formal matrix on the right-hand side provides the components of the vector product that appears in relation (3.4), while u, v, w indicate the components along the axes X, Y, Z of the elementary displacements, and φ_x, φ_y, φ_z indicate the components of the rotation vector.

Relation (3.5) can alternatively be presented in the form of a product of an antisymmetric matrix, called the **rotation matrix,** for the position vector of the point P with respect to the point O,

$$\{ds_P\} - \{ds_O\} = \begin{bmatrix} 0 & -\varphi_z & \varphi_y \\ \varphi_z & 0 & -\varphi_x \\ -\varphi_y & \varphi_x & 0 \end{bmatrix} \begin{bmatrix} x_P - x_O \\ y_P - y_O \\ z_P - z_O \end{bmatrix} \tag{3.6}$$

In the particular case of an elementary rotation of a two-dimensional rigid body in its XY plane (Figure 3.8), Equation 3.5 is particularized as follows:

$$(u_P - u_O)\bar{i} + (v_P - v_O)\bar{j} = \begin{bmatrix} \bar{i} & \bar{j} & \bar{k} \\ 0 & 0 & \varphi_z \\ (x_P - x_O) & (y_P - y_O) & 0 \end{bmatrix} \tag{3.7}$$

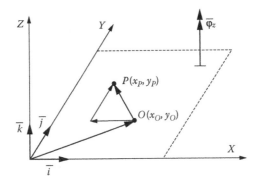

Figure 3.8

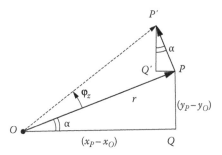

Figure 3.9

Evaluating the determinant of the symbolic matrix on the right-hand side of the equation, we obtain

$$u_P - u_O = -(y_P - y_O)\varphi_z \tag{3.8a}$$

$$\upsilon_P - \upsilon_O = (x_P - x_O)\varphi_z \tag{3.8b}$$

Equations 3.8 are linear relations that describe the concept of linearization of the constraint algebraically, already introduced previously on a more intuitive basis. A geometrical interpretation (Figure 3.9) of Equation 3.8 may then be given, considering the elementary rotation φ_z of P about O. The point P will move to P' at a distance equal to $r\varphi_z$ (except for infinitesimals of a higher order). The horizontal relative displacement, represented in Figure 3.9 by segment PQ', thus equals

$$u_P - u_O = PQ' = -r\varphi_z \sin\alpha \tag{3.9}$$

Since triangles OQP and $P'Q'P$ are similar, it follows that $r \sin\alpha = (y_P - y_O)$, so that from Equation 3.9 we obtain again Equation 3.8a. In the same way, it is possible to verify also the meaning of the linear relation (3.8b).

3.3 KINEMATIC DEFINITION OF PLANE CONSTRAINTS

The constraints that we shall hereafter assume to be connecting the plane rigid body to the fixed reference system are referred to as **external constraints**. These can be classified on the basis of the elementary movements of the constrained point P that can be prevented, these movements consisting of the two translations u_P and υ_P and of the elementary rotation φ_P. The subscript P has been applied to the latter quantity, even though this is a characteristic of the act of rigid motion and thus of each point of the body.

There is thus created a hierarchy of constraints, from those that restrain the body more weakly (single constraints) to those that more effectively inhibit its movements (triple constraints or fixed joints).

The simplest kind of constraint, frequently used in technical applications, is the **roller support** (Figure 3.10a) or the **connecting rod** (Figure 3.10b). This constraint imposes a movement along the straight line p on the point P. It should be noted that in the case of the

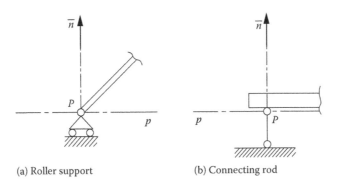

(a) Roller support   (b) Connecting rod

Figure 3.10

connecting rod of Figure 3.10b, the constraint is linearized, as was clarified in the previous section. In mathematical terms, the symbols of Figure 3.10 impose the following condition:

$$\{ds_P\}^T\{n\} = 0 \qquad (3.10)$$

i.e., the elementary displacement of the point P cannot present components different from zero on the straight line n, perpendicular to the straight line p. The elementary rotation φ_z can, on the other hand, be different from zero. Since Equation 3.10 is a scalar relation (it represents, in fact, the scalar product of the displacement and the unit vector of the straight line n), it can be stated that the roller support or the connecting rod is a single constraint. Single constraints require that any centre of instantaneous rotation must lie on a straight line. In the case of the roller support or the connecting rod, the centre of instantaneous rotation must lie on the straight line n, since this is perpendicular to the instantaneous trajectory p.

Another constraint that has a wide application is the **hinge** (Figure 3.11a). This imposes on the constrained point P to remain fixed in the plane, while the elementary rotation φ_z can be about the same point P, which thus coincides with the centre of instantaneous rotation. In mathematical terms,

$$\{ds_P\} = \{0\} \qquad (3.11)$$

The foregoing condition is of a vector nature and thus the hinge can be classified as a double constraint. We may then consider that, by suitably combining two single constraints, the result for the body can be a double constraint. The most typical case is that of two non-parallel connecting rods (Figure 3.11b), the axes of which come together in the centre of

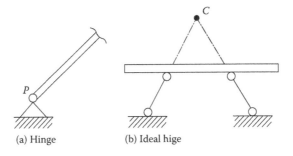

(a) Hinge   (b) Ideal hige

Figure 3.11

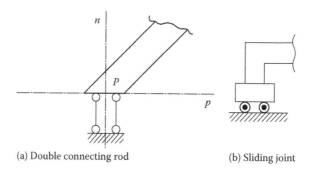

(a) Double connecting rod　　　　　(b) Sliding joint

Figure 3.12

instantaneous rotation C. Each connecting rod in fact conditions the centre to belong to its axis, and the two conditions are compatible, except when the connecting rods are parallel, in which case the centre would be at infinity. The centre C is said to constitute an **ideal hinge**.

When the two connecting rods are parallel, the centre of instantaneous rotation coincides with the point at infinity of the axes of the connecting rods (Figure 3.12a). This means that the rigid motion becomes one of pure translation in the p direction perpendicular to the axis n. In mathematical terms, we have the following two scalar conditions:

$$\{ds_P\}^T \{n\} = 0 \tag{3.12a}$$

$$\varphi_z = 0 \tag{3.12b}$$

The **double connecting rod** is thus a double constraint and can be represented, in an altogether equivalent way, by a **sliding joint** (Figure 3.12b).

The only triple constraint of a local type is the **fixed joint** or **built-in support** (Figure 3.13a), which, by definition, prevents all three movements (hence, no centre of rotation exists):

$$\{ds_P\} = \{0\} \tag{3.13a}$$

$$\varphi_z = 0 \tag{3.13b}$$

Obviously, the triple constraint may be obtained by suitably combining a double constraint with a single one (Figure 3.13b).

In the hierarchy of plane constraints outlined earlier, just one case is missing, one that, as a rule, is rarely applied in building practice, but may be defined mathematically and may prove useful in applications of the principle of virtual work. This constraint is that which allows motion of translation in all directions, while it inhibits rotation

$$\varphi_z = 0 \tag{3.14}$$

This can be represented by a **double articulated parallelogram** (Figure 3.14) and is a single constraint that obliges the centre of instantaneous rotation to remain on the straight line at infinity.

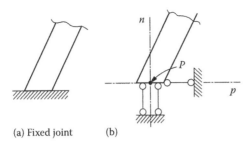

(a) Fixed joint   (b)

Figure 3.13

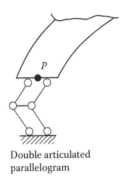

Double articulated
parallelogram

Figure 3.14

So far, only point constraints have been introduced, i.e., constraints concentrated in one point or, at the most, acting within an infinitesimal area of the body. The only exception is that of the ideal hinge (Figure 3.11b) in which two connecting rods are applied, at a finite distance from one another. On the other hand, in building practice, the constraints are arranged in different points of the structural element, in such a way as to prevent its movement. When the constraints are insufficient to fix the position of the rigid body in the plane, the constraint condition is said to be **hypostatic**. In the case of Figure 3.15a, for instance, a single hinge allows rotation of the plate constrained by it about its axis. In the case of Figure 3.15b, a hinge and a roller support prove to constitute a strictly sufficient condition to hold the body in position; the body is then said to be constrained **isostatically**. While, in fact, the hinge A forces any rotation to be centred in A, the roller support B forces the centre to lie along the normal n. These two conditions are evidently mutually incompatible, as the point A does not lie along the line n. Likewise, it may be observed how the hinge A imposes the trajectory n on the point B, while the roller support B imposes the trajectory p on the same point B. These two trajectories are again incompatible, in that they have in common only the point B. In the case of Figure 3.15c, finally, the two hinges constitute a redundant condition of constraint. The body is then said to be **hyperstatically constrained**.

On the other hand, it should at once be pointed out that, if the constraints are not suitably disposed, they can lose their effectiveness. Thus, it is those bodies that are apparently isostatic or even hyperstatic may then in fact prove to be hypostatic. The hinge and the roller support of Figure 3.16a, e.g., are **ill-disposed** constraints, because the centre of rotation

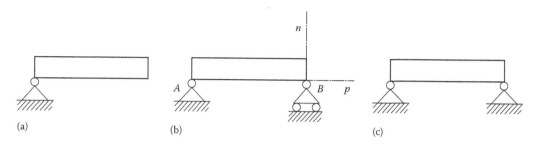

Figure 3.15

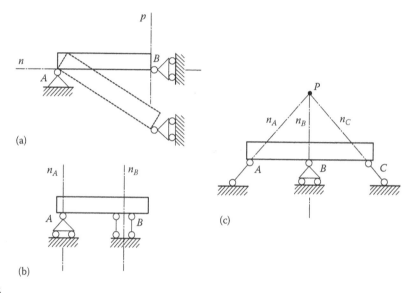

Figure 3.16

allowed by the roller support B can come to find itself on the straight line n, which passes through the hinge A. Hence, it is possible for the centre of rotation to be in A.

In other words, the trajectories imposed on the point B by both constraints coincide in the straight line p. Likewise, the roller support and the double connecting rod of Figure 3.16b are ill-disposed, because both of them permit horizontal translations of the body. The centre of rotation will thus be at infinity, as the meeting point of the normals n_A and n_B. Finally, the case illustrated in Figure 3.16c is yet another example of ill-disposition of the constraints in that the three normals, n_A, n_B and n_c, all meet at the point P, i.e., there is a centre of instantaneous rotation at the point where all three lines meet, since this centre must belong to each one of them.

Just as external constraints impose particular conditions on absolute elementary displacements of points held to the foundation, **internal constraints** impose similar conditions on the relative elementary displacements of points belonging to different rigid bodies. Consider the case of the **internal connecting rod** (Figure 3.17). This prevents discontinuity of displacement in the direction of its axis n:

$$\{ds_A - ds_B\}^T\{n\} = 0 \tag{3.15a}$$

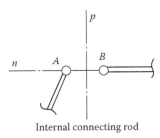

Internal connecting rod

Figure 3.17

while it allows discontinuity of displacement in the perpendicular direction p:

$$\{ds_A - ds_B\}^T\{p\} \neq 0 \tag{3.15b}$$

just as it allows discontinuity of rotation, i.e., relative rotation:

$$\varphi_A - \varphi_B \neq 0 \tag{3.15c}$$

Any centre of relative rotation must lie on the line n. Internal constraints may also be considered, in a complementary way, as **disconnections** of degree $(3 - g)$, where g is the degree of constraint; i.e., it is possible to start from a single rigid body that has two parts that are firmly joined and to reduce it, *via* elementary disconnections, to the case under examination. In this context, the connecting rod is a single constraint and, at the same time, a double disconnection.

The **internal hinge** eliminates any relative displacement of the two points that it connects, while it allows relative rotation (Figure 3.18):

$$\{ds_A - ds_B\} = \{0\} \tag{3.16a}$$

$$\varphi_A - \varphi_B \neq 0 \tag{3.16b}$$

Any centre of relative rotation will coincide with the hinge, which is in fact a double constraint and, at the same time, a single disconnection.

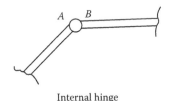

Internal hinge

Figure 3.18

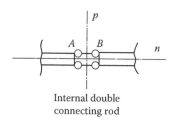

Internal double
connecting rod

Figure 3.19

The **double internal rod** (Figure 3.19) eliminates relative displacement in the direction n as well as relative rotation:

$$\{ds_A - ds_B\}^T \{n\} = 0 \tag{3.17a}$$

$$\varphi_A - \varphi_B = 0 \tag{3.17b}$$

while it allows the relative displacement in the perpendicular direction p:

$$\{ds_A - ds_B\}^T \{p\} \neq 0 \tag{3.17c}$$

Any centre of relative rotation will be the point at infinity of the straight line n, and thus the only relative motion allowed will be the translational one in the p direction. The double rod is a double constraint or can be considered as a single disconnection. The **internal fixed joint** (Figure 3.20) firmly joins one portion of the body with the other and is a triple constraint. In other words, there is no disconnection.

To bring our discussion of internal constraints to a close, let us consider the **internal double articulated parallelogram** (Figure 3.21), which allows any relative translation but

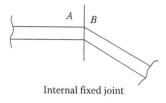

Internal fixed joint

Figure 3.20

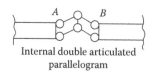

Internal double articulated
parallelogram

Figure 3.21

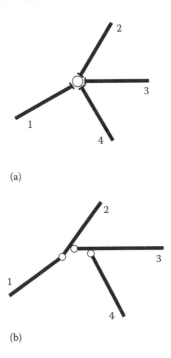

(a)

(b)

Figure 3.22

not relative rotation. It is thus a single constraint or a double disconnection. The relative centre lies on the straight line at infinity and can be any point on that line.

The internal hinge often connects more than two elements (Figure 3.22a), in such a way that the relative centres all come to coincide with it. In these cases, it is logical to consider $(n - 1)$ mutual connections, if n is the number of the connected elements (Figure 3.22b). In the case, therefore, of four bars mutually connected as in Figure 3.22a, the residual degrees of freedom are $g = (4 \times 3) - (3 \times 2) = 6$. We have, in fact, 12 original degrees of freedom and 6 degrees of constraint. More synthetically, it is possible to arrive at the same result again by considering the Cartesian coordinates of the hinge plus the four angles of orientation of the bars as generalized coordinates of the system. In the same way, the roller support of Figure 3.23, with three-hinged bars, forms a mechanical system having four degrees of freedom. A less immediate calculation again gives $g = (3 \times 3) - (2 \times 2) - 1 = 4$, as there are nine original degrees of freedom, and five degrees suppressed by two double (internal) constraints and one single (external) constraint. In the case where the bars are built into one another so as to form one or more closed configurations, the system is said to be internally hyperstatic,

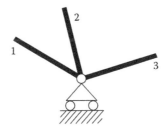

Figure 3.23

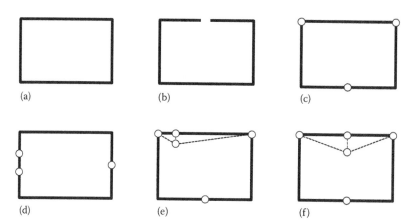

Figure 3.24

and there will be three degrees of hyperstaticity for each closed configuration. For example, for the rectangle of Figure 3.24a, formed by four bars built into one another, the calculation $g = (4 \times 3) - (4 \times 3) = 0$ does not hold good, as it would suggest, quite falsely, that the system is isostatic. In actual fact, the doubly connected body presents three external degrees of freedom (it is not externally constrained), and three degrees of internal indeterminacy, since the various parts would remain firmly joined even if a triple disconnection (i.e. a cut) were made in any point of the axis (Figure 3.24b). The triple disconnection could, on the other hand, be distributed in two or three points (Figure 3.24c and d). It is necessary, however, to take care in arranging such disconnections. The hinge on the axis of the connecting rod or the three aligned hinges, e.g., would be ill-disposed, with the consequent development of an internal mechanism (Figure 3.24e and f).

Two bars connected both to the foundation and to one another by three hinges constitute a fundamental isostatic scheme called a **three-hinged arch** (Figure 3.25). Naturally, for the same reasons already seen previously, the three hinges must not be collinear; otherwise, a mechanism is created. By inserting a fourth hinge (single disconnection), a mechanism with one degree of freedom is obtained (Figure 3.26); this is known as an **articulated parallelogram** ($g = 9 - 8 = 1$). Alternatively, yet another simple mechanism is obtained by

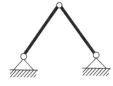

Figure 3.25

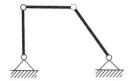

Figure 3.26

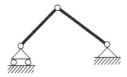

Figure 3.27

disconnecting, with respect to the horizontal translation, one of the two hinges connected to the foundation ($g = 6 - 5 = 1$). In this case, the **crank mechanism** of Figure 3.27 is obtained.

3.4 ALGEBRAIC STUDY OF KINEMATICS OF RIGID BODY SYSTEMS

We shall now approach, from the algebraic point of view, the problem of the kinematics of rigid body systems, thus giving a rigorous interpretation to the degenerate case of the ill-disposition of constraints. Progressively more complex examples will be introduced, with an increasing number of bodies.

As a first case, consider the L-shaped beam of Figure 3.28a. This is constrained by a roller support hinged in A and by a hinge in B. The unknowns of the problem will be the displacements of a representative point of the rigid body, chosen arbitrarily, referred to as the **pole** or **centre of reduction**. This point is chosen to fall at the convergence O of the two bars. A Cartesian system is chosen with centre in O and axes parallel to the two bars. The unknowns are then u_O, v_O, φ_O, i.e., the two elementary translations of the point O with reference to the axes X and Y and the elementary rotation of the entire rigid body. On the other hand, there are also three constraint equations:

$$v_A = 0 \tag{3.18a}$$

$$u_B = 0 \tag{3.18b}$$

$$v_B = 0 \tag{3.18c}$$

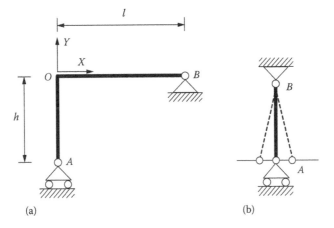

(a)   (b)

Figure 3.28

The task, therefore, will be to render explicit Equations 3.18 as a function of the elementary displacements of the pole. Applying Equations 3.8, we have

$$v_A = v_O + (x_A - x_O)\varphi_O = v_O \tag{3.19a}$$

$$u_B = u_O - (y_B - y_O)\varphi_O = u_O \tag{3.19b}$$

$$v_B = v_O + (x_B - x_O)\varphi_O = v_O + l\varphi_O \tag{3.19c}$$

and thus the constraint equations become

$$v_O = 0 \tag{3.20a}$$

$$u_O = 0 \tag{3.20b}$$

$$v_O + l\varphi_O = 0 \tag{3.20c}$$

The system of linear algebraic equations (3.20) may be resolved very quickly by substitution, thus affording the obvious or trivial solution. The displacements of the pole O are all zero, which means that the constraints are well disposed. Since, however, it is our purpose to reason in terms that are more general and that can be more readily extrapolated to cases of greater complexity, it is necessary to introduce the matrix notation

$$\begin{bmatrix} 0 & 1 & 0 \\ 1 & 0 & 0 \\ 0 & 1 & l \end{bmatrix} \begin{bmatrix} u_O \\ v_O \\ \varphi_O \end{bmatrix} = \begin{bmatrix} 0 \\ 0 \\ 0 \end{bmatrix} \tag{3.21}$$

It is well known from linear algebra that a necessary and sufficient condition for a homogeneous system of linear algebraic equations to admit of the obvious solution is that the determinant D of the matrix of the coefficients should be different from zero. In the case of the system of Equations 3.21, we have $D = -l$. When $l > 0$, as appears in Figure 3.28a, the solution is thus the trivial one. On the other hand, carrying out a parametric study on the variable l, we find that, for $l = 0$, the kinematic solutions become ∞^1 and the system becomes a mechanism, since an ill-disposition of constraints is produced. The normal n to the plane in which the roller support moves, in fact, in this case contains the hinge B (Figure 3.28b), so that a centre of instantaneous rotation arises in the same point B.

As a second example, consider again the L-shaped beam met with earlier, this time, however, restrained in A with a roller support having its plane of movement inclined at the angle α with respect to the vertical direction (Figure 3.29a). The first of Equations 3.18 will then be substituted by the following:

$$\{ds_A\}^T\{n\} = 0 \tag{3.22a}$$

or more explicitly by

$$u_A \cos\alpha + v_A \sin\alpha = 0 \tag{3.22b}$$

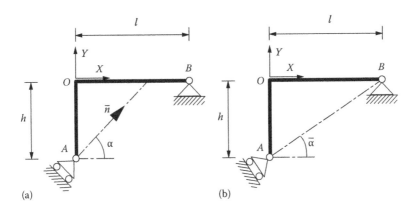

Figure 3.29

Writing the left-hand side expression as a function of the displacements of the pole O, we have

$$(u_O + h\varphi_O)\cos\alpha + v_O \sin\alpha = 0 \tag{3.22c}$$

In the matrix of the coefficients that appears in Equation 3.21, only the first line will vary with respect to the previous case:

$$\begin{bmatrix} \cos\alpha & \sin\alpha & h\cos\alpha \\ 1 & 0 & 0 \\ 0 & 1 & l \end{bmatrix} \begin{bmatrix} u_O \\ v_O \\ \varphi_O \end{bmatrix} = \begin{bmatrix} 0 \\ 0 \\ 0 \end{bmatrix} \tag{3.23}$$

The determinant of the matrix of the coefficients has the value $D = -(l \sin\alpha - h \cos\alpha)$ and is generally different from zero. Assuming as parameter the angle α, we find that D vanishes for $\alpha = \bar{\alpha} = \arctan(h/l)$. In this case, in fact, the plane in which the roller support moves is set perpendicular to the line joining A and B (Figure 3.29b).

We now examine the case of a portal frame made up of two L-shaped beams, connected together by a hinge and to the foundation by a hinge and by a horizontal double rod, respectively (Figure 3.30a). The points A and C of external constraint are assumed as poles.

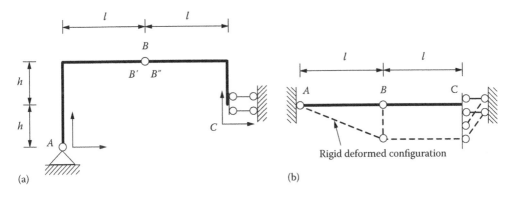

Figure 3.30

The six kinematic unknowns will thus consist of the elementary displacements of the two poles: u_A, v_A, φ_A, u_C, v_C and φ_C. On the other hand, the six constraint equations will impose

$$u_A = 0 \tag{3.24a}$$

$$v_A = 0 \tag{3.24b}$$

$$u_{B'} - u_{B''} = 0 \tag{3.24c}$$

$$v_{B'} - v_{B''} = 0 \tag{3.24d}$$

$$u_C = 0 \tag{3.24e}$$

$$\varphi_C = 0 \tag{3.24f}$$

While the unknowns of the problem appear directly in the first and last pairs of equations, the intermediate ones, which exclude relative displacements in B, must be expressed in terms of the generalized displacements of the poles A and C. Applying Equations 3.8, the displacements of the central ends B' and B'' of the two beams are given by

$$u_{B'} = u_A - 2h\varphi_A \tag{3.25a}$$

$$v_{B'} = v_A + l\varphi_A \tag{3.25b}$$

$$u_{B''} = u_C - h\varphi_C \tag{3.25c}$$

$$v_{B''} = v_C - l\varphi_C \tag{3.25d}$$

Substituting Equations 3.25 into Equations 3.24c and 3.24d), we obtain

$$u_A = 0 \tag{3.26a}$$

$$v_A = 0 \tag{3.26b}$$

$$u_A - 2h\varphi_A - u_C + h\varphi_C = 0 \tag{3.26c}$$

$$v_A + l\varphi_A - v_C + l\varphi_C = 0 \tag{3.26d}$$

$$u_C = 0 \tag{3.26e}$$

$$\varphi_C = 0 \tag{3.26f}$$

and, in matrix form

$$
\begin{bmatrix}
1 & 0 & 0 & 0 & 0 & 0 \\
0 & 1 & 0 & 0 & 0 & 0 \\
1 & 0 & -2h & -1 & 0 & h \\
0 & 1 & l & 0 & -1 & l \\
0 & 0 & 0 & 1 & 0 & 0 \\
0 & 0 & 0 & 0 & 0 & 1
\end{bmatrix}
\begin{bmatrix}
u_A \\
v_A \\
\varphi_A \\
u_C \\
v_C \\
\varphi_C
\end{bmatrix}
=
\begin{bmatrix}
0 \\
0 \\
0 \\
0 \\
0 \\
0
\end{bmatrix}
\tag{3.27}
$$

The determinant of the matrix of the coefficients has the value $D = -2h$. For $h = 0$, this vanishes and we end up with an ill-disposition of the constraints, as emerges from the rigid deformed configuration of Figure 3.30b. In this case, it may be noted how the two absolute centres and the relative one are aligned.

If the portal frame just considered is further constrained in C, by adding a vertical connecting rod, the constraint equations become seven, while the kinematic unknowns remain the six previously introduced. In the case of $h > 0$ (Figure 3.31a), the solution will still be the obvious one. On the other hand, even when $h = 0$ (Figure 3.31b), the solution is the obvious one, because, from the matrix of the coefficients (7×6), it will be possible to extract a nonzero minor of order 6, since the added line is not linearly dependent on the others.

If, instead, one of the two rods in C of the portal frame of Figure 3.30a is suppressed, the constraint equations are reduced to five. In the case of $h > 0$ (Figure 3.32a), it will be possible to extract a nonzero minor of order 5, and thus the solutions will be ∞^1 and the system will have one degree of freedom. On the other hand, when $h = 0$ (Figure 3.32b), it will be possible to extract a nonzero minor of order 4, and thus the solutions will be ∞^2, which means that the mechanical system is hypostatic to the second degree. In fact, two angles of rotation will be necessary, e.g., the absolute angle ϑ and the relative angle φ (Figure 3.32b), to describe its deformed configuration.

Now imagine parameterizing, in the portal frame of Figure 3.33, the angle of orientation α of the double connecting rod in C. Thus, only Equation 3.26e is altered:

$$
u_C \cos\alpha + v_C \sin\alpha = 0
\tag{3.28}
$$

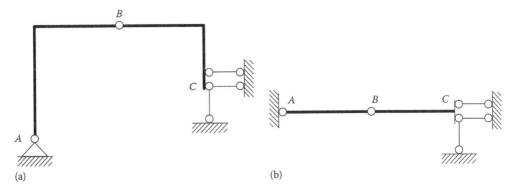

(a)   (b)

Figure 3.31

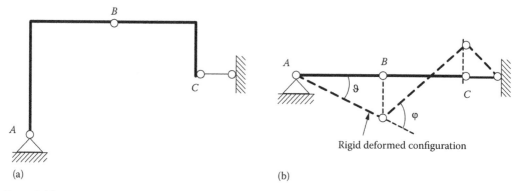

Figure 3.32

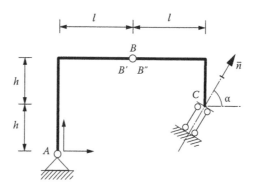

Figure 3.33

and, consequently, in the matrix of the coefficients (3.27), only the fifth row varies:

$$
\begin{bmatrix}
1 & 0 & 0 & 0 & 0 & 0 \\
0 & 1 & 0 & 0 & 0 & 0 \\
1 & 0 & -2h & -1 & 0 & h \\
0 & 1 & l & 0 & -1 & l \\
0 & 0 & 0 & \cos\alpha & \sin\alpha & 0 \\
0 & 0 & 0 & 0 & 0 & 1
\end{bmatrix}
\begin{bmatrix}
u_A \\
v_A \\
\varphi_A \\
u_C \\
v_C \\
\varphi_C
\end{bmatrix}
=
\begin{bmatrix}
0 \\
0 \\
0 \\
0 \\
0 \\
0
\end{bmatrix}
\tag{3.29}
$$

The determinant of the matrix of the coefficients has the value $D = -2h \cos \alpha + l \sin \alpha$ and vanishes for $\alpha = \bar{\alpha} = \arctan(2h/l)$, i.e., when the double connecting rod is parallel to the line joining A and B. In this case, the constraints are ill-disposed, since, e.g., the hinge A on the one hand, and the double rod C on the other, allows the point B the same trajectory about the original configuration, *viz.* the normal for B to the line joining A and B. Also in this example, when a condition of ill-disposition of the constraints and consequently of freedom is reached, the absolute centres and the relative centre are aligned.

Finally, consider a structure made up of three rigid parts, connected together with hinges and attached to the foundation by a vertical double rod, a hinge and a roller support (Figure 3.34). There are, in all, nine original degrees of freedom and nine degrees of constraints (four double constraints and one single constraint). We shall study the system,

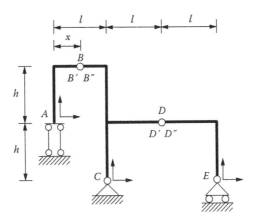

Figure 3.34

assuming as variable parameter the distance x of the internal hinge B from the left upper vertex of the structure. It is convenient to choose the point constrained to the foundation as pole of each rigid part, so that the nine kinematic unknowns will be

$$u_A, v_A, \varphi_A, u_C, v_C, \varphi_C, u_E, v_E, \varphi_E$$

The constraint conditions are the following:

$$v_A = 0 \tag{3.30a}$$

$$\varphi_A = 0 \tag{3.30b}$$

$$u_{B'} - u_{B''} = 0 \tag{3.30c}$$

$$v_{B'} - v_{B''} = 0 \tag{3.30d}$$

$$u_C = 0 \tag{3.30e}$$

$$v_C = 0 \tag{3.30f}$$

$$u_{D'} - u_{D''} = 0 \tag{3.30g}$$

$$v_{D'} - v_{D''} = 0 \tag{3.30h}$$

$$v_E = 0 \tag{3.30i}$$

In the third, fourth, seventh and eighth, the kinematic parameters of the poles do not appear directly, and thus it will be necessary to have recourse to Equations 3.8:

$$u_{B'} = u_A - h\varphi_A \tag{3.31a}$$

$$v_{B'} = v_A + x\varphi_A \tag{3.31b}$$

$$u_{B''} = u_C - 2h\varphi_C \tag{3.31c}$$

$$\upsilon_{B''} = \upsilon_C + (x - l)\varphi_C \tag{3.31d}$$

$$u_{D'} = u_C - h\varphi_C \tag{3.31e}$$

$$\upsilon_{D'} = \upsilon_C + l\varphi_C \tag{3.31f}$$

$$u_{D''} = u_E - h\varphi_E \tag{3.31g}$$

$$\upsilon_{D'} = \upsilon_E - l\varphi_E \tag{3.31h}$$

The equations of the problem appear, then, as follows:

$$\upsilon_A = 0 \tag{3.32a}$$

$$\varphi_A = 0 \tag{3.32b}$$

$$u_A - h\varphi_A - u_C + 2h\varphi_C = 0 \tag{3.32c}$$

$$\upsilon_A + x\varphi_A - \upsilon_C + (l - x)\varphi_C = 0 \tag{3.32d}$$

$$u_C = 0 \tag{3.32e}$$

$$\upsilon_C = 0 \tag{3.32f}$$

$$u_C - h\varphi_C - u_E + h\varphi_E = 0 \tag{3.32g}$$

$$\upsilon_C + l\varphi_C - \upsilon_E + l\varphi_E = 0 \tag{3.32h}$$

$$\upsilon_E = 0 \tag{3.32i}$$

and in matrix form, we have

$$
\begin{bmatrix}
0 & 1 & 0 & 0 & 0 & 0 & 0 & 0 & 0 \\
0 & 0 & 1 & 0 & 0 & 0 & 0 & 0 & 0 \\
1 & 0 & -h & -1 & 0 & 2h & 0 & 0 & 0 \\
0 & 1 & x & 0 & -1 & l-x & 0 & 0 & 0 \\
0 & 0 & 0 & 1 & 0 & 0 & 0 & 0 & 0 \\
0 & 0 & 0 & 0 & 1 & 0 & 0 & 0 & 0 \\
0 & 0 & 0 & 1 & 0 & -h & -1 & 0 & h \\
0 & 0 & 0 & 0 & 1 & l & 0 & -1 & l \\
0 & 0 & 0 & 0 & 0 & 0 & 0 & 1 & 0
\end{bmatrix}
\begin{bmatrix}
u_A \\
\upsilon_A \\
\varphi_A \\
u_C \\
\upsilon_C \\
\varphi_C \\
u_E \\
\upsilon_E \\
\varphi_E
\end{bmatrix}
=
\begin{bmatrix}
0 \\
0 \\
0 \\
0 \\
0 \\
0 \\
0 \\
0 \\
0
\end{bmatrix}
\tag{3.33}
$$

The determinant of the matrix of the coefficients has the value $D = (l - x)l$ and vanishes for $x = l$, when the hinge B is at the right end of the upper horizontal beam. In the next section, we shall discuss the type of mechanism that is created in this last particular case (Figure 3.39).

3.5 GRAPHICAL STUDY OF KINEMATICS OF SYSTEMS HAVING ONE DEGREE OF FREEDOM

The kinematic study of systems having one degree of freedom, herein also referred to as **kinematic chains,** will be proposed in this section from the graphical point of view. Also certain cases of the foregoing section will be taken up again, with the aim of defining the displacement vector field and the deformed configuration of the rigid bodies (here all assumed as being one-dimensional) that make up the chain.

The graphical study of kinematic chains is founded on two theorems, which, here for reasons of brevity, will not be demonstrated but only stated.

First theorem of kinematic chains (applicable when the chain is made up of at least two rigid bodies): A necessary and sufficient condition for the mechanical system to be hypostatic is that, for each pair of bodies i and j, the absolute centres of rotation C_i and C_j and the relative centre C_{ij} should be aligned.

Second theorem of kinematic chains (applicable when the chain is made up of at least three rigid bodies): A necessary and sufficient condition for the mechanical system to be hypostatic is that, for each group of three bodies i, j and k, the three relative centres C_{ij}, C_{jk} and C_{ki} should be aligned.

Let us consider the crank mechanism of Figure 3.35a in its initial configuration and let us proceed to study its elementary movements about that configuration. The absolute centre of rotation C_1 of the connecting rod I coincides with the point hinged to the foundation, just as the relative centre C_{12} will coincide with the internal hinge. The absolute centre of the bar II is not known *a priori*. Since one end of this bar is hinged to a roller support, the straight line n on which that centre must lie is, however, known. By the first theorem of kinematic chains,

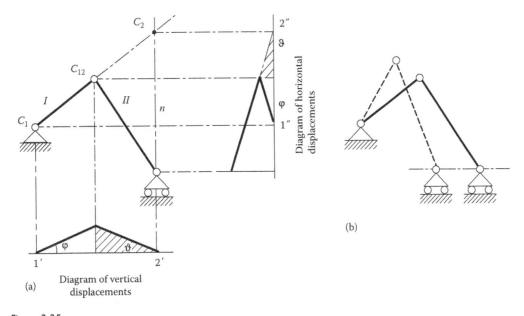

(a) Diagram of vertical displacements

(b)

Figure 3.35

the centre C_2 must then be collinear to centres C_1 and C_{12}. We are thus provided with the information indicating that centre C_2 belongs to two different straight lines:

$$C_2 \in n, C_2 \in (C_1, C_{12})$$

so that it must consequently lie at the intersection of these two lines (Figure 3.35a).

Once the centres of instantaneous rotation have been identified, these and the other notable points of the system will be projected onto two orthogonal lines (Figure 3.35a). It will thus be possible to draw the diagrams of the horizontal and vertical displacements, respectively, on the vertical and horizontal fundamental lines. To be able to do this, we shall, then, have to take into consideration the linear equations (3.8), the points where the displacement functions vanish (coinciding with the projections of the absolute centres) and the conditions of continuity imposed by the internal constraints on the displacements and the rotations.

The actual procedure will be to draw a segment of a line, inclined at an arbitrary angle φ with respect to the horizontal fundamental line (Figure 3.35a), which will represent the vertical displacements of the corresponding points projected from the connecting rod *I*. As regards the bar *II*, the projection of the absolute centre is known, as is also the vertical displacement of the end hinged to the connecting rod *I*. The internal hinge, in fact, prevents relative displacements in that point, i.e., it imposes the continuity of the vertical (and horizontal) displacement function. It may be stated, therefore, that the vertical displacements of the points of the bar *II* will be represented by the segment of straight line that joins the projection of the centre C_2 to the right end of the previous linear diagram. This segment is thus rotated clockwise by an angle ϑ, while the rod *I* is rotated counterclockwise by an angle φ.

The horizontal displacements can then be read in reference to the vertical fundamental line (Figure 3.35a). Draw a segment rotated counterclockwise by an angle φ about the projection of the centre C_1. Next, consider the line that joins the projection of centre C_2 to the point representative of the horizontal displacement of the internal hinge. The horizontal projection of bar *II* on this line represents the horizontal displacements of the points of the bar itself. Note that the rotations on the two displacement diagrams must be the same for each rigid body. It can be demonstrated in fact, by exploiting the similitude of the hatched triangles in Figure 3.35a, that also on the diagram of the horizontal displacements the bar *II* rotates clockwise by the angle ϑ.

To conclude, it may be of interest for a verification of the results to draw the deformed configuration of the system on the basis of the diagrams obtained. The procedure is to compose horizontal and vertical displacements of the notable points of the chain and to reconnect these points, transformed by the movement, with straight line segments. In the case of the crank mechanism of Figure 3.35a, the internal hinge is displaced leftwards and upwards (by amounts that can be deduced from the diagrams described earlier), while the roller support moves towards the left, dragged by the rotation of the connecting rod *I*. Of course, when we come to draw the deformed configuration, we must bear in mind that the linearization of the constraints, which is justified in a context of infinitesimal displacements, instead deforms rigid bodies when these displacements are amplified to meet the requirements of graphical clarity. In the crank mechanism, e.g., the connecting rod *I* appears dilated owing to the movement (Figure 3.35b).

As a second example, consider the hypostatic arch of Figure 3.36a. This time the roller support moves along a vertical plane. The absolute centre C_2 is identifiable as the intersection of the normal to the plane in which the roller support moves with the line joining the centres C_1 and C_{12}. With the absolute and relative centres projected on the two fundamental lines, and taking into account the conditions of continuity imposed on the displacements by the internal constraint, we can proceed to draw the diagrams of the horizontal and vertical

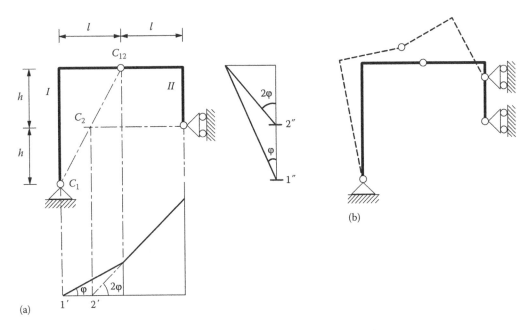

Figure 3.36

displacements. It may be noted (Figure 3.36a) how the absolute centres of rotation undergo null displacements (this is true by definition), just as the extreme points of the two rigid bodies connected by the internal hinge are displaced by the same amounts, both vertically and horizontally. Both bodies rotate counterclockwise, the first by an arbitrary (and infinitesimal) angle φ and the second by an angle 2φ, twice as much on account of the particular geometrical proportions of the system (the upright on the left is twice as long as the one on the right). These angles emerge from both diagrams of horizontal and vertical displacements. While, therefore, the body *I* rotates about the external hinge, the body *II* is dragged in such a way as to rotate twice as much, pulling the roller support upwards with a displacement equal to $3\varphi l$ (Figure 3.36b).

We are now confronted with the case of a hypostatic arch (Figure 3.37) consisting of two rigid L-shaped bodies, connected together by a horizontal double rod and to the foundation with a hinge and a roller support, respectively. Also in this case, the centre C_2 is obtained as the intersection of the normal to the plane of movement of the roller support with the line joining C_1 and $C_{12}(\infty)$. Since $C_{12}(\infty)$ is the point at infinity of the horizontal lines, the line joining C_1 and C_{12} is the horizontal line passing through C_1. With the absolute centres projected on the two fundamentals, the two rigid bodies must rotate by the same angle (since the double rod does not allow relative rotations), while the horizontal displacement cannot present discontinuity in correspondence with the double rod. There is, instead, a discontinuity in the diagram of the vertical displacement equal to $2\varphi l$, so that the end to the left of the double rod, belonging to body *I*, will be displaced upwards (as well as leftwards), while the end to the right, belonging to body *II*, will be displaced downwards (as well as leftwards). The roller support will, meanwhile, be displaced by an amount φh towards the right.

So far we have considered systems that are hypostatic for manifest insufficiency of constraints. We shall now consider two cases (already met with in the previous section) that concern structures having ill-disposed constraints. The difference, therefore, in this case lies in the fact that the initial configuration is not arbitrary, as instead occurs in the case of crank mechanisms and, in general, for all mechanisms having one degree of freedom.

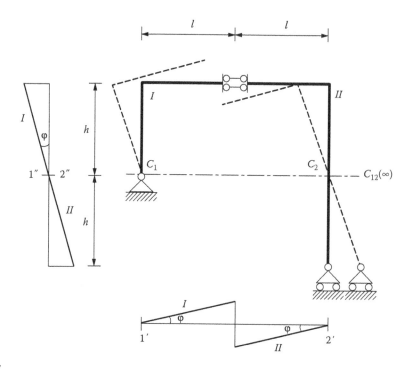

Figure 3.37

Consider the arch with ill-disposed constraints of Figure 3.38, already studied algebraically earlier. The three centres, two of which absolute, C_1 and $C_2(\infty)$, and one relative, C_{12}, as stated earlier, are aligned and, thus, according to the first theorem of kinematic chains, the structure is hypostatic. The projections of the centre $C_2(\infty)$ fall at the points at infinity of the two fundamentals, so that the diagrams of the horizontal and vertical displacements of the points of body II are two segments parallel to their respective fundamentals. These points translate downwards by the amount φl and rightwards by $2\varphi h$, and thus, globally, translate in a direction orthogonal to that of the axes of the rods. At the same time, body I rotates clockwise by an angle φ, so that the internal hinge finds itself translated rightwards by $2\varphi h$ and downwards by φl.

To conclude, reconsider the structure made up of three rigid bodies, in the case of ill-disposition of the constraints (Figure 3.39). The absolute centre $C_1(\infty)$ is the point at infinity of the vertical straight lines. The centre C_2 is found on the external hinge, while the centre C_3 is unknown a priori. The relative centres C_{12} and C_{23} coincide with the internal hinges.

The centre C_3 is found at the intersection of the normal to the plane of movement of the roller support with the line joining C_2 and C_{23}. On the other hand, the third relative centre C_{13} is found at the intersection of the line joining C_{12} and C_{23} (second theorem of kinematic chains) with the line joining $C_1(\infty)$ and C_3, which is the vertical line through C_3 (first theorem of kinematic chains). In short, C_{13} is found to coincide with the hinge of the roller support.

Once all the centres are known, both absolute and relative, it is possible to project them on the two fundamentals and to draw the linearized functions of displacement graphically, turning, so to speak, on these projections. As is natural, the body I translates only horizontally, while the bodies II and III both rotate by the same angle, the former, however, counterclockwise, the latter clockwise. The roller support moves towards the left by the amount $2\varphi h$, as does the upper internal hinge. The lower internal hinge translates leftwards by φh and upwards by φl.

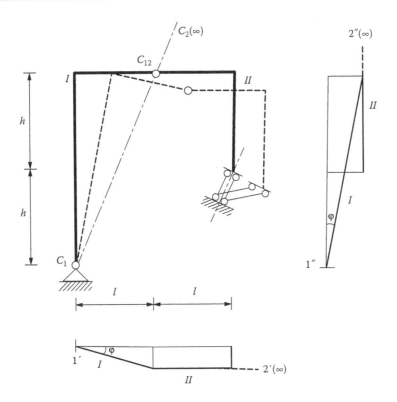

Figure 3.38

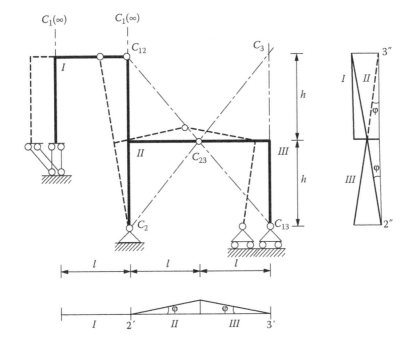

Figure 3.39

3.6 CARDINAL EQUATIONS OF STATICS

Consider a plane rigid body subjected to the action of n concentrated forces F_i and of m concentrated moments (or couples) M_i (Figure 3.40). A necessary and sufficient condition for this body to be in equilibrium is that the system of loads should satisfy the cardinal equations of statics:

$$\{R\} = \sum_{i=1}^{n} \{F_i\} = \{0\} \tag{3.34a}$$

$$M(O) = \sum_{i=1}^{m} M_i + \sum_{i=1}^{n} \left(\{r_i\} \wedge \{F_i\}\right)^{T} \{k\} = 0 \tag{3.34b}$$

where
$\{R\}$ is the resultant force
$M(O)$ is the resultant moment (scalar because the system is plane) with respect to an arbitrary pole O of the plane

The arbitrariness of the pole is permitted by the condition whereby the resultant force is zero. In fact, the resultant moment with respect to a different pole O' is linked to the foregoing one by the following relation:

$$M(O') = M(O) + \left((O - O') \wedge \{R\}\right)^{T} \{k\} \tag{3.35}$$

A system of loads that satisfies the conditions (3.34) is said to be **balanced** or **equivalent to zero** (two systems being equivalent when they possess equal resultant forces and equal resultant moments). On the other hand, two systems of loads are said to be one the **equilibrant** of the other when their sum is a balanced system. It follows from this that an equilibrant of a system of loads is the opposite of an equivalent system. It will be seen later how the system of external loads and the system of constraint reactions balance one another, their sum necessarily constituting a system equivalent to zero.

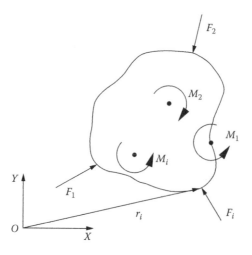

Figure 3.40

3.7 STATIC DEFINITION OF PLANE CONSTRAINTS

Plane constraints, both internal and external, have already been introduced and defined from the kinematic point of view in Section 3.3. We shall now give a static definition, which means that we shall specify the reactions that the constraints themselves are able to exert. There is a perfect correspondence, known as **duality**, between the kinematic definition and the static one. To each elementary movement (or generalized displacement) prevented by the constraint, there corresponds a generalized force exerted by the constraint on the body (and *vice versa*) in the case of external constraint and mutually between two connected bodies in the case of internal constraint.

Recalling that for the **roller support** or **connecting rod** the kinematic conditions are (Figure 3.41)

$$\{ds_P\}^T\{p\}\neq0 \tag{3.36a}$$

$$\{ds_P\}^T\{n\}=0 \tag{3.36b}$$

$$\varphi_z\neq0 \tag{3.36c}$$

to these they will correspond in a complementary (or dual) fashion the following elementary reactions:

$$\{R_P\}^T\{p\}=0 \tag{3.37a}$$

$$\{R_P\}^T\{n\}\neq0 \tag{3.37b}$$

$$M_P = 0 \tag{3.37c}$$

Hence, only the reaction orthogonal to the plane of movement of the roller support is different from zero, while the reaction parallel to this plane and the reaction moment are zero. Absence of friction (smooth constraint) is therefore assumed. Note that, as a consequence of the duality, the total work of the constraint reactions is zero:

$$\{R_P\}^T\{ds_P\} + M_P\varphi_z = 0 \tag{3.38}$$

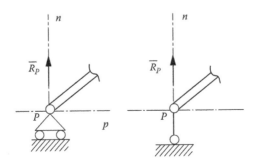

Figure 3.41

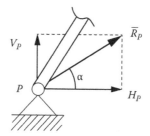

Figure 3.42

This relation is valid, however, only when the constraint is ideal, i.e., when the constraint is **rigid** and **smooth,** the conditions whereby displacements and reactions cancel rigorously holding good.

The **hinge,** which is a double constraint, reacts with a force whose line of action passes through the hinge itself (Figure 3.42). The kinematic and static conditions are

$$\{ds_P\} = \{0\} \Rightarrow \{R_P\} \neq \{0\} \tag{3.39a}$$

$$\varphi_z \neq 0 \Rightarrow M_P = 0 \tag{3.39b}$$

The parameters that identify the reaction force are in any case two: the two orthogonal components, or magnitude and angle of orientation.

Also the **double rod** is a double constraint, which reacts with a force parallel to the axes of the rods. The kinematic and static conditions are (Figure 3.43)

$$\{ds_P\}^T\{p\} \neq 0 \Rightarrow \{R_P\}^T\{p\} = 0 \tag{3.40a}$$

$$\{ds_P\}^T\{n\} = 0 \Rightarrow \{R_P\}^T\{n\} \neq 0 \tag{3.40b}$$

$$\varphi_z = 0 \Rightarrow M_P \neq 0 \tag{3.40c}$$

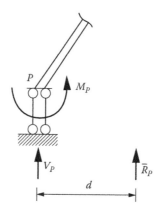

Figure 3.43

There are two parameters that identify the reaction: force and reaction moment, or translated force (after its composition with the reaction moment) and distance of the line of action from the constraint (Figure 3.43). Note that, considering the double rod as an ideal hinge at infinity, it may be stated that the reaction is a force whose line of action passes through this ideal hinge.

The triple constraint, or **fixed joint**, inhibits all elementary movements, and thus all three elementary reactions can be different from zero (Figure 3.44):

$$\{ds_P\} = \{0\} \Rightarrow \{R_P\} \neq \{0\} \tag{3.41a}$$

$$\varphi_z = 0 \Rightarrow M_P \neq 0 \tag{3.41b}$$

There are three parameters that identify the reaction: two orthogonal components of the reaction force and the reaction moment, or magnitude and angle of orientation of the translated force and distance of the line of action from the constraint (Figure 3.44).

Finally, the last external constraint that remains to be considered is the **double articulated parallelogram** (Figure 3.45), which is a single constraint that inhibits rotation and that will thus react only with the moment:

$$\{ds_P\} \neq \{0\} \Rightarrow \{R_P\} = \{0\} \tag{3.42a}$$

$$\varphi_z = 0 \Rightarrow M_P \neq 0 \tag{3.42b}$$

For **internal constraints**, the considerations are altogether similar to those just set forth for external constraints. On the other hand, just as for external constraints the reactions are understood as mutual actions (i.e. equal and opposite) exerted by the foundation on the

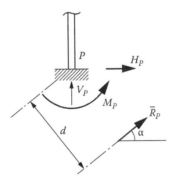

Figure 3.44

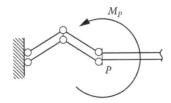

Figure 3.45

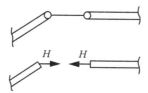

Figure 3.46

body and *vice versa*, so for internal constraints the mutual action is to be understood as being exerted between the two bodies connected by the constraint.

The **connecting rod** exerts an equal and opposite force on the two bodies (Figure 3.46). This force has the line of action coinciding with the axis of the connecting rod, and thus the only static parameter involved is its magnitude (in addition to the sense). The force perpendicular to its axis and the reaction moment are not transmitted by the connecting rod.

The **hinge** exerts a force that passes through its own centre (Figure 3.47). Since it is a double constraint, there are two static parameters involved: the two orthogonal components, or the magnitude and the angle of orientation. The reaction moment is zero since relative rotations are allowed. The hinge is usually said not to react to moment.

The **double rod** transmits a force parallel to the axes of the rods, of which the magnitude and the distance from the constraint are to be defined (Figure 3.48). In an equivalent manner, it may be stated that the double rod transmits a force, with line of action coincident with the axes of the rods (which are at an infinitesimal distance), and a reaction moment.

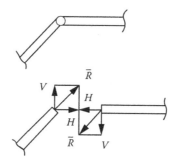

Figure 3.47

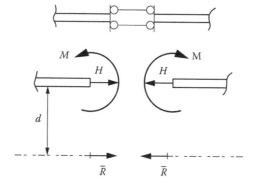

Figure 3.48

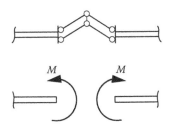

Figure 3.49

Considering the double rod as an ideal hinge at infinity, the reaction is a force passing through this ideal hinge.

The **double articulated parallelogram** transmits the reaction moment only (Figure 3.49).

The **internal joint**, since it is a triple constraint, transmits all three elementary reactions: horizontal and vertical components of the force and the reaction moment (Figure 3.50). It is equivalent, on the other hand, to consider the global reaction obtained by summing up the single components: this can be any force, however oriented in the plane and however distant from the constraint.

It is important to note at this point that a rigid beam can be considered as a succession of infinite triple constraints that connect the infinite elementary segments that make it up (Figure 3.51). In this case, the elementary reactions assume peculiar structural meanings, as will be seen in the sequel, and particular denominations. The axial reaction is also referred to as **normal reaction**; the reaction perpendicular to the axis is known as **transverse** or **shear reaction**; reaction moment is called **bending moment**. We shall hereafter refer to these elementary reactions in general as **characteristics of internal reaction**.

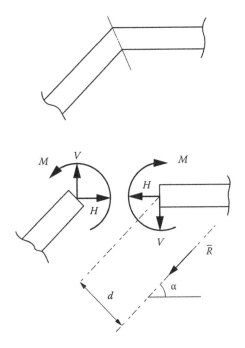

Figure 3.50

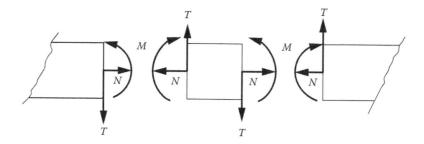

Figure 3.51

3.8 ALGEBRAIC STUDY OF STATICS OF RIGID BODY SYSTEMS

In this section, we shall show how hypostaticity, isostaticity and hyperstaticity of rigid systems also find a highly meaningful algebraic interpretation in the field of statics. Moreover, we shall see that, whereas in the kinematic field the condition of the system is an intrinsic property of the system itself, in the static field it is also a function of the external load.

As an introduction to the subject, consider again the L-shaped beam of Figure 3.52, loaded by a vertical force concentrated in the centre of the horizontal part. The external constraint reactions consist of the force R_A, perpendicular to the plane of movement of the roller support A, passing through the hinge of the roller support, and assumed to be acting upwards, and of the components H_B and V_B of the reaction exerted by the hinge B, assumed to be acting rightwards and upwards, respectively. These three parameters, R_A, H_B, V_B, represent the unknowns of the static problem, while the equations for resolving them are the cardinal equations of statics. If, once the solution has been obtained, a parameter is found to have a negative value, this means that the corresponding elementary reaction has a sense contrary to the one initially assumed.

To bring out the correspondence, or duality, between statics and kinematics, it is necessary to choose the static pole about which equilibrium to rotation is to be expressed, corresponding to the same kinematic pole that characterized the displacements of the individual rigid body (Figure 3.29a). This point O is thus chosen, also in this case, at the point of convergence of the two sections, horizontal and vertical. We must then express, in order,

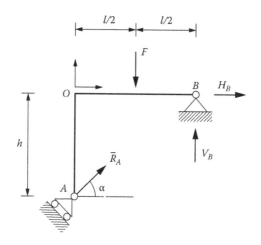

Figure 3.52

equilibrium to horizontal translation, equilibrium to vertical translation and equilibrium to rotation about the pole (or centre of reduction) O:

$$R_A \cos\alpha + H_B = 0 \tag{3.43a}$$

$$R_A \sin\alpha + V_B - F = 0 \tag{3.43b}$$

$$R_A h \cos\alpha + V_B l - F\frac{l}{2} = 0 \tag{3.43c}$$

Note that the vertical component $R_A \sin\alpha$ of the reaction of the roller support and the horizontal component H_B of the reaction of the hinge do not appear in the third equation, as these elementary reactions have a zero arm with respect to the pole O. The term $R_A h \cos\alpha$ may alternatively be read as the product of the magnitude R_A of the roller support reaction and the corresponding moment arm (Figure 3.52).

A matrix version of Equations 3.43 may be given as

$$\begin{bmatrix} \cos\alpha & 1 & 0 \\ \sin\alpha & 0 & 1 \\ h\cos\alpha & 0 & l \end{bmatrix} \begin{bmatrix} R_A \\ H_B \\ V_B \end{bmatrix} = - \begin{bmatrix} 0 \\ -F \\ -F\dfrac{l}{2} \end{bmatrix} \tag{3.44}$$

The vector of the known terms is the opposite of the so-called **vector of reduced external forces**. The latter represents a system of loads equivalent to the system of external forces and acting precisely at the pole. In the case in point, it is the vertical force F translated at O, plus the moment of translation $-Fl/2$ (negative as it is clockwise).

Note that the matrix of the coefficients of relation (3.44) is exactly the transpose of the matrix in Equation 3.23. This, as shall be seen more clearly later, is a property altogether general, known as **static–kinematic duality of rigid body systems**. The matrix of the static coefficients is therefore the transpose of the matrix of the kinematic coefficients.

To discuss the system (3.44), it is necessary to refer to the well-known Rouché–Capelli theorem, valid for systems of nonhomogeneous linear algebraic equations. It is useful here to recall the statement of this theorem.

Rouché–Capelli theorem
A necessary and sufficient condition for a system of m linear equations in n unknowns to possess a solution is that the matrix of the coefficients and the matrix made up of the coefficients and the known terms, the so-called **augmented matrix**, should have the same rank.

We recall that the rank of a matrix is the integer expressing the maximum order of its nonzero minors.

The determinant of the matrix that appears in system (3.44) vanishes, as has already been seen, for $\alpha = \bar{\alpha} = \arctan(h/l)$. The determinant of a square matrix is in fact equal to that of its transpose. When, therefore, the line joining A and B is perpendicular to the plane of movement of the roller support, the matrix of the coefficients presents rank 2 whereas the augmented matrix admits of rank 3. On the basis of the Rouché–Capelli theorem, the system, therefore, does not possess a solution. This means that, in a condition of ill-disposition of the constraints and hence of hypostaticity, calculation of the constraint reactions is **impossible**.

As a second example, consider the arch made up of two rigid bodies, already analysed from the kinematic point of view in Section 3.3 (Figure 3.53a). Let this be loaded by an oblique force F. In this case, the unknowns consist of the constraint reactions of the external hinge, H_A, V_A, of the internal hinge, H_B, V_B, and of the double rod, R_C, M_C. The centres

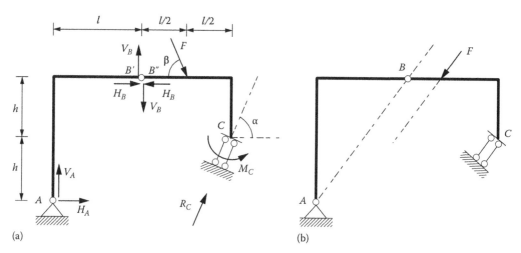

Figure 3.53

of reduction are chosen coincident with the external constraints, at A and C, respectively, while, as is customary, the horizontal forces directed towards the right, the vertical forces directed upwards and the counterclockwise moments are considered positive. These choices are, in actual fact, altogether arbitrary and conventional.

The six equilibrium equations, the first three relative to the left-hand body and the following three relative to the right-hand body, are presented as follows:

$$H_A + H_B = 0 \tag{3.45a}$$

$$V_A + V_B = 0 \tag{3.45b}$$

$$-2hH_B + V_Bl = 0 \tag{3.45c}$$

$$R_C \cos\alpha - H_B + F\cos\beta = 0 \tag{3.45d}$$

$$R_C \sin\alpha - V_B - F\sin\beta = 0 \tag{3.45e}$$

$$H_Bh + V_Bl - hF\cos\beta + \frac{l}{2}F\sin\beta + M_C = 0 \tag{3.45f}$$

In matrix form, we obtain

$$
\begin{bmatrix}
1 & 0 & 1 & 0 & 0 & 0 \\
0 & 1 & 0 & 1 & 0 & 0 \\
0 & 0 & -2h & l & 0 & 0 \\
0 & 0 & -1 & 0 & \cos\alpha & 0 \\
0 & 0 & 0 & -1 & \sin\alpha & 0 \\
0 & 0 & h & l & 0 & 1
\end{bmatrix}
\begin{bmatrix}
H_A \\
V_A \\
H_B \\
V_B \\
R_C \\
M_C
\end{bmatrix}
= -
\begin{bmatrix}
0 \\
0 \\
0 \\
F\cos\beta \\
-F\sin\beta \\
-hF\cos\beta + \dfrac{l}{2}F\sin\beta
\end{bmatrix}
\tag{3.46}
$$

The opposite of the vector of the known terms represents the vector of the external forces reduced at the poles A and C. The first three terms are zero because the left-hand body is not subjected to external loads.

The static matrix of system (3.46) is the transpose of the kinematic matrix of system (3.29). When the determinant is different from zero, both the matrix of the coefficients and its augmented matrix evidently present the same rank 6. By the Rouché–Capelli theorem, the algebraic system then possesses only one solution, i.e., an orderly set of six values. In this case, the mechanical system is said to be **statically determinate** or **isostatic**.

The determinant, on the other hand, vanishes for $\alpha = \bar{\alpha} = \arctan(2h/l)$. In this case, the matrix of the coefficients is of rank 5 while its augmented matrix in general is of rank 6. The algebraic system is thus impossible and does not possess any solution, while the mechanical system is **hypostatic** and loaded by external forces that cannot be balanced in any way.

An even more particular case is where $\alpha = \bar{\alpha} = \arctan(2h/l)$, $\beta = \pi - \bar{\alpha} = \arctan(-2h/l)$. Also the external force becomes parallel to the line joining A and B, as well as the double rod at C (Figure 3.53b). In this case, the augmented matrix will be as follows:

$$
\begin{bmatrix}
1 & 0 & 1 & 0 & 0 & 0 & 0 \\
0 & 1 & 0 & 1 & 0 & 0 & 0 \\
0 & 0 & -2h & l & 0 & 0 & 0 \\
0 & 0 & -1 & 0 & \cos\bar{\alpha} & 0 & F\cos\bar{\alpha} \\
0 & 0 & 0 & -1 & \sin\bar{\alpha} & 0 & F\sin\bar{\alpha} \\
0 & 0 & h & l & 0 & 1 & -hF\cos\bar{\alpha} - \left(\dfrac{l}{2}\right)F\sin\bar{\alpha}
\end{bmatrix}
\tag{3.47}
$$

The seventh column, of the known terms, is a linear combination of the fifth (multiplied by F) and the sixth (multiplied by $-hF\cos\bar{\alpha} - (l/2)F\sin\bar{\alpha}$). Consequently, both the matrix of the coefficients and the augmented matrix present rank 5. By the Rouché–Capelli theorem, there exist then ∞^1 solutions. The mechanical system is thus intrinsically hypostatic, but is loaded by an external force that can be balanced in ∞^1 different ways (Figure 3.53b). The mechanical system is thus in equilibrium on account of the particular load condition. But this equilibrium presents itself as **statically indeterminate** or **hyperstatic**.

To conclude, let us re-examine the double portal frame of Figure 3.54a, loaded by a horizontal force F_1 and by a vertical force F_2. In this case, we have nine equilibrium equations in nine unknowns, which, in a matrix form, are presented as follows:

$$
\begin{bmatrix}
0 & 0 & 1 & 0 & 0 & 0 & 0 & 0 & 0 \\
1 & 0 & 0 & 1 & 0 & 0 & 0 & 0 & 0 \\
0 & 1 & -h & x & 0 & 0 & 0 & 0 & 0 \\
0 & 0 & -1 & 0 & 1 & 0 & 1 & 0 & 0 \\
0 & 0 & 0 & -1 & 0 & 1 & 0 & 1 & 0 \\
0 & 0 & 2h & l-x & 0 & 0 & -h & l & 0 \\
0 & 0 & 0 & 0 & 0 & 0 & -1 & 0 & 0 \\
0 & 0 & 0 & 0 & 0 & 0 & 0 & -1 & 1 \\
0 & 0 & 0 & 0 & 0 & 0 & h & l & 0
\end{bmatrix}
\begin{bmatrix}
V_A \\ M_A \\ H_B \\ V_B \\ H_C \\ V_C \\ H_D \\ V_D \\ V_E
\end{bmatrix}
= -
\begin{bmatrix}
F_1 \\ 0 \\ -F_1 h \\ 0 \\ -F_2 \\ 0 \\ 0 \\ 0 \\ 0
\end{bmatrix}
\tag{3.48}
$$

The matrix of the coefficients is the transpose of Equation 3.33.

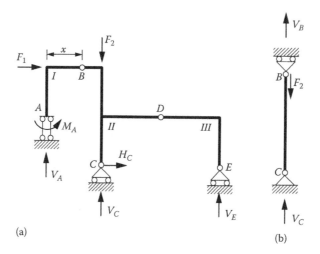

Figure 3.54

When $0 \le x < l$, the matrix of the coefficients is of rank 9, as is the augmented matrix. The algebraic system thus possesses a solution and the mechanical system is said to be **statically determinate**.

When $x = l$ and $F_2 = 0$, the matrix of the coefficients is of rank 8, while the augmented matrix is again of rank 9. The algebraic system is thus impossible, just as the mechanical system is hypostatic and loaded by an unbalanceable force.

Finally, when $x = l$ and $F_1 = 0$, the matrix of the coefficients is again of rank 8, as is the augmented matrix. The column of the known terms is in fact proportional to the sixth column of the static matrix. The algebraic system thus possesses ∞^1 solutions, the rigid system being hypostatic but in equilibrium on account of the particular load condition. This equilibrium can be obtained in ∞^1 different ways. This can be immediately justified if the body I is replaced by a roller support that moves horizontally in B, and the body III is eliminated since it does not react (Figure 3.54b). In this way, having reduced the system to its essential elements, we may verify how the force F_2 can be balanced by an infinite number of pairs of vertical reactions V_C and V_B, such that $V_B + V_C - F_2 = 0$.

3.9 STATIC–KINEMATIC DUALITY

We close this chapter with a summary of its contents, underlining the aspect of duality between kinematics and statics. A rigorous demonstration will therefore be given of the fact that the **static matrix** is the transpose of the **kinematic matrix** and *vice versa*.

The **kinematic equations** of a rigid body system, with g original degrees of freedom and v degrees of constraint, can be represented in a compact form as follows:

$$\underset{v \times g}{[C]} \underset{g \times 1}{\{s_O\}} = \underset{v \times 1}{\{s_X\}} = \underset{v \times 1}{\{0\}}$$

(3.49)

where
  $[C]$ is the kinematic matrix
  $\{s_O\}$ is the vector of displacements of poles
  $\{s_X\}$ is the vector of displacements (absolute or relative) of constraint points

When $v < g$, the algebraic system possesses at least ∞^{g-v} solutions and the mechanical system is at least $(g - v)$ times **hypostatic**.

When $v = g$, the algebraic system possesses at least the obvious solution, and the mechanical system is **isostatic**, if there is no ill-disposition of the constraints.

When $v > g$, the algebraic system possesses at least the obvious solution, and the mechanical system is $(v - g)$ times **hyperstatic**, if there is no ill-disposition of the constraints.

On the other hand, the **static equations** of the same rigid system can be represented as follows:

$$[A]\{X\} = -\{F\} \tag{3.50}$$
$$\underset{g \times v}{} \underset{v \times 1}{} \quad \underset{g \times 1}{}$$

where
    $[A]$ is the static matrix
    $\{X\}$ is the vector of constraint reactions
    $\{F\}$ is the vector of reduced external forces

When $v < g$, the algebraic system is generally impossible and the system of external forces cannot be balanced in any way.

When $v = g$, the algebraic system generally possesses one solution and the mechanical system is said to be **statically determinate** or **isostatic**.

When $v > g$, the algebraic system generally possesses ∞^{v-g} solutions and the mechanical system is said to be **statically indeterminate** or **hyperstatic**.

The principle of virtual work can be applied to a rigid system in equilibrium, subjected to external loads and constraint reactions. If the external forces are replaced, as is permissible, with the forces reduced to the poles, we have

$$\{s_O\}^T\{F\} + \{s_X\}^T\{X\} = 0 \tag{3.51}$$

On the other hand, using the matrix relations (3.49) and (3.50), Equation 3.51 is transformed as follows:

$$-\{s_O\}^T[A]\{X\} + \{s_O\}^T[C]^T\{X\} = 0 \tag{3.52}$$

from which we obtain

$$\{s_O\}^T([A] - [C]^T)\{X\} = 0 \tag{3.53}$$

This equation, on account of the arbitrariness of $\{s_O\}$ and $\{X\}$, is satisfied if and only if

$$[A] = [C]^T \tag{3.54}$$

The fundamental relation (3.54), which arose inductively in this chapter, can thus be deemed to be rationally demonstrated.

Chapter 4

Determination of constraint reactions

4.1 INTRODUCTION

This chapter presents some methods for determining constraint reactions in the case of statically determinate structures, i.e. structures constrained in a nonredundant manner. In addition to the algebraic method of auxiliary equations, which helps us to split the general solution system into two or more systems of smaller dimensions, the method based on the principle of virtual work, as well as the classical graphical method using pressure lines, is also proposed.

In this chapter, particular attention is again drawn to the problem of ill-disposition of constraints for which the potential centres of rotation, both absolute and relative, fall on a straight line, just as, from the static viewpoint, the force polygons do not close except at infinity.

In the case of continuous distributions of forces acting in the same direction, the differential equation of the pressure line is obtained, so revealing how this line represents, save for one factor, the diagram of the bending moment. The cases of arch bridges and suspension bridges are considered as typical examples in which, since the geometrical axis of the load-bearing member coincides with the pressure line, the member itself is either only compressed (arch) or else only stretched (cable).

4.2 AUXILIARY EQUATIONS

As we have already seen in the previous chapter, if the number of rigid bodies making up an isostatic, or statically determinate, structure is n, it will be possible to write $3n$ equations of partial equilibrium in $3n$ unknown elementary reactions. On the other hand, this leads to systems of equations that are unwieldy to resolve even with a low n.

An alternative way that can be followed is that of the **method of auxiliary equations**. In this case, we consider the three equations of global equilibrium, i.e. of the entire structure, with the addition of s auxiliary equations of partial equilibrium, if s is the degree of internal disconnection of the structure. We thus have $(3 + s)$ equilibrium equations in $(3 + s)$ unknown external reactions. The s auxiliary equations are chosen in such a way that the internal reactions are not involved in the system of resolution.

To clarify the process outlined earlier, let us examine once more the arch of Figure 4.1, consisting of two rigid parts hinged at B and connected to the foundation with a hinge and a double rod, respectively. In this case, the degree of internal disconnection s is equal to one (the relative rotation allowed by the hinge B). We can, therefore, write three equations of global equilibrium plus an auxiliary equation of partial equilibrium with regard to rotation of the body AB about point B, in the four unknown external reactions, H_A, V_A, R_C and M_C.

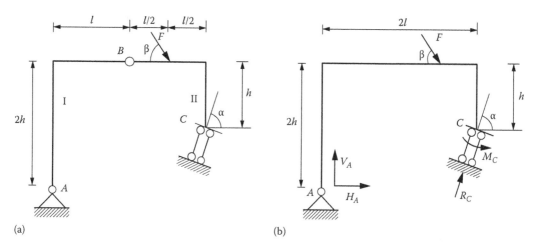

Figure 4.1

The auxiliary equation of partial equilibrium with regard to rotation of the body BC about point B is equivalent to the foregoing auxiliary equation, it being a linear combination of the four equations so far considered. It is, however, more complex. It is therefore more convenient to write

$$H_A + R_C \cos\alpha + F\cos\beta = 0 \tag{4.1a}$$

$$V_A + R_C \sin\alpha - F\sin\beta = 0 \tag{4.1b}$$

$$H_A h - 2V_A l - hF\cos\beta + \frac{l}{2}F\sin\beta + M_C = 0 \tag{4.1c}$$

$$2H_A h - V_A l = 0 \tag{4.1d}$$

Whereas, then, the first three are the equations of equilibrium of the **braced** structure, i.e. of the hyperstatic structure obtained by replacing the internal hinge with an internal built-in constraint (Figure 4.1b), the fourth is the auxiliary equation that expresses equilibrium with regard to rotation of the left-hand body about the hinge. In this way, the internal unknowns H_B and V_B do not, for the moment, enter into the balance, and the system (4.1) consists of four equations in four unknowns, as against the six equations in six unknowns of the system (3.45).

It is, however, possible subsequently and once the external reactions have been determined, also, to determine the internal reactions. These may be deduced using the equations of partial equilibrium with regard to translation, of the body AB or, alternatively, of the body BC. Since the body AB is not subjected to external loads, the first way is the more convenient:

$$H_A + H_B = 0 \tag{4.1e}$$

$$V_A + V_B = 0 \tag{4.1f}$$

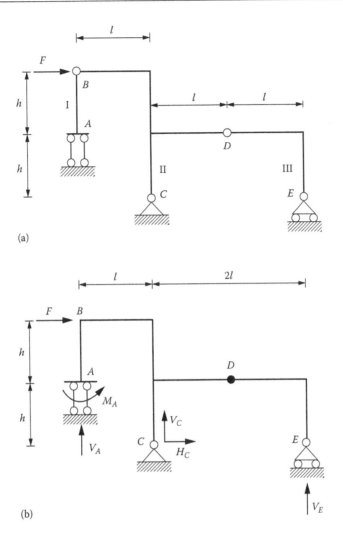

Figure 4.2

Now consider the case, already introduced in the last chapter, of three rigid bodies (Figure 4.2a). For this structure, we have $s = 2$, since two rotational single disconnections in B and D are present. There will then be five equations (three global equilibrium equations and two auxiliary equations) in the five unknown external reactions: V_A, M_A, H_C, V_C and V_E. The first three equations refer to the braced structure (twice hyperstatic) of Figure 4.2b:

$$H_C + F = 0 \tag{4.2a}$$

$$V_A + V_C + V_E = 0 \tag{4.2b}$$

$$-V_A l + M_A + 2V_E l - 2Fh = 0 \tag{4.2c}$$

Of these, Equation (4.2c) expresses equilibrium with regard to rotation about point C. It is then necessary to provide the information that there exists a hinge at B, by writing the equation of partial equilibrium with regard to rotation of the body AB about B (in this way, we avoid introducing additional unknowns),

$$M_A = 0 \tag{4.2d}$$

and, finally, that there also exists a hinge at point D, expressing the partial equilibrium with regard to rotation of the body DE about D,

$$V_E l = 0 \tag{4.2e}$$

Once we have obtained the five external unknowns, it will also be possible to deduce the internal ones, by resolving the equations of equilibrium with regard to translation of the two end bodies:

$$F + H_B = 0 \tag{4.2f}$$

$$V_A + V_B = 0 \tag{4.2g}$$

$$-V_D + V_E = 0 \tag{4.2h}$$

$$-H_D = 0 \tag{4.2i}$$

More generally, it is then possible to consider structures having a generic number s of internal hinges and thus $(3 + s)$ external elementary reactions. The procedure will be altogether analogous to that presented previously. There are, however, auxiliary equations to be considered relative to partial sections, having one end internally constrained and the other externally constrained. These sections can contain possible internal disconnections.

On the other hand, the forms of internal disconnection that may be considered are not limited to the hinge. For instance, the portal frame of Figure 4.3 contains an internal disconnection to the vertical translation, i.e. a horizontal double rod. The three global equations

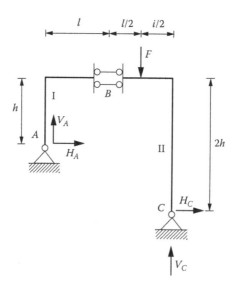

Figure 4.3

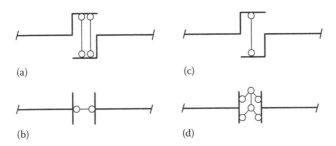

(a)

(c)

(b)

(d)

Figure 4.4

are those of equilibrium with regard to horizontal and vertical translations and the equation of equilibrium with regard to rotation about point C:

$$H_A + H_C = 0 \qquad\qquad (4.3a)$$

$$V_A + V_C - F = 0 \qquad\qquad (4.3b)$$

$$-2V_A l - H_A h + F\frac{l}{2} = 0 \qquad\qquad (4.3c)$$

while the auxiliary equation to be considered is that of partial equilibrium with regard to vertical translation of the left-hand section:

$$V_A = 0 \qquad\qquad (4.3d)$$

In the case of the double rod being vertically oriented (Figure 4.4a), the auxiliary equation would then be that of equilibrium with regard to horizontal translation of one of the two rigid sections.

When, instead, there is a double internal disconnection, i.e. $s = 2$, which is concentrated in the same point, there are two auxiliary equations corresponding to it, and they must be formulated so that the only elementary reaction transmitted does not appear. In the case of a horizontal connecting rod, e.g. it will be necessary to consider the partial equilibrium with regard to rotation about the constraint and with regard to vertical translation of one of the two sections into which the structure is separated by the connecting rod itself (Figure 4.4b). When, instead, the connecting rod is vertical (Figure 4.4c), the partial equilibrium with regard to rotation and with regard to horizontal translation will be considered. Finally, in the case of the double articulated parallelogram (Figure 4.4d), the two auxiliary equations will both be equations of equilibrium with regard to translation.

4.3 PRINCIPLE OF VIRTUAL WORK

We have previously described two algebraic methods for determining constraint reactions: (1) the **general method,** according to which each single rigid body is set in equilibrium, writing $3n$ equations in $3n$ unknowns (n is the number of rigid bodies of the system); (2) the **method of auxiliary equations,** according to which the global equilibrium is considered and at the same time the information is provided that there exist s internal disconnections, by writing $(3 + s)$ equations in $(3 + s)$ unknowns.

A semi-graphical method will now be introduced that is based on the principle of virtual work and on the theory of mechanisms, which is, on the other hand, able to provide a single elementary reaction each time. This is thus a method that can be used to advantage when we wish to determine a specific reaction, necessary, e.g. for dimensioning the constraint supporting it.

As an introduction to this method, consider the bar system of Figure 4.5a, subjected to the horizontal force F. We intend to define the value of the horizontal reaction H_C exerted by the hinge C. It will then be necessary to effect a disconnection in such a way that, apart from the external force, only the reaction sought will be able to perform work. The hinge C will then be degraded by being transformed into a horizontally moving roller support (from a double constraint to a single constraint) and the horizontal force H_C, exerted by the hinge C,

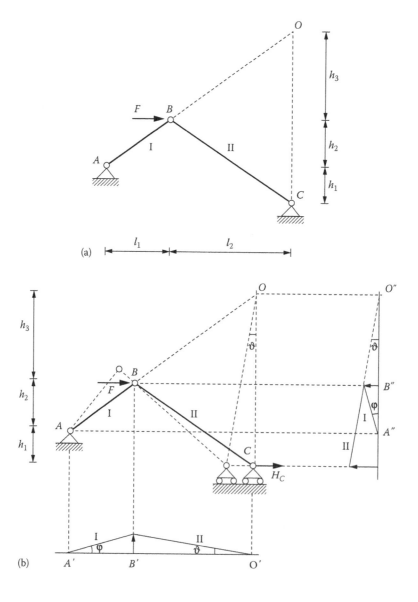

(a)

(b)

Figure 4.5

will be applied, assumed to be acting towards the right. It is evident that the assigned structure has now been transformed into the mechanism already studied in Section 3.5 (Figure 4.5b). The reactions H_A and V_A of the external hinge do not perform work because their point of application is not displaced. The internal reactions H_B and V_B perform equal and opposite work on the two bars that make up the bar system. Finally, the reaction V_C of the roller support does not perform work, since it is displaced in a direction perpendicular to that of its line of action. There thus remain to be taken into account the external force F, which is displaced by the quantity ϑh_3, and the reaction H_C (which in this scheme has the role of an external force), which is displaced by the amount $\vartheta (h_1 + h_2 + h_3)$. Since these two forces undergo displacements opposite to their direction, the two contributions will both be negative (Figure 4.5b):

$$\text{Work} = -F\vartheta h_3 - H_C\vartheta(h_1 + h_2 + h_3) = 0 \tag{4.4}$$

from which

$$H_C = -F\frac{h_3}{h_1 + h_2 + h_3} \tag{4.5}$$

It may be noted that the angle of virtual rotation ϑ of the bar BC is cancelled by both the contributions and therefore does not enter into the expression (4.5). Nor do the abscissae l_1 and l_2 enter into this expression (Figure 4.5). The mechanism thus remains in equilibrium on account of the particular load condition, and H_C is negative because it is, in actual fact, acting in the opposite direction to the one assumed.

As a second example, consider the asymmetrical portal frame of Figure 4.6a, subjected to a constant distributed load q. We intend to determine the vertical reaction of the hinge C. The procedure, then, is to reduce the constraint C in a dual manner with respect to the reaction that we are seeking. The hinge C will thus be transformed into a vertically moving roller support. In this case, only the diagram of the vertical displacements defined already in Figure 3.36a will be used. With the aim of applying the principle of virtual work, we shall consider the two partial resultants of the external load, each acting on one rigid section, plus the reaction V_C:

$$-(ql)\left(\varphi\frac{l}{2}\right) - (ql)(2\varphi l) + V_C(3\varphi l) = 0 \tag{4.6}$$

from which

$$V_C = \frac{5}{6}ql \tag{4.7}$$

In this case, the direction assumed proves to be the actual one.

If we intend then to determine the vertical reaction T (shearing force) transmitted by the internal hinge B (Figure 4.6b), we have to reduce the hinge itself and to transform it into a disconnection of a higher order that allows the vertical relative translations. If, then, a horizontal connecting rod is introduced in place of the hinge, the mechanism will undergo the horizontal and vertical displacements shown in Figure 4.6b. The two portions both turn in the same direction, the one on the right through an angle twice as large. Note how, in correspondence with the relative centre C_{12}, relative displacements of the two bodies do not occur.

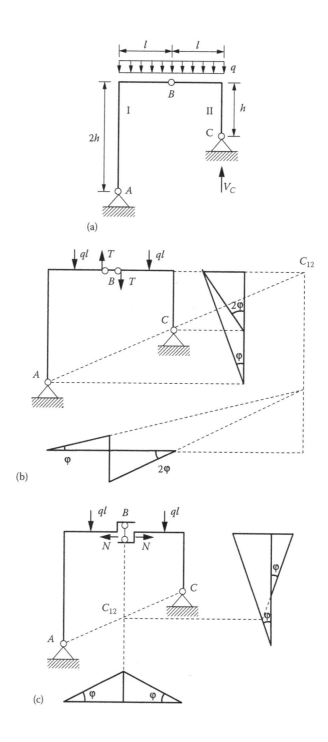

Figure 4.6

While the force T, acting on the left-hand portion, performs the work $T\varphi l$, that on the right-hand portion performs twice as much work. The same applies to the external partial resultants, even though in this case the amounts of work performed are of opposite algebraic sign. Altogether, we have

$$T\varphi l + 2T\varphi l - ql\left(\varphi\frac{l}{2}\right) + ql\left(2\varphi\frac{l}{2}\right) = 0 \tag{4.8}$$

from which

$$T = -\frac{ql}{6} \tag{4.9}$$

If, instead, we wish to obtain the horizontal reaction N (axial force) transmitted by the internal hinge B (Figure 4.6a), we must replace the hinge with a vertical connecting rod (Figure 4.6c), which allows horizontal relative translations. This time the two sections turn through the same angle but in opposite directions. Also in this case, in correspondence with the relative centre C_{12}, the two bodies are displaced by the same amounts. The total balance works out as follows:

$$-ql\left(\varphi\frac{l}{2}\right) - ql\left(\varphi\frac{l}{2}\right) + N(2\varphi h) + N(\varphi h) = 0 \tag{4.10}$$

whence

$$N = \frac{ql^2}{3h} \tag{4.11}$$

It is possible to check the results obtained using the principle of virtual work, by applying the method of auxiliary equations:

$$H_A + H_C = 0 \tag{4.12a}$$

$$V_A + V_C - 2ql = 0 \tag{4.12b}$$

$$-2V_A l + H_A h + 2ql^2 = 0 \tag{4.12c}$$

$$-V_A l + 2H_A h + q\frac{l^2}{2} = 0 \tag{4.12d}$$

where the last equation is the auxiliary one (equilibrium to rotation of the section AB about B). Resolving the system (4.12), we obtain

$$H_A = \frac{ql^2}{3h} \tag{4.13a}$$

$$V_A = \frac{7}{6}ql \tag{4.13b}$$

$$H_C = -\frac{ql^2}{3h}$$

(4.13c)

$$V_C = \frac{5}{6}ql$$

(4.13d)

Considering equilibrium with regard to translation of the left-hand section, we have then

$$H_A - N = 0$$

(4.14a)

$$V_A + T - ql = 0$$

(4.14b)

from which we obtain the axial and shearing forces transmitted by the hinge *B*:

$$N = \frac{ql^2}{3h}, \quad T = -\frac{ql}{6}$$

(4.15)

As regards the asymmetrical portal frame of Figure 4.7a, obtained from the foregoing one by replacing the internal hinge with a horizontal double rod, the determination of the two elementary reactions transmitted by the internal constraint *B* can be arrived at, applying the principle of virtual work and using the two mechanisms shown in Figure 4.7b and c.

4.4 GRAPHICAL METHOD

The graphical method for determining the constraint reactions is based on the cardinal equations of statics. In the case where we have three forces in equilibrium in the plane, these must form a triangle if laid out one after the other. This derives from the well-known **parallelogram law** and hence from the first cardinal equation of equilibrium with regard to translation. At the same time, the lines of action of the three forces must all pass through the same point of the plane. The moment of the three forces must in fact be zero with respect to any point in the plane, and thus also with respect to the intersection of each pair of lines of action: the third line must then pass through that point. This latter requirement follows directly from the second cardinal equation of equilibrium with regard to rotation.

For greater clarity, let us take an example of application of the graphical method. Examine the L-shaped beam of Figure 4.8a, already considered from both the static and the kinematic viewpoints. We know *a priori* two of the three lines of action for the forces involved: the line of action of the external force *F* and the line of action of the reaction R_A of the roller support. These two lines are concurrent in point *P*, and hence, for equilibrium with regard to rotation, also the reaction R_B of the hinge must pass through *P*. Thus we have defined the direction of the reaction R_B, which is that of the line joining *B* and *P*. The magnitudes, which were previously unknown, of the two constraint reactions can then be determined graphically, laying off to scale the given force *F* and drawing through its ends two lines oriented as the lines of action of R_A and R_B (Figure 4.8a). The triangle that is thus obtained represents half of the parallelogram of the forces R_A and R_B, which will assume a direction so that they follow the given force *F*. In this way, the force *F* is equilibrant of R_A and R_B (while the opposite force is their resultant).

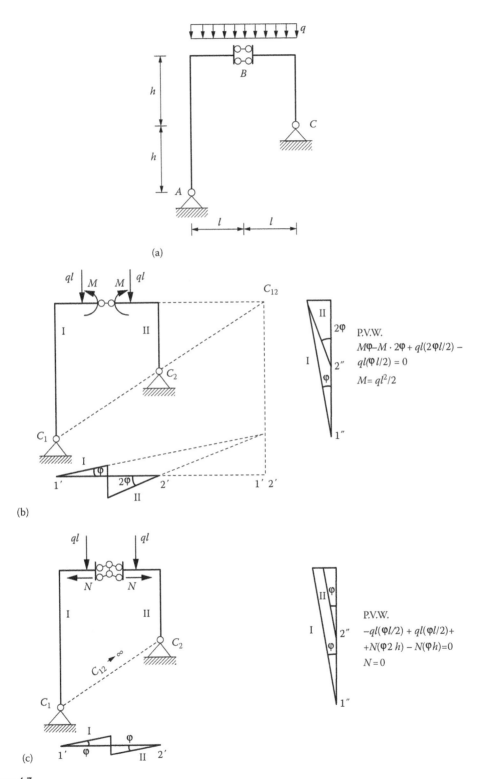

Figure 4.7

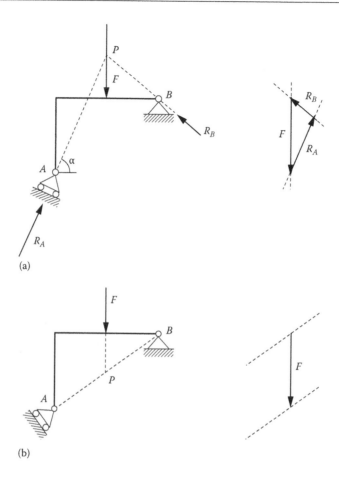

Figure 4.8

When the constraints are ill-disposed (Figure 4.8b), the lines of action of the reactions R_A and R_B both coincide with the line joining A and B and the triangle of forces does not close, other than at infinity. The reactions thus tend to infinity and equilibrium is impossible.

However, there are not always only three forces involved in ensuring equilibrium of a rigid body. On the other hand, if we compose the forces suitably, it is always possible to reduce them to three partial resultants. For example, Figure 4.9a shows the case of a beam constrained by

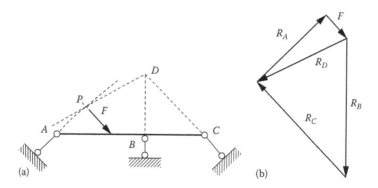

Figure 4.9

three well-disposed connecting rods, i.e. one not passing through the same point and loaded by an external force F. Composing the connecting rods B and C, we can assume that we have an ideal hinge in D, and thus the graphical resolution will closely resemble the previous one (Figure 4.9b). We identify, in fact, point P, the intersection of the line of action of the external force with the axis of the rod A, and we join P with the ideal hinge D. The line PD is the line of action of the composite reaction R_D. Once we have found the magnitude of R_D using the triangle of forces (Figure 4.9b), we can resolve this reaction into its two elementary components R_B and R_C by drawing a new triangle of forces, which in Figure 4.9b appears adjacent to the foregoing one. Note, however, that while the first triangle resolves a problem of equilibrium (the arrows follow one another round), the second one resolves a problem of equivalence (parallelogram law).

In the case where the external load consists of a concentrated couple or moment m (Figure 4.10), we must consider the straight line passing through the ideal hinge D and parallel to the axis of the rod A. The moment m will, in fact, be balanced by a couple of equal and opposite forces R_A and R_D, the magnitude of which is $R_A = R_D = m/d$, where d is the distance between the two lines of action. The reaction R_D can then be resolved into its two components R_B and R_C.

The graphical method can be used also when the structure consists of more than one rigid body. As regards three-hinged arches, a rapid and convenient application of the method requires, however, that the external load should act only on one of the two rigid sections, or else on the internal hinge.

In the case of the arch of Figure 4.11a, the vertical external force is applied to the hinge C, which, being considered in this case as a material point, is found in equilibrium with regard to translation under the action of F, R_A and R_B (Figure 4.11b). The triangle of forces will therefore have one side vertical and the other two in the directions of the lines joining A and C, and B and C, respectively.

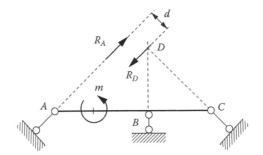

Figure 4.10

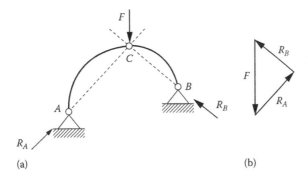

(a)   (b)

Figure 4.11

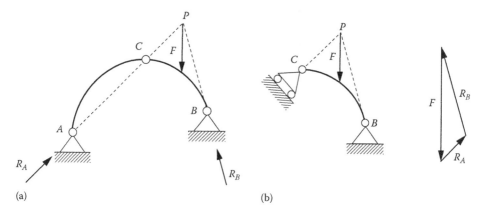

Figure 4.12

If the external force acts on the section BC (Figure 4.12a), the section AC performs the function of a connecting rod and we find ourselves once again in a case similar to those that we have already seen previously and that correspond to a single rigid body (Figure 4.12b).

In the event then that the three hinges are aligned and hence ill-disposed, also the external reactions R_A and R_C become collinear and the triangle of forces does not close, thus ruling out the possibility of any static solution.

So far only hinges and roller supports have been considered, but also the double rod can be involved in a solution of a graphical type. This may, in fact, be considered as an ideal hinge at infinity and thus be treated as we have by now already seen more than once. In the case, for instance, of an L-shaped beam constrained by a double rod and a roller support (Figure 4.13a) and loaded by an oblique force F, the pole P is given by the intersection of the line of action of F with the normal to the plane of movement of the roller support. The double rod will react with a horizontal force passing through P. The triangle of forces will have the hypotenuse parallel to the external force and the two catheti parallel, respectively, to the two rectilinear sections of the beam (Figure 4.13b).

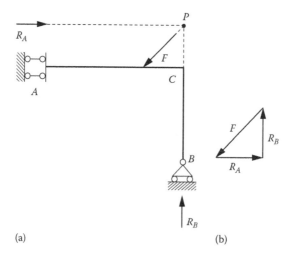

Figure 4.13

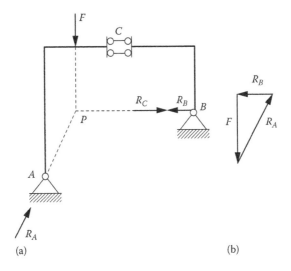

Figure 4.14

In the case of generalized three-hinged arches, i.e. arches where double rods are also involved (Figure 4.14), our analysis will once more be based on the interpretation of the double rod as an ideal hinge and on the partial equilibrium of the portion that is not externally loaded. The portion BC, in fact, is in equilibrium under the action of two equal and opposite collinear forces. These two forces must be horizontal and must pass through the hinge B. In other words, their line of action will be the line joining the two hinges, the real one B and the ideal one consisting of the point at infinity of the horizontal lines. Given this, the pole P is furnished by the intersection of the line of action of F with the horizontal line through B. The third force acting on the structure is the reaction R_A, whose line of action is given by the line joining A and P.

In the case where the portal frame is symmetrical, i.e. where it has uprights of equal height (Figure 4.15), the pole P falls on the line joining A and B, so that the triangle of forces does not close. The ill-disposition of the constraints thus renders the equilibrium of the system impossible, once this is subjected to the vertical force F.

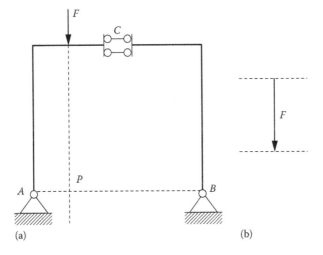

Figure 4.15

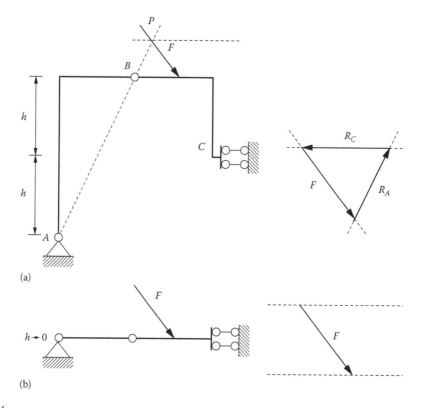

Figure 4.16

Now reconsider the arch of Figure 4.16a, already studied in the previous chapter. The equilibrium of the portion AB is achieved with two equal and opposite forces on the line joining A and B. The intersection of the line AB with the line of action of the external force gives the pole P, through which also the horizontal reaction R_C of the double rod must pass. When the parameter h vanishes, the system becomes hypostatic and equilibrium is impossible (Figure 4.16b).

Another example, already amply studied, is the portal frame of Figure 3.53a. When the double rod is disposed parallel to the line joining A and B (Figure 4.17a), the system is hypostatic and the generic force F cannot be balanced in any way, since the triangle of forces does not close (Figure 4.17b). When, moreover, also the external force is directed as the straight line AB and the double rod, the pole P comes to coincide with the point at infinity of this direction, so that the line of action of the reaction R_C of the double rod remains indeterminate. On the other hand, in this case, the triangle of forces is degenerate, with all three sides collinear (Figure 4.17c), and there exist ∞^1 pairs of vectors R_A and R_C that satisfy equilibrium. As was already seen algebraically in Chapter 3, the mechanical system is hypostatic but in equilibrium for the particular condition of load. Since, however, there exist ∞^1 solutions, both on account of the indetermination of the line of action of the reaction R_C of the double rod and on account of the possibility of balancing the force F with two reactions parallel to it (of which one is arbitrary), the system is said to be statically indeterminate (or once hyperstatic).

Another case that has already been amply discussed is that of the double portal frame of Figure 4.18a. Imagine that the hinge B is in the centre of the upper horizontal beam ($x = l/2$) and that the system is loaded by the horizontal force F. To start with, it should be noted that

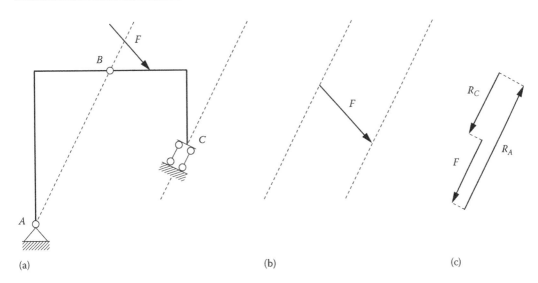

(a)   (b)   (c)

Figure 4.17

the vertical reaction V_E of the roller support must be zero. If we were to assume *ab absurdo* that this were not so, it ought to be balanced by the vertical reaction V_D of the hinge D, and thus a couple would be formed that cannot be balanced by the only reaction H_D remaining to be considered as acting on the section III. This section is thus completely inactive or, to use the term normally applied, unloaded. The remaining sections, I and II, constitute a generalized three-hinged arch of the same type as those already introduced. The section II is not loaded externally and hence its equilibrium develops on the line joining B and C. The hinge B, therefore, coincides with the pole P, and the vertical reaction of the double rod A must pass through this point. Since the triangle of forces (Figure 4.18b) is geometrically similar to the triangle $BB'C$ (Figure 4.18a), it is immediately evident that

$$R_A = \frac{4Fh}{l}, \quad R_C = F\left(1 + \frac{16h^2}{l^2}\right)^{1/2} \tag{4.16}$$

and the structure is completely resolved.

When the hinge B is at the right end of the upper horizontal beam, the beam system is transformed into a mechanism (Figure 3.39). The vertical reaction R_A of the double rod is reduced to being collinear to the reaction R_C of the hinge so that, if the external force is horizontal, the triangle of forces cannot close and equilibrium is impossible (Figure 4.18c). If the external force is, instead, vertical (Figure 4.18d), the beam system reveals itself to be (once) hyperstatic owing to the particular load condition.

Consequently, with extreme economy, we have been able to arrive back at the results that are already known to us, which in Chapter 3 were brought to light algebraically, using the Rouché–Capelli theorem.

As a final case, consider that of an internal constraint made up of two nonparallel connecting rods (Figure 4.19a). The system of bars can be considered as a three-hinged arch, where the internal hinge consists of the ideal hinge H, the point of concurrence of the axes of the two connecting rods, and the two rigid bodies are the bars AE and BG. The pole P is the intersection of the line of action of the external force F with the line joining B and H.

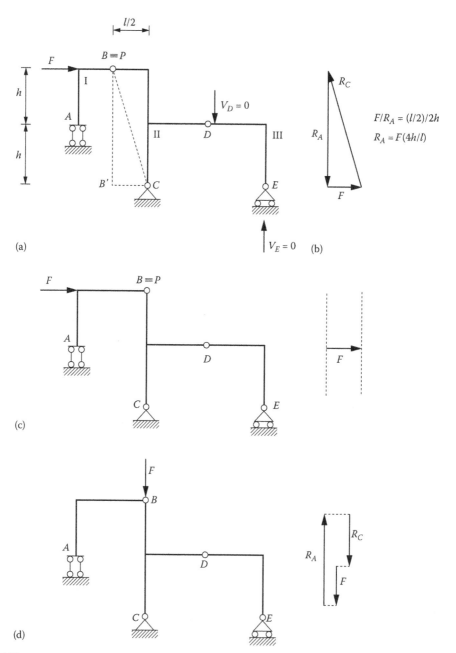

Figure 4.18

The reaction R_A of the external hinge A must pass through the pole P, as well as through A. This is determined by the triangle of forces, together with the reaction R_B. The latter is equal and opposite to the resultant reaction transmitted by the two connecting rods to the section BG, while it is equivalent to the resultant reaction transmitted by the two connecting rods to the section AE. Figure 4.19b shows the triangles of forces for global equilibrium, for equilibrium of the section BG and for the equivalence of the reaction R_B with the sum of the two reactions R_E and R_C transmitted by the two connecting rods to the section AE.

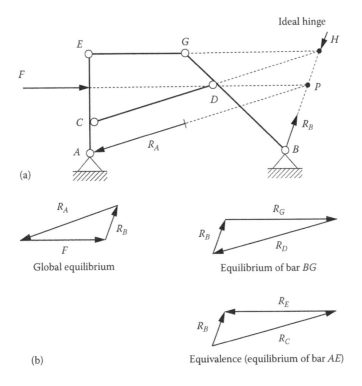

(a)

R_A R_B F

Global equilibrium

R_G R_B R_D

Equilibrium of bar BG

R_E R_B R_C

Equivalence (equilibrium of bar AE)

(b)

Figure 4.19

We could equally have resolved the exercise with the method of auxiliary equations. The primary unknowns would have been the external ones, H_A, V_A, H_B, V_B, while the four resolving equations would have been those of global equilibrium plus the auxiliary equation of equilibrium with regard to the partial rotation of the section BG about the ideal hinge H.

4.5 LINE OF PRESSURE

The set of lines of action of the successive resultant forces acting on a structure, or rather, that act as internal constraint reactions, proceeding from one end to the other of the structure itself, is called a **line of pressure**.

To illustrate this definition more clearly, consider the isostatic arch of Figure 4.20a, subjected to four external forces, of which the resultant R is known, and to the two constraint reactions R_A and R_B, identified *via* the triangle of forces of Figure 4.20b. Imagine then composing the constraint reaction R_A with the first external force F_1: the first partial resultant R_1 will then be given by the corresponding triangle of forces of Figure 4.20b and its line of action will pass through point P_1 of intersection of the lines of action of R_A and F_1. Compose then the first partial resultant R_1 with the second force F_2. The second partial resultant R_2 passes through the pole P_2 and is obtained from the triangle $R_1 F_2 R_2$ of Figure 4.20b. We then proceed by adding each partial resultant vectorially with the following external force, thus obtaining the next partial resultant. Finally, by adding R_3 and F_4, we obtain the last resultant R_4, which must be a vector

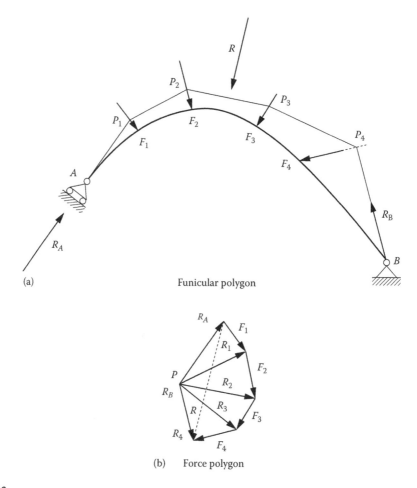

(a)   Funicular polygon

(b)   Force polygon

Figure 4.20

equal and opposite to the reaction R_B. In this way, the **polygon** (in this case, a hexagon) **of forces** closes (Figure 4.20b). It is made up of four triangular sectors, each of which represents a vector sum (i.e. a problem of equivalence). Note once again that the vector that has as its foot that of F_1 and as its head that of F_4 represents the resultant R. The partial resultants R_i, $i = 1, ..., 4$, are thus the internal fixed joint reactions, which each portion to the left of a generic section contained between F_i and F_{i+1} transmits to the complementary portion to the right of the same section. If in the triangular sectors of Figure 4.20b we invert the sense of the vectors R_i, $i = 1, ..., 4$, this means to consider a problem of equilibrium (instead of equivalence), and these vectors would represent, in this case, the internal fixed joint reactions that each right-hand portion transmits to the complementary left-hand portion.

It may now be understood how the **funicular polygon** or **pressure line**, being the set of the lines of action of the successive resultants, is represented in Figure 4.20a by the broken line $AP_1P_2P_3P_4B$. The sides of the funicular polygon are parallel to the rays of the **polygon of forces**, represented in Figure 4.20b by the polygonal line $R_AF_1F_2F_3F_4R_B$.

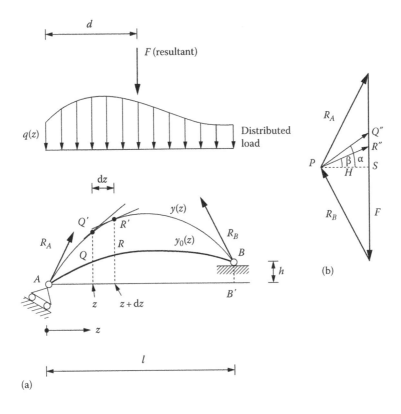

Figure 4.21

Now examine the case of a continuous system of forces $q(z)$ having equal direction and acting on the arch AB (Figure 4.21a). Integrating, we can obtain the resultant

$$F = \int_0^l q(z)\,dz \tag{4.17}$$

and its arm d with respect to the straight line $z = 0$

$$Fd = \int_0^l q(z)z\,dz \tag{4.18}$$

It will then be possible to obtain the constraint reactions R_A and R_B *via* the global triangle of forces (Figure 4.21b). In this case, the polygon of forces reduces to a triangle, because the distributed forces are all acting in the same direction. This is made up of an infinite number of infinitesimal triangular sectors, each of which represents the vector sum of the partial resultant with the subsequent increment of load $q(z)dz$.

Let us then take the pressure line to be known *a priori* and assume that it may be described analytically *via* the function $y(z)$. If Q and R represent two points of the arch infinitely close to one another of abscissae, respectively, z and $z + dz$, the corresponding partial resultants will be directed along the respective tangents to the pressure line (this being the envelope of the infinite lines of action of the subsequent resultants). These partial resultants will then be definable in their magnitudes *via* the rays PQ'' and PR'' of the triangle of forces of Figure 4.21b. Their angles of orientation α and β will differ by an infinitesimal amount and will be given by the respective tangents to the pressure line.

The segment $Q''R''$ in Figure 4.21b represents the increment of distributed load

$$Q''R'' = q(z)\,dz \tag{4.19}$$

On the other hand, some elementary geometrical considerations give

$$Q''R'' = Q''S - R''S = H(\tan\alpha - \tan\beta)$$

$$= -H[y'(R) - y'(Q)] \tag{4.20}$$

where H indicates the magnitude of the horizontal component of R_A and R_B while the prime indicates the first derivation with respect to the coordinate z. From Equations (4.19) and (4.20), by the transitive law, we have

$$q(z)\,dz = -H\,dy' \tag{4.21}$$

where the difference has been replaced by the differential, as the quantities are extremely close and as a pressure line is considered devoid of cusps (in fact there are no concentrated forces).

Finally, from relation (4.21), we obtain the differential equation of the pressure line for distributed loads acting in the same direction:

$$\frac{d^2y}{dz^2} = -\frac{q(z)}{H} \tag{4.22}$$

This is a second-order differential equation and hence is to be combined with two boundary conditions, which are, in this case $y(0) = 0$, $y(l) = h$, the pressure line passing through the two end hinges. The first partial resultants that we meet going around the structure from the left or from the right are the reaction R_A or the reaction R_B, respectively. Hence the pressure line, besides passing through A and B, must be tangential at these points to the relative external reactions.

When the distributed load q is constant, the funicular curve is parabolic. We shall see various examples of this hereafter.

It is interesting to note that the pressure line, of equation $y(z)$, represents, but for one factor, the diagram of the bending moment. The partial resultant, PQ'', in fact, is made up of the horizontal force PS and the vertical force SQ'' (Figure 4.21b) and is applied in Q' (Figure 4.21a). Thus, it is clear that the moment of internal reaction PQ'' is equal to the sum of the moments of PS and SQ''. While the former is equal to $H \times \overline{QQ'}$, the latter is zero because the corresponding arm vanishes.

Finally,

$$M = H(y - y_0) \tag{4.23}$$

where

    M is the so-called **bending moment**

    H is the projection of the external reactions R_A and R_B on the normal to the direction of the external forces

    y is the distance of the pressure line from the fundamental straight line AB', while y_0 indicates the distance of the axis of the arch from that fundamental

The segment intercepted between the pressure line and the axis of the beam thus represents, but for the factor H, the corresponding bending moment.

From relation (4.22), we then obtain

$$\frac{d^2 M}{dz^2} = -q(z) - H\frac{d^2 y_0}{dz^2} \tag{4.24}$$

which is the differential equation of the bending moment in the case of a beam with a curvilinear axis. When the beam is rectilinear, we have $y_0 = hz/l$, and thus

$$\frac{d^2 M}{dz^2} = -q(z) \tag{4.25}$$

If we imagine constructing an arch that presents exactly the form of the pressure line, we have $y = y_0$, and thus the bending moment vanishes at each point of the arch. Between one section and another, only a compressive force would then be transmitted, as the internal reaction is always tangential to the axis of the curved beam. This is the situation that tends to occur when incoherent materials are used, i.e. those without tensile strength.

If all the forces acting on the arch, and thus also the reactions R_A and R_B, were inverted, only tensile internal reactions would be obtained. A string of length equal to that of the pressure line would, in fact, be disposed according to the configuration of this line, it not being able to support other than tensile stresses; hence, the name **funicular curve**, from the Latin *funis* meaning cable or rope.

Consider a three-hinged parabolic arch, subjected to a uniform vertically distributed load q (Figure 4.22). The pressure line is parabolic and passes through the three hinges, where the bending moment vanishes. Recalling that only one parabola may pass through three given points, the pressure line must necessarily coincide with the axis of the arch, which will be found to be entirely in compression and devoid of bending moment.

Wide use has traditionally been, and still is, made of **arches** for buildings having wide spans, such as bridges. Usually, it is the deck that transmits the vertical loads, which consist of its own weight and any live loads, to the supporting arch by means of connecting bars that can all be in compression (Figure 4.23a), all in tension (Figure 4.23b) or partly in compression and partly in tension (Figure 4.23c), according to the level at which the deck is disposed. Then there are **suspension bridges**, where the static scheme is inverted and the supporting element is represented by a cable in tension in a parabolic configuration (Figure 4.23d). In this case, the elements of transmission are all tie rods.

We recall that, in the case where the vertical distributed load is not constant per unit of span, but constant per unit length of the arch, i.e. in practice it represents the weight of the arch itself assumed to be of uniform section, the pressure line is no longer exactly parabolic but assumes the form of a **catenary**.

Let us now examine another notable case in which the pressure line coincides with the axis of the arch. Let the three-hinged semicircular arch of Figure 4.24a be subjected to a

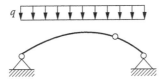

Figure 4.22

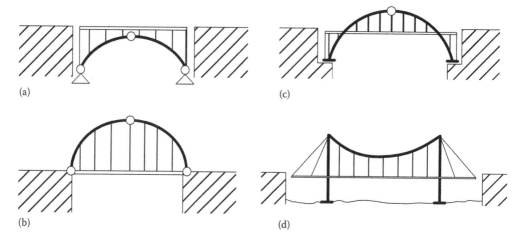

Figure 4.23

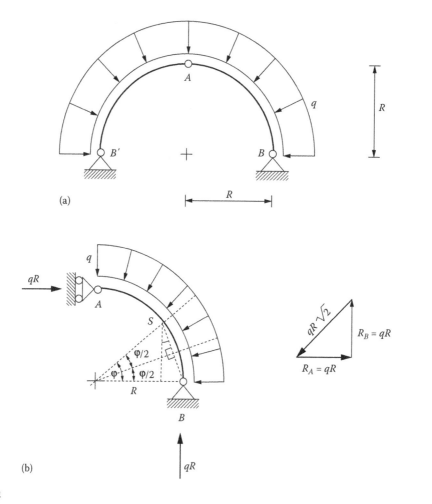

Figure 4.24

constant radial distributed load q. For reasons of symmetry, it is possible to consider only one half of the arch, if it is noted that the vertical reaction transmitted by the hinge A must vanish (Figure 4.24b). On the basis of the triangle of forces, the reactions R_A and R_B are thus of magnitude equal to qR. The moment transmitted by a generic section S of the arch, identified by the polar angle φ, may be calculated as the algebraic sum of the moment of the reaction R_B and the moment of forces distributed between 0 and φ. The first is counter-clockwise and equal to

$$M_S(R_B) = qR^2(1 - \cos\varphi) \tag{4.26}$$

while the second is clockwise and equal to

$$M_S(q) = \left(2qR\sin\left(\frac{\varphi}{2}\right)\right)\left(R\sin\left(\frac{\varphi}{2}\right)\right) \tag{4.27}$$

The resultant of a constant radial distribution q of forces is given by the product of q and the chord subtended by the arc on which the forces act. Since then from trigonometry we have

$$\sin\left(\frac{\varphi}{2}\right) = \left(\frac{1 - \cos\varphi}{2}\right)^{\frac{1}{2}} \tag{4.28}$$

it follows that the expressions (4.26) and (4.27) are equal and that, therefore, the moment M_S in any generic section is zero.

To demonstrate rigorously that the pressure line coincides with the circular axis of the beam, it is necessary to demonstrate that, in addition to M_S, in each section S the internal radial reaction T_S also vanishes.

The radial reaction transmitted by the reaction R_B is equal to (Figure 4.24b)

$$T_S(R_B) = qR\sin\varphi \tag{4.29}$$

while the radial reaction transmitted by the forces distributed between 0 and φ is equal to

$$T_S(q) = -\int_0^\varphi qR\cos(\varphi - \omega)d\omega \tag{4.30}$$

$qRd\omega$ being the elementary contribution, acting at an inclined polar angle at ω on the horizontal (Figure 4.25). Setting $x = \varphi - \omega$ and integrating Equation (4.30), we obtain

$$T_S(q) = -qR\int_0^\varphi \cos x\, dx$$

$$= -qR[\sin x]_0^\varphi = -qR\sin\varphi \tag{4.31}$$

The contributions (4.29) and (4.31) cancel each other out, and thus the pressure line coincides with the semi-circumference of radius R.

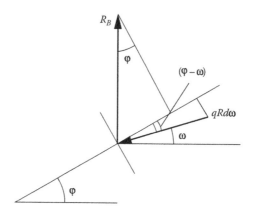

Figure 4.25

The only reaction transmitted by the internal built-in constraints is the tangential one N_S. The tangential reaction transmitted by R_B is equal to (Figure 4.24b)

$$N_S(R_B) = qR\cos\varphi \tag{4.32}$$

while the tangential reaction transmitted by the radial forces between 0 and φ is equal to (Figure 4.25)

$$N_S(q) = \int_0^\varphi qR\sin(\varphi - \omega)\,d\omega \tag{4.33}$$

Setting again $x = \varphi - \omega$, we obtain

$$N_S(q) = qR\int_0^\varphi \sin x\,dx$$

$$= qR[-\cos x]_0^\varphi = qR(1 - \cos\varphi) \tag{4.34}$$

Summing up the axial forces (4.32) and (4.34), we have finally

$$N_S = qR \tag{4.35}$$

The axial force is thus constant and compressive over the whole arch. Its absolute value coincides with that of reactions R_A and R_B.

In the simple cases considered in the previous chapter, consisting of one, two or, at the most, three rigid bodies, loaded by a concentrated force, the pressure line reduces to the set of a finite number of straight lines. In the case of the L-shaped beam of Figure 4.8a, the pressure line consists of the line of action of the reaction R_A for all the points contained between A and the point of application of the force, and by the line of action of the reaction R_B for all the points contained between the point of application of the force and the hinge B. In other words, it is a pair of straight lines, and we pass discontinuously from one to the other, passing over the point of application of the concentrated force.

For the bar system of Figure 4.5a, the pressure line consists of the axes of the two bars, while for the three-hinged arch of Figure 4.11, the pressure line consists of the line joining A

and C for all the points of the section AC, and of the line joining B and C for all the points of the section BC. In the case of the arch of Figure 4.12, the pressure line is the straight line BP for the points to the right of the external force, while it is the straight line AP for the points contained between A and the force.

For the beam constrained by the three connecting rods of Figure 4.9a, the pressure line consists of the line joining A and P for the points between A and the point of application of the force, of the straight line DP for the points contained between the point of application of the force and the rod B, and finally of the axis of the rod C for the points contained between the last two right-hand connecting rods, B and C.

When the same beam is subjected to a concentrated moment m (Figure 4.10), the pressure line is represented by the axis of the rod A for the points contained between A and the loaded point, by a parallel straight line passing through point D for the points contained between m and B, and finally by the axis of the rod C for the remaining points.

Now consider the three-hinged portal frame of Figure 4.26a, subjected to a vertical distributed load q on the right-hand section. The external reaction R_A has as its line of action the line joining A and B, which intersects the line of action of the resultant ql in the pole P. The second external reaction R_C passes through points C and P and can be determined graphically, together with R_A, by means of the triangle of forces. The pressure line for the left-hand section will again be the line joining A and B, while for the section BC, this will be composed of an infinite number of straight lines that have a parabolic envelope with a vertical axis. This parabola passes, of course, through the hinges B and C and admits, in those points, of the lines of action of the reactions R_A and R_C as its tangents. At this point,

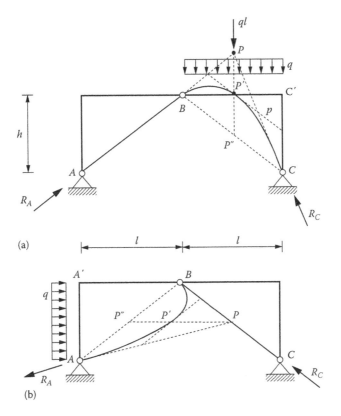

Figure 4.26

the arc of parabola between B and C is defined, since three data are sufficient to identify a second-order parabola. Even though two points of the parabola are already known with their corresponding tangents, to make the graphical construction easier, a third point will be identified along with its tangent. Indicate by P'' the intersection of the vertical through P with the line joining the extreme points B and C, and by P' the midpoint of the segment PP''. Then draw through P' a straight line p parallel to the line joining B and C. It is possible to demonstrate that P' and the latter line p constitute the third point and the third tangent that we had set out to obtain. It is now extremely easy to draw the arc of parabola, since this must pass through B, P' and C, and it is inscribed in the polygonal line made up of the straight lines AP, p and PC (Figure 4.26a). In the section CC', the pressure line is represented by the straight line CP.

Consider the case of a horizontal distributed load, acting on the left-hand section of the portal frame previously studied (Figure 4.26b). While for the section CA', which is not externally loaded, the pressure line is represented by the line joining the two hinges B and C, for the section AA', the pressure line is represented by an arc of parabola with horizontal axis, which has as its extreme points A and B and as its tangents at those points the lines of action of the reactions R_A and R_B. It will then be possible, as shown previously, to identify a third point P' and a third tangent parallel to the line joining the extreme points. Note that, if the axis of the structure were now conceived with the same form as the funicular curve found, the section AB would be found to be completely in tension and the section BC would act as a strut (connecting bar in compression).

As our last example, consider the asymmetrical portal frame of Figure 4.27, already studied more than once, loaded by a constant distribution of vertical forces on the right-hand section. The horizontal reaction of the double rod passes through the pole P and constitutes, with its line of action, one of the two extreme tangents to the arc of parabola with a vertical

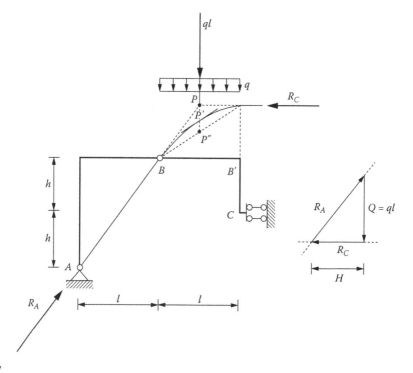

Figure 4.27

axis, which represents the pressure line for the section BB'. This arc must be contained in the vertical strip of the plane containing also the vertical distribution q. The second tangent to the arc of parabola is, of course, given by the straight line AB. We therefore find, finally, that the pressure line for the section CB' is represented by the line of action of the reaction R_C, for the section AB, it is represented by the line joining A and B, while for the loaded section BB', it is represented by the arc of parabola connecting the previous two straight lines.

Note that, in the case of a concentrated load equal to the resultant of the distributed load under examination, the pressure line would be made up of the two earlier-mentioned extreme straight lines only, passing from one to the other with discontinuity in crossing the point of application of the force. Hence there derives from this a sort of rounding off brought about by the continuous, rather than discrete, distribution of the external forces.

Chapter 5

Internal beam reactions

5.1 INTRODUCTION

The **characteristics of internal reaction** for a beam section are, as stated in the earlier chapters, the elementary internal reactions transmitted by the section itself. In the case of a plane beam (line of axis contained in the plane), there are three characteristics of internal reaction (Figure 5.1):

1. **Axial force,** which is the component of the force tangential to the axis of the beam
2. **Shearing force** (or, more simply, **shear**), which is the component of the force perpendicular to the axis of the beam
3. **Bending moment,** which is the moment of the force that the two portions of beam transmit to one another, with respect to the section being considered

The usual conventions regarding the signs of the characteristics in the plane are the following:

1. Axial force is taken to be positive when it is tensile.
2. Shearing force is taken to be positive when it tends to rotate the segment of the beam on which it acts in a clockwise direction.
3. Bending moment is taken to be positive when it stretches the lower fibres and compresses the upper fibres of the beam.

It is at once evident that these conventions are purely arbitrary and, except in the case of axial force, relative to the observer's orientation (bending moment) or half-space of observation (shear).

In the case of a beam with skewed axis, it becomes necessary to establish a direction in which we consider the axis of the beam, and to establish, for each section, an intrinsic reference system consisting of the tangent, the normal and the binormal to the curve (Figure 5.2). More precisely, the reference system will be right-handed, with the Z axis (middle finger) oriented according to the tangent and in the direction chosen for considering the axis, the Y axis (index) oriented according to the normal and the X axis (thumb) according to the binormal. The force transmitted by the lower section to the upper section must be projected on these axes and the respective components are the **axial force N**, the **shearing force T_y** and the **shearing force T_x** (Figure 5.2a). On the other hand, the opposite force transmitted by the upper section to the lower section must be projected on to the left-handed reference system, opposite to the system previously considered, in order to obtain characteristics of the same sign as the previous ones. As regards the moment vector mutually exchanged between the two beam portions (Figure 5.2b), conventions and considerations altogether analogous

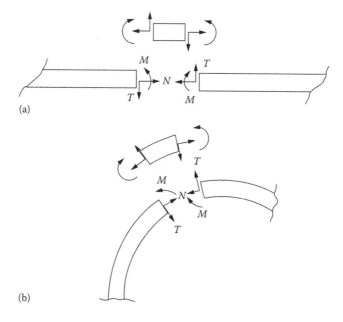

Figure 5.1

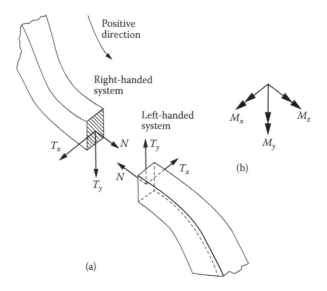

Figure 5.2

to those set out earlier hold good also here. The moment vector comprises three components. The one along the Z axis is called the **twisting moment** M_z while the remaining two are the bending moments M_y and M_x. In the case, therefore, of the beam in 3D space, there are six characteristics of the internal reaction: axial force, two shearing forces, twisting moment and two bending moments. Also in the 3D case, the conventions are arbitrary and the signs of the characteristics come to depend on the direction in which we consider the axis of the beam.

5.2 INDEFINITE EQUATIONS OF EQUILIBRIUM
FOR PLANE BEAMS

We shall now deduce the differential equations that govern the equilibrium of a plane beam. For this purpose, consider an infinitesimal element of the beam, bounded by two right sections, i.e. perpendicular to the axis of the beam (Figure 5.3). Indicate by O the centre of curvature of the axis of the beam in relation to the element considered. If r is the radius of curvature, ds the increment of the curvilinear coordinate in the direction of the axis of the beam and dϑ the angle formed by the two end sections of the element (positive if the upper section is superposed on the lower section rotating in a counterclockwise direction), we have by definition

$$\mathrm{d}s = r\,\mathrm{d}\vartheta \tag{5.1}$$

In this way, the radius of curvature r, as well as the increments, assumes an algebraic sign (in Figure 5.3, r is positive). We point out that, if instead the element were observed from the opposite half-space, the radius r would be negative.

The infinitesimal element of the beam is in general subjected to an axial distributed load $p(s)\mathrm{d}s$, a transverse distributed load $q(s)\mathrm{d}s$ and a distributed moment $m(s)\mathrm{d}s$, as well as to the characteristics N, T, M, at the upward end, and to the incremented characteristics $N + \mathrm{d}N$, $T + \mathrm{d}T$, $M + \mathrm{d}M$, at the downward end (Figure 5.3). Noting that dϑ/2 represents both the angle between the median radial line and the end sections, and the angle between the median tangent and the extreme tangents, it is possible to impose equilibrium with regard to translation of the element in the directions of the median tangent and the median radial line, respectively:

$$p\mathrm{d}s - N\cos\frac{\mathrm{d}\vartheta}{2} + (N + \mathrm{d}N)\cos\frac{\mathrm{d}\vartheta}{2} + T\sin\frac{\mathrm{d}\vartheta}{2} + (T + \mathrm{d}T)\sin\frac{\mathrm{d}\vartheta}{2} = 0 \tag{5.2a}$$

$$q\mathrm{d}s - N\sin\frac{\mathrm{d}\vartheta}{2} - (N + \mathrm{d}N)\sin\frac{\mathrm{d}\vartheta}{2} - T\cos\frac{\mathrm{d}\vartheta}{2} + (T + \mathrm{d}T)\cos\frac{\mathrm{d}\vartheta}{2} = 0 \tag{5.2b}$$

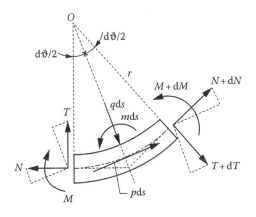

Figure 5.3

As regards rotational equilibrium, it is expedient to choose as centre of reduction the point of intersection of the extreme tangents, i.e. of the lines of action of the axial forces on the end sections:

$$m\,ds + dM - p\,ds\left(\frac{r}{\cos\left(\dfrac{d\vartheta}{2}\right)} - r\right) - Tr\tan\frac{d\vartheta}{2} - (T + dT)r\tan\frac{d\vartheta}{2} = 0 \qquad (5.2c)$$

Since the angle $d\vartheta/2$ is assumed to be infinitesimal, it is legitimate to take this angle as equal to its sine or its tangent, and put $\cos(d\vartheta/2) \simeq 1$. By so doing, Equations 5.2 are transformed as follows:

$$p\,ds + dN + T\,d\vartheta = 0 \qquad (5.3a)$$

$$q\,ds - N\,d\vartheta + dT = 0 \qquad (5.3b)$$

$$m\,ds + dM - Tr\,d\vartheta = 0 \qquad (5.3c)$$

Dividing the foregoing equations by ds and applying relation (5.1), we obtain the **indefinite equations of equilibrium** for the beam:

$$\frac{dN}{ds} + \frac{T}{r} + p = 0 \qquad (5.4a)$$

$$\frac{dT}{ds} - \frac{N}{r} + q = 0 \qquad (5.4b)$$

$$\frac{dM}{ds} - T + m = 0 \qquad (5.4c)$$

In the case where the radius of curvature r is not a function of the curvilinear coordinate s, namely that of **circular arches** and **rings**, Equations 5.4 can be presented in the following form:

$$\frac{dN}{d\vartheta} + T + pR = 0 \qquad (5.5a)$$

$$\frac{dT}{d\vartheta} - N + qR = 0 \qquad (5.5b)$$

$$\frac{dM}{d\vartheta} - TR + mR = 0 \qquad (5.5c)$$

where the independent variable is represented by the angular coordinate ϑ and R is the radius of the circular axis of the beam.

Equations 5.5, like Equations 5.4, form a system of three linear differential equations of the first order in the three unknown functions N, T, M. It is possible to decouple the function M from the other two and obtain a third-order differential equation, where only the unknown $M(\vartheta)$ appears.

From Equation 5.5c, we have

$$T = m + \frac{1}{R}\frac{dM}{d\vartheta} \tag{5.6}$$

so that Equations 5.5a and b are transformed as follows:

$$\frac{dN}{d\vartheta} + m + \frac{1}{R}\frac{dM}{d\vartheta} + pR = 0 \tag{5.7a}$$

$$\frac{dm}{d\vartheta} + \frac{1}{R}\frac{d^2M}{d\vartheta^2} - N + qR = 0 \tag{5.7b}$$

From Equation 5.7b, then, we obtain

$$N = qR + \frac{dm}{d\vartheta} + \frac{1}{R}\frac{d^2M}{d\vartheta^2} \tag{5.8}$$

and hence from Equation 5.7a there follows

$$\frac{d^3M}{d\vartheta^3} + \frac{dM}{d\vartheta} = -R^2\left(p + \frac{dq}{d\vartheta}\right) - R\left(m + \frac{d^2m}{d\vartheta^2}\right) \tag{5.9}$$

Equation 5.9 is a nonhomogeneous third-order differential equation, which has the following complete integral:

$$M(\vartheta) = M_0(\vartheta) + C_1 \sin\vartheta + C_2 \cos\vartheta + C_3 \tag{5.10}$$

where $M_0(\vartheta)$ indicates the particular solution. In the case, for instance, of the circular arch of Figure 5.4, subjected to a hydrostatic load, we have

$$p(\vartheta) = m(\vartheta) = 0 \tag{5.11a}$$

$$q(\vartheta) = -\gamma R(1 - \cos\vartheta) \tag{5.11b}$$

where γ indicates the specific weight of the fluid exerting pressure. The three boundary conditions are those which cause the corresponding characteristics to vanish at the two ends: $M(A) = M(B) = T(A) = 0$. In order to resolve the earlier analytical problem, it is possible to apply the method of variation of the arbitrary constants (Appendix A).

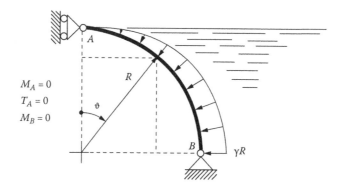

Figure 5.4

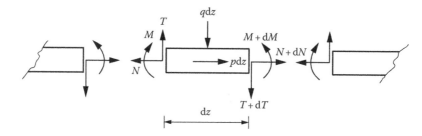

Figure 5.5

In the case where the radius of curvature r tends to infinity, namely that of **rectilinear beams** (Figure 5.5), and in the absence of distributed moments ($m = 0$), the indefinite equations of equilibrium (5.4) reduce to the following:

$$\frac{dN}{dz} = -p(z) \tag{5.12a}$$

$$\frac{dT}{dz} = -q(z) \tag{5.12b}$$

$$\frac{dM}{dz} = T \tag{5.12c}$$

The first of these tells us that the derivative of the axial force with respect to the axial coordinate is equal to the opposite of the axial distributed load; the second that the derivative of the shearing force is equal to the opposite of the distributed load perpendicular to the axis; and finally, the third that the derivative of the bending moment is equal to the shearing force. In addition, deriving both sides of Equation 5.12c and taking into account Equation 5.12b, we again obtain Equation 4.25, which in the previous chapter was obtained on the basis of considerations involving the pressure line.

In the absence of distributed loads, then, both the axial force and the shearing force are constant, while the bending moment is a linear function of the z coordinate. The diagram of the characteristics is to be studied for the portions of beam contained between one concentrated load and another. At the points of application of the concentrated loads, there emerges, instead, the discontinuity of the corresponding characteristics. When the distributed loads are constant, on the other hand, both the axial force and the shearing force are linear functions of the z coordinate, while the bending moment is a second-order parabolic function. More generally, when the distributed loads are polynomial functions of order n, the axial force and the shearing force are polynomial functions of order $(n + 1)$, while the bending moment is a polynomial function of order $(n + 2)$.

We shall now apply the indefinite equations of equilibrium (5.12) to determine analytically the functions M, T, N, for the inclined rectilinear beam of Figure 5.6a, which is subjected to a uniform vertical distributed load $q_0 = F/l$, where F is the resultant force and l the projection of the beam on the horizontal. The triangle of equilibrium gives the constraint reactions $R_A = q_0 l/\tan \beta$, $R_B = q_0 l/\sin \beta$, β being the angle that the line joining B and P forms with the horizontal (Figure 5.6b). The pressure line is a parabola with vertical axis that passes through the extreme points of the beam and presents as extreme tangents the lines of action of the two external reactions. Later we shall verify how this represents, but for a factor of proportionality, the diagram of the bending moment is shown.

The vertical distributed load per unit length of the beam is

$$q^* = \frac{q_0 l}{l/\cos \alpha} = q_0 \cos \alpha \tag{5.13}$$

where
 α is the angle of inclination of the beam on the horizontal
 $l/\cos \alpha$ is the length of the beam

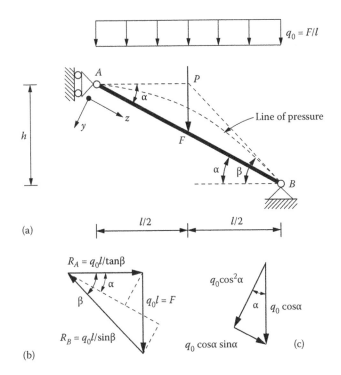

Figure 5.6

The axial component and the component perpendicular to the axis of the distributed load q^* are then equal to (Figure 5.6c)

$$p(z) = q_0 \cos\alpha \sin\alpha \tag{5.14a}$$

$$q(z) = q_0 \cos^2\alpha \tag{5.14b}$$

The differential equation that governs the bending moment will thus be the following:

$$\frac{\mathrm{d}^2 M}{\mathrm{d}z^2} = -q_0 \cos^2\alpha \tag{5.15}$$

with the boundary conditions

$$M(0) = M\left(\frac{l}{\cos\alpha}\right) = 0 \tag{5.16}$$

The complete integral of the function $M(z)$ contains two constants that depend on the foregoing boundary conditions:

$$M(z) = -q_0 \cos^2\alpha \frac{z^2}{2} + C_1 z + C_2 \tag{5.17}$$

Applying the two conditions (5.16), we obtain two algebraic equations in the two unknowns C_1 and C_2:

$$M(0) = C_2 = 0 \tag{5.18a}$$

$$M\left(\frac{l}{\cos\alpha}\right) = -\frac{1}{2}q_0 l^2 + C_1 \frac{l}{\cos\alpha} = 0 \tag{5.18b}$$

whence

$$C_1 = \frac{1}{2}q_0 l \cos\alpha, \quad C_2 = 0 \tag{5.19}$$

The moment function is then given by

$$M(z) = \frac{1}{2}q_0 lz \cos\alpha\left(1 - \frac{z}{l}\cos\alpha\right) \tag{5.20}$$

for $0 \leq z \leq l/\cos\alpha$.

The diagram of the moment function, given in Figure 5.7a on the side of the fibres in tension, represents a parabola having the axis perpendicular to the beam. Note that the distribution is symmetrical with respect to the centre of the beam and that the maximum that is reached in the centre is independent of the inclination α of the beam:

$$M_{\max} = M\left(\frac{l}{2\cos\alpha}\right) = \frac{1}{8}q_0 l^2 \tag{5.21}$$

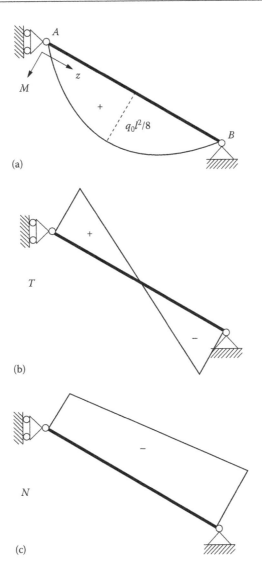

(a)

(b)

(c)

Figure 5.7

The shearing force can be found by derivation of the moment function (5.20):

$$T(z) = \frac{dM}{dz} = \frac{1}{2} q_0 l \cos \alpha \left(1 - \frac{2z}{l} \cos \alpha \right) \tag{5.22}$$

The diagram of the shear function (Figure 5.7b) is linear and skew-symmetrical with respect to the centre of the beam, where it vanishes (stationary point of the bending moment).

The axial force, finally, may be obtained from the first of Equations 5.12:

$$\frac{dN}{dz} = -q_0 \cos \alpha \sin \alpha \tag{5.23}$$

which gives

$$N(z) = -q_0 z \cos\alpha \sin\alpha + C \tag{5.24}$$

where the constant C is determined by imposing a suitable boundary condition. It is possible, e.g. to consider that, at the end A, the axial compressive force coincides with the component of the reaction R_A along the axis of the beam (Figure 5.6b):

$$N(0) = C = -q_0 l \frac{\cos\alpha}{\tan\beta} \tag{5.25}$$

hence we have

$$N(z) = -q_0 \cos\alpha \left(z\sin\alpha + \frac{l}{\tan\beta} \right) \tag{5.26}$$

It is possible, on the other hand and in an equivalent manner, to assume that the axial force at the end B is compressive and coincides with the component of the reaction R_B along the axis of the beam (Figure 5.6b):

$$N\left(\frac{l}{\cos\alpha} \right) = -q_0 l \sin\alpha + C = -q_0 l \frac{\cos(\beta - \alpha)}{\sin\beta} \tag{5.27}$$

From Equation 5.27, we obtain

$$C = -q_0 l \left(\frac{\sin\alpha \sin\beta + \cos\alpha \cos\beta}{\sin\beta} - \sin\alpha \right) = -q_0 l \frac{\cos\beta}{\tan\beta} \tag{5.28}$$

which confirms Equation 5.25.

The trapezoidal diagram of the axial force appearing in Figure 5.7c shows how the beam is entirely in compression and how the maximum of this force is reached at the end B.

The distinction between distributed load per unit of horizontal projection and distributed load per unit length of the beam is necessary, e.g. also in the case of the circular arch of Figure 5.8, subjected to a vertical distributed load q_0, which is uniform if considered per

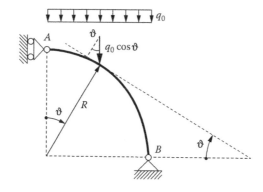

Figure 5.8

unit length of horizontal span. Just as was seen in the case of the inclined rectilinear beam (cf. Equation 5.14), the elementary components of the distributed load are

$$p(\vartheta) = q_0 \cos \vartheta \sin \vartheta \tag{5.29a}$$

$$q(\vartheta) = -q_0 \cos^2 \vartheta \tag{5.29b}$$

The full solution for the earlier example is given in Appendix B.

5.3 DIAGRAMS OF CHARACTERISTICS OF INTERNAL REACTION: DIRECT METHOD AND GRAPHICAL METHOD

In the last section, we introduced the characteristics of internal reaction and the differential equations governing them. It has thus been possible to note how the problem of the determination of these functions can be set and resolved in a purely analytical manner. In this section, we shall see how it is possible to approach the same problem from a direct point of view, i.e. imposing equilibrium on finite portions of beam, subjected to external loads, to known constraint reactions and to unknown characteristics. On these bases, it is possible in many cases to draw the diagrams of the characteristics M, T, N, using purely graphical procedures.

To start with, let us consider the so-called **elementary schemes**, which, on account of their simplicity, recur very frequently, also being inserted within more complex structural schemes. A built-in rectilinear beam, known as a **cantilever beam**, is subjected to a force F perpendicular to its axis and with the point of application in the end B (Figure 5.9a). The built-in support A reacts with a force equal and opposite to the external one, so that the pressure line coincides with the line of action of F for all the points of the beam. On the other hand, the reaction R_A can be thought of as acting in point A together with the counterclockwise moment of transport $M_A = R_A l = Fl$. The built-in support A will react, therefore, transmitting a positive shear F and a negative bending moment $-Fl$ to the beam. The diagram of the bending moment is linear, owing to the absence of distributed loads $q(z)$, and vanishes at point B, with respect to which the external force has a zero arm (Figure 5.9b). To draw it, it will be sufficient to perform a simple graphical operation, joining point B with the upper end of the segment that represents the moment in the built-in support $-Fl$. This operation, however commonplace it may be, involves a series of logical steps that we shall endeavour to illustrate.

The fact that, in a generic section of the cantilever beam, the bending moment is $M(z) = -F(l - z)$ and the shear is $T(z) = F$ (Figure 5.9c) means that the beam portion ZB is found to be in equilibrium under the action of the external force F, the positive shear F and the counterclockwise bending moment $F(l - z)$ (Figure 5.10a). The last two loads are the internal reactions transmitted by the left-hand portion AZ to the one being considered, ZB. On the other hand, it also means that the portion AZ (Figure 5.10b) is in equilibrium under the action of the counterclockwise fixed-end moment Fl, the vertical fixed-end reaction F, the positive shear F and clockwise bending moment $F(l - z)$. To determine, section by section, the internal characteristics M and T, it will suffice, therefore, to consider a problem of equilibrium for either of the two portions into which the beam is divided by the section under examination. Similarly, it is possible, instead, to consider a problem of equivalence and to transfer into the section under examination all the forces acting upwards of the section itself, applying this equivalent system to the complementary portion of the beam.

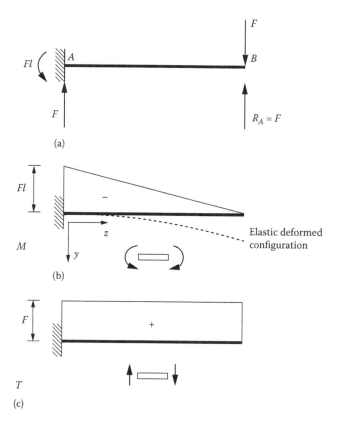

Figure 5.9

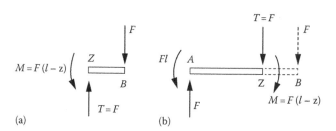

Figure 5.10

If the cantilever beam is loaded by a concentrated moment at the end B (Figure 5.11a), the built-in support reacts with an equal and opposite moment, so that each partial portion of the beam, whether finite or infinitesimal, is in equilibrium under the action of two opposite moments. The line of pressure is the straight line at infinity, since the resultant force of a couple is zero and tends to act at infinity. The moment diagram is constant and negative (Figure 5.11b), while the shear is identically zero (Figure 5.11c), being equal to the derivative of a constant function and there being no vertical forces involved.

The last elementary scheme for the cantilever beam is that of the uniform distributed load q (Figure 5.12a). The reaction of the built-in support A is equal and opposite to the resultant ql of the distributed load. The pressure line is thus degenerate and consists of the sheaf of

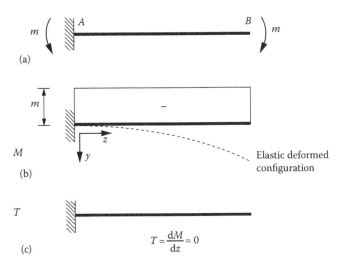

Figure 5.11

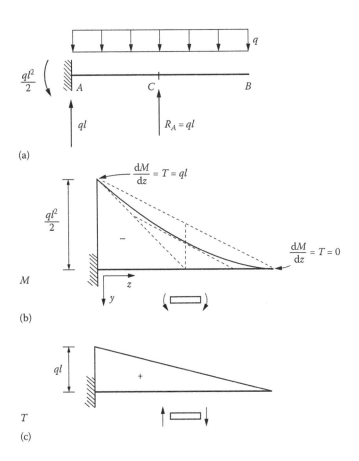

Figure 5.12

vertical straight lines contained between the end B and the midpoint C (Figure 5.12a). The bending moment in one generic section of abscissa z is thus (Figure 5.12b)

$$M(z) = -q\frac{(l-z)^2}{2} \tag{5.30}$$

while the shear is (Figure 5.12c)

$$T(z) = q(l-z) \tag{5.31}$$

These two functions have been obtained by reducing, in the section under consideration, the system of forces acting to the right of the section. This reduction of the forces that precedes a generic section is referred to as the **direct method.**

It is, on the other hand, possible to proceed in a completely graphical manner, using the moment diagram for a concentrated force equal to the resultant of the system and acting in the middle of the beam. This diagram is linear and presents the zero value in the centre C and the maximum absolute value $ql^2/2$ at the built-in support A (Figure 5.12b). In actual fact, the moment diagram for the distribution of forces q is parabolic and presents a zero value at the end B and a maximum absolute value equal to the previous one of $ql^2/2$ at the built-in support A, while its tangent in A coincides with the linear diagram described formerly, the shear being transmitted by the built-in support in a like manner in both cases. The tangent at B is then horizontal, the shear vanishing at that point. We thus have the two extreme points with the corresponding tangents to the arc of parabola that is sought. It is then simple, by applying the graphical construction already illustrated in the previous chapter, to identify a third point with its corresponding tangent and to draw the diagram of $M(z)$ precisely.

As regards the shear (Figure 5.12c), the graphical construction of the diagram is immediate, if we join the end of the cantilever beam with the upper end of the segment that represents the vertical reaction of the built-in support ql. In this case, it is also possible to verify that the function $T(z)$ is the derivative of the function $M(z)$.

Now consider the L-shaped cantilever beam of Figure 5.13a, subjected to a horizontal concentrated load at the end C. The reaction of the built-in support R_A is a force equal and opposite to the external one, so that, reduced in point A, it is equivalent to a horizontal reaction F and to a fixed-end moment equal to the moment of translation Fh. The axial force (Figure 5.13b) is zero on the portion CB, while it is compressive and of an absolute value equal to F on the portion BA. The shearing force (Figure 5.13c) is, *vice versa*, zero on the portion BA and equal to F on the portion CB. Finally, the bending moment (Figure 5.13d) increases linearly in absolute value proceeding from the end C to the knee B. From B to A, the absolute value remains constant and equal to the product of the force and the arm h. The algebraic sign of the bending moment depends on the conventions (taking those of Figure 5.13d, it is negative). However, what is independent of the reference system and physically important is the part (or edge) of the beam in which the longitudinal fibres are in tension. As has already been said, it is customary to draw the moment diagram (whatever sign it may have) on the side of the fibres in tension. This is illustrated by the elastic deformed configuration of Figure 5.13e. From the graphical point of view, we use the term **overturning** of the value of the moment at B, implying by this the equilibrium to rotation of the built-in node B (Figure 5.13d).

If the cantilever beam has a skewed axis, the characteristics that could be present total six. An example is shown in Figure 5.14, where the intrinsic reference system has been highlighted for each rectilinear portion of the beam. In the portion AB, only two characteristics are different from zero: $T_y = F$, $M_x = Fz$. In the portion BC, we have $N = -F$,

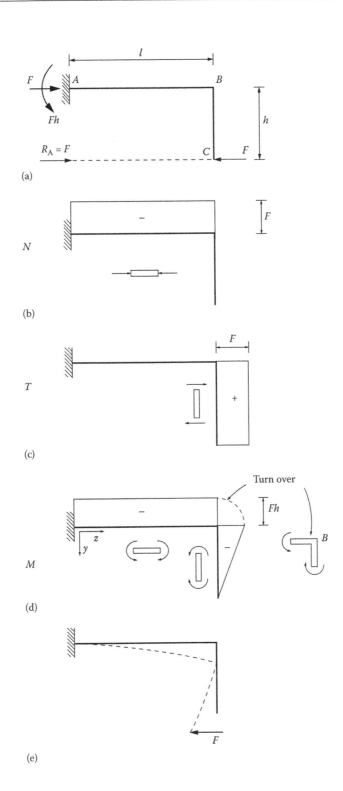

Figure 5.13

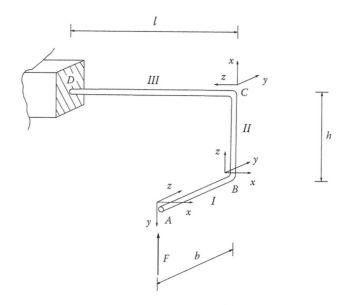

Figure 5.14

$M_x = Fb$. In the last portion CD, there also emerges the internal reaction of twisting moment: $T_x = -F$, $M_y = Fz$, $M_z = -Fb$.

Also in plane cantilever beams, if these are loaded with forces not contained in the plane, there is present the internal characteristic M_z. In the case, e.g. of the semicircular cantilever beam of Figure 5.15a, loaded with a force F perpendicular to the plane and applied at the end A (the direction assumed looking into the plane of the diagram), we have

$$N = 0 \tag{5.32a}$$

$$T_x = 0 \tag{5.32b}$$

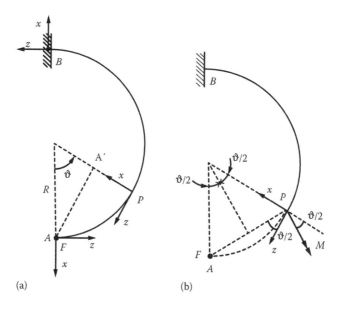

Figure 5.15

$$T_y = -F \tag{5.32c}$$

$$M_z = FR(1 - \cos \vartheta) \tag{5.32d}$$

$$M_x = -FR \sin \vartheta \tag{5.32e}$$

$$M_y = 0 \tag{5.32f}$$

The moments M_z and M_x can be determined in two different ways. The first consists of transferring the force F at A' (Figure 5.15a), adding the first moment of translation M_x, and subsequently in P, adding the second moment of translation M_z. The second way consists of considering the moment vector of the force F with respect to the generic point P (double-headed arrow in Figure 5.15b) and projecting this vector onto the left-handed XYZ reference system. The total moment has the magnitude

$$M = 2FR \sin\left(\frac{\vartheta}{2}\right) \tag{5.33}$$

and thus

$$M_z = M \sin\left(\frac{\vartheta}{2}\right) = FR(1 - \cos \vartheta) \tag{5.34a}$$

$$M_x = -M \cos\left(\frac{\vartheta}{2}\right) = -FR \sin \vartheta \tag{5.34b}$$

Now consider a rectilinear beam, hinged at one end and constrained with a horizontally moving roller support at the other (Figure 5.16a). This elementary scheme is referred to as the **supported beam** and will be studied in the various cases of external loading. Let the supported beam be subjected to the concentrated force F acting on its centre. The constraint reactions consist of two vertical forces, each equal to $F/2$. The horizontal reaction of the hinge is, in fact, zero by equilibrium to horizontal translation. The pressure line will thus consist of the line of action of the corresponding constraint reaction for each of the two portions AB and BC. The moment diagram will consist of two linear functions, symmetrical with respect to the centre (Figure 5.16b):

$$M(z) = \frac{F}{2}z, \quad \text{for } 0 \leq z \leq \frac{l}{2} \tag{5.35a}$$

$$M(z) = \frac{F}{2}z - F\left(z - \frac{l}{2}\right), \quad \text{for } \frac{l}{2} \leq z \leq l \tag{5.35b}$$

The function (5.35b), which emphasizes the sum of the two contributions, may be rewritten as follows:

$$M(z) = -\frac{F}{2}z + F\frac{l}{2} \tag{5.36}$$

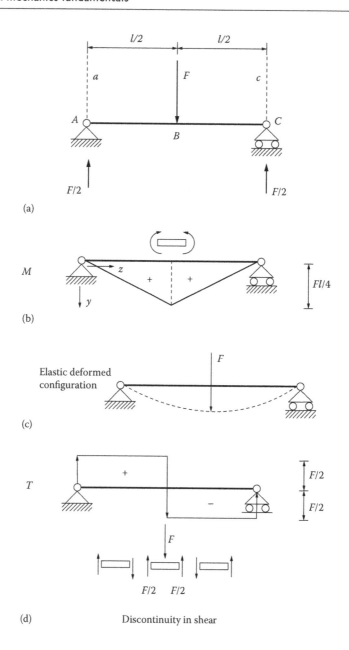

Figure 5.16

The maximum of the function is obtained in the centre and equals *Fl*/4. This maximum is not a stationary point but a cusp of the function, i.e. the left-hand derivative appears different from the right-hand one. The bitriangular diagram *M(z)* can be drawn graphically, referring to the symmetry of the problem and joining the end points *A* and *C* of the beam with the lower end of the vertical segment that represents the moment in the centre. The diagram is drawn from the side of the lower longitudinal fibres, which are the ones physically in tension (Figure 5.16c).

The shear diagram, on the other hand, is birectangular and skew-symmetrical with respect to the centre (Figure 5.16d). It represents exactly the derivative of the moment function.

Where the function $M(z)$ presents a cusp, its derivative $T(z)$ presents a discontinuity of the first kind, i.e. a negative jump equal to the applied force. The infinitesimal element of the beam straddling the centre is, in fact, in equilibrium under the action of the external force F and the two shearing forces $F/2$ both directed upwards (Figure 5.16d).

Let the supported beam be subjected to the concentrated moment m acting in the centre (Figure 5.17a). The two constraint reactions in this case will be opposite and equal to m/l so as to form a couple equal and opposite to the one applied externally. The pressure line, as before, consists of two vertical straight lines passing through A and through C. The moment diagram consists of two linear functions, which are skew-symmetrical with respect to the centre (Figure 5.17b):

$$M(z) = \frac{m}{l} z, \qquad \text{for } 0 \le z \le \frac{l}{2} \tag{5.37a}$$

$$M(z) = \frac{m}{l} z - m, \quad \text{for } \frac{l}{2} \le z \le l \tag{5.37b}$$

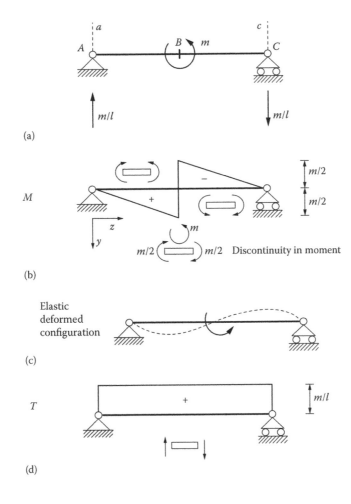

Figure 5.17

In the centre, a discontinuity of the first kind is thus created, i.e. a positive jump, equal to the concentrated moment applied there. The infinitesimal element straddling the centre will thus be in equilibrium with regard to rotation under the action of the counterclockwise external moment m and the two bending moments $m/2$, both clockwise. The elastic deformed configuration presents an inflection in the centre, so that the fibres in tension appear below in the portion AB and above in the portion BC (Figure 5.17c).

The shear diagram is constant, positive and equal to the magnitude of the constraint reactions (Figure 5.17d). In fact, going along the axis of the beam and encountering the moment m, no contribution is added to the vertical force. On the other hand, the derivative of the function $M(z)$ of Figure 5.17b is defined and is equal to m/l in each section. It would not be analytically defined only in the centre, where, instead, physically it is defined and is equal to the left-hand and right-hand derivatives.

Let us imagine applying the moment m in a generic section, other than that of the centre (Figure 5.18). The constraint reactions are the same as in the previous case, so that the moment diagram appears still made up of two linear segments of equal inclination (Figure 5.18):

$$M(z) = \frac{m}{l}z, \qquad \text{for } 0 \leq z \leq a \tag{5.38a}$$

$$M(z) = \frac{m}{l}z - m, \quad \text{for } a \leq z \leq l \tag{5.38b}$$

The shear diagram is obviously identical to the previous one (Figure 5.17d).

When, finally, the concentrated moment is applied to one of the two ends of the beam (Figure 5.19a), the moment diagram reduces to a single linear function that has as its maximum the value of the applied moment itself (Figure 5.19b). In this case, the pressure line consists of the single straight line a, and the inflection of the elastic deformed configuration disappears, leaving the fibres in tension always underneath. The shear diagram, of course, coincides with that of Figure 5.17d.

As a last elementary scheme, consider the supported beam subjected to the distributed load q (Figure 5.20a). The constraint reactions are two vertical forces having the same direction equal to $ql/2$, so that the pressure line is made up of the sheaf of vertical straight lines external to the beam. Imagine going along the axis of the beam, starting from the end A. In the end section A, the pressure line is the line of action of the constraint reaction. In a section immediately to the right of A, the reaction R_A will have to be composed with the partial resultant of the distributed forces qdz. Since the two forces have opposite direction, their resultant will be a vertical force ($R_A - qdz$) passing to the

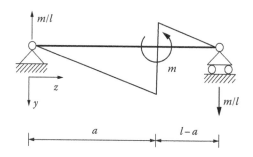

Figure 5.18

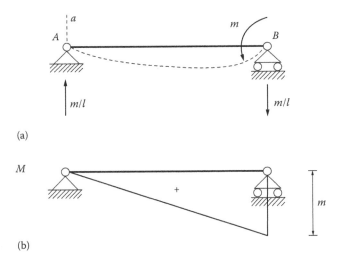

Figure 5.19

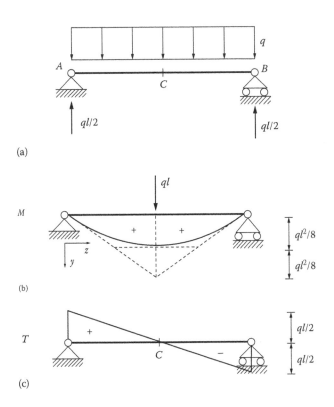

Figure 5.20

left of A. If we increase the coordinate z, the line of action of the subsequent resultants remains vertical but departs more and more from point A, until it reaches infinity for $z = l/2$. The system of the forces acting to the left of the centre is, in fact, equivalent to a couple of moment $ql^2/8$.

Note that, if the roller support B were not moving horizontally (Figure 5.21), the pressure line would be made up of all the tangents enveloping the arc of parabola that has as

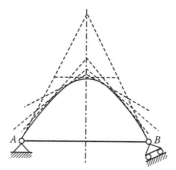

Figure 5.21

its extreme points A and B and as its extreme tangents the lines of action of the constraint reactions. Considering the foregoing case as a limiting case, we can understand then how the parabola becomes degenerate and the infinite tangents are transformed into the sheaf of vertical straight lines described earlier.

The moment diagram is parabolic and may be determined using the direct method, taking into account all the contributions upwards (or downwards) of a generic section of coordinate z (Figure 5.20b):

$$M(z) = q\frac{l}{2}z - (qz)\frac{z}{2} = q\frac{z}{2}(l - z) \tag{5.39}$$

This function vanishes for $z = 0$ and for $z = l$ and is symmetrical with respect to the centre, where it presents a maximum equal to $ql^2/8$.

The same diagram may be determined also using a purely graphical procedure, considering the auxiliary diagram of the moment relative to a concentrated force acting in the centre and equal to the total resultant ql. The auxiliary diagram is bitriangular with the maximum equal to $ql^2/4$. The diagram sought presents the same extreme points and the same extreme tangents, the constraint reactions, and hence the shear value at the ends, as unchanging. On the other hand, according to the by now familiar graphical construction, the third point and the third tangent correspond to the stationary point in the centre.

The shear diagram is linear and skew-symmetrical with respect to the centre (Figure 5.20c). This vanishes at the central point, where the moment diagram is stationary. It is also possible to proceed on the basis of merely graphical considerations, joining the ends of the vertical segments that represent the values that the shear assumes at the ends A and B of the beam. On the other hand, the direct method applied to a generic section of coordinate z gives

$$T(z) = q\frac{l}{2} - qz \tag{5.40}$$

where the first term is the contribution of the reaction R_A, while the second is the contribution of the partial distribution of external forces that extends from A to the section under examination. It may thus be verified that the function (5.40) is the derivative of the function (5.39).

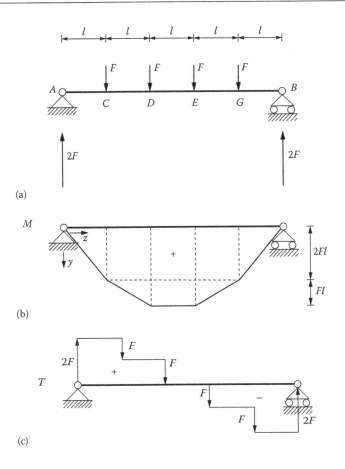

Figure 5.22

In the case of the distribution of constant concentrated forces of Figure 5.22a, the moment diagram is represented by a polygonal line with the sides contained between each pair of consecutive forces (Figure 5.22b):

$$M(z) = 2Fz, \qquad\qquad \text{for } 0 \leq z \leq l \qquad\qquad (5.41\text{a})$$

$$M(z) = 2Fz - F(z - l) = Fz + Fl, \qquad\qquad \text{for } l \leq z \leq 2l \qquad\qquad (5.41\text{b})$$

$$M(z) = Fz + Fl - F(z - 2l) = 3Fl, \qquad\qquad \text{for } 2l \leq z \leq 3l \qquad\qquad (5.41\text{c})$$

$$M(z) = 3Fl - F(z - 3l) = 6Fl - Fz, \qquad\qquad \text{for } 3l \leq z \leq 4l \qquad\qquad (5.41\text{d})$$

$$M(z) = 6Fl - Fz - F(z - 4l) = 10Fl - 2Fz, \quad \text{for } 4l \leq z \leq 5l \qquad\qquad (5.41\text{e})$$

The shear diagram is represented by a step function, with discontinuity of the first kind in each point in which a force is applied (Figure 5.22c). This may be derived analytically from the moment diagram, or it may be obtained using the direct method and summing up algebraically all the contributions that precede a section.

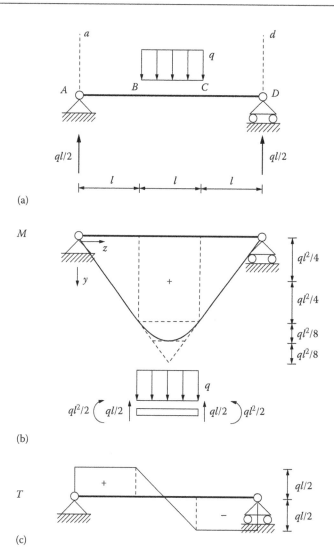

Figure 5.23

In the case of the supported beam of Figure 5.23a, with a constant distributed load only on the intermediate portion, the constraint reactions are symmetrical, vertical and equal to $ql/2$. The pressure line is then represented by the vertical straight line a for the portion AB, by the vertical straight line d for the portion CD and by the sheaf of vertical straight lines external to the beam for the portion BC.

The moment diagram is obtained using the direct method (Figure 5.23b):

$$M(z) = q\frac{l}{2}z, \quad \text{for } 0 \leq z \leq l \tag{5.42a}$$

$$M(z) = q\frac{l}{2}z - q\frac{(z-l)^2}{2}$$

$$= -\frac{1}{2}qz^2 + \frac{3}{2}qlz - \frac{1}{2}ql^2, \quad \text{for } l \leq z \leq 2l \tag{5.42b}$$

$$M(z) = q\frac{l}{2}z - ql\left(z - \frac{3}{2}l\right)$$

$$= -\frac{1}{2}qlz + \frac{3}{2}ql^2, \quad \text{for } 2l \le z \le 3l \tag{5.42c}$$

The stationary point of the function $M(z)$ is obtained by equating its derivative in the portion $l \le z \le 2l$ to zero

$$\frac{dM}{dz} = T(z) = -qz + \frac{3}{2}ql = 0 \tag{5.43}$$

for $z = \frac{3}{2}l$.

The moment diagram will then be symmetrical, with the maximum at the centre, which is equal to

$$M_{max} = M\left(\frac{3}{2}l\right) = \frac{5}{8}ql^2$$

This will be made up of two linear external segments and the parabolic intermediate segment (Figure 5.23b).

It is also possible, however, to draw the moment diagram graphically, using the auxiliary diagram corresponding to the resultant. This diagram is bitriangular and symmetrical and its maximum in the cusp is equal to $3/4ql^2$ (Figure 5.23b). In the outermost portions AB and CD, the auxiliary diagram coincides with the one sought. In the intermediate portion BC, the diagram for the distribution q will present the same extreme values and the same extreme tangents as the auxiliary diagram. It is therefore easy to draw an arc of parabola that corresponds to these conditions and thus to refind the solution described earlier.

The shear diagram, as usual, may be obtained by derivation of the moment diagram or rather, directly, considering the successive contributions of the forces acting perpendicularly to the beam (Figure 5.23c). In the outermost portions, the function $T(z)$ is constant, while it varies linearly where the distributed load is applied; in fact, the differential equation (5.12b) must hold good at all points. The cusps in the shear diagram reflect the discontinuity of the distributed load, which passes sharply from zero to q and *vice versa*. The point of zero shear corresponds, of course, to the stationary point of $M(z)$ (Figure 5.23b).

Now consider a supported beam not symmetrically loaded (Figure 5.24a). The load weighs only on the left-hand half, so that the reaction V_A will be greater than the other reaction V_B. The respective values are $V_A = 3/4ql$, $V_B = 1/4ql$. The pressure line is made up of the sheaf of vertical straight lines external to the beam, for the section AC, and by the line of action of the reaction V_B, for the section CB.

The moment diagram, as has already been seen, may be obtained using the direct method or using the graphical method. The direct method gives two functions, one parabolic and the other linear (Figure 5.24b):

$$M(z) = \frac{3}{4}qlz - q\frac{z^2}{2}, \quad \text{for } 0 \le z \le l \tag{5.44a}$$

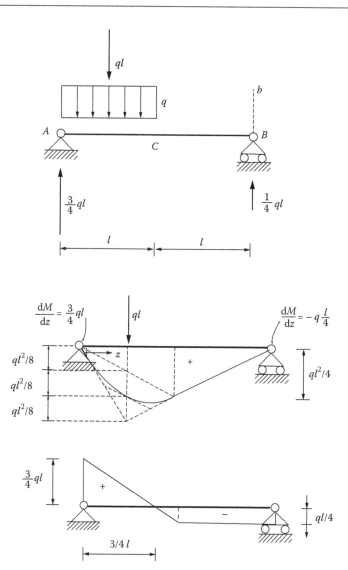

Figure 5.24

$$M(z) = \frac{3}{4}qlz - ql\left(z - \frac{l}{2}\right)$$

$$= -\frac{1}{4}qlz + \frac{1}{2}ql^2, \quad \text{for } l \le z \le 2l \tag{5.44b}$$

We have the stationary point in the left-hand half when

$$\frac{dM}{dz} = T(z) = \frac{3}{4}ql - qz = 0 \tag{5.45}$$

i.e. for $z = 3/4\ l$. The value of the moment in the centre may be obtained more simply by going along the beam from B leftwards:

$$M(l) = V_B l = \frac{1}{4} q l^2 \tag{5.46}$$

The graphical method is applied using the auxiliary diagram for the resultant. This diagram is bitriangular and presents a maximum in the cusp equal to $3/8 q l^2$. The linear segment between B and C coincides with the diagram for the distributed load, since between B and C no external loads act. On the other hand, the two linear segments of the auxiliary diagram constitute the extreme tangents of the arc of parabola that represents the diagram $M(z)$ between A and C. Also in this case, the real diagram follows the course of the auxiliary one, the cusp rounding off considerably.

The shear diagram is linear between A and C and constant between C and B (Figure 5.24c). There are no discontinuities in $T(z)$, as there are no concentrated forces apart from the constraint reactions, but there is a discontinuity in the derivative of $T(z)$ that reflects the discontinuity that the distributed load undergoes in the centre. The shear vanishes where the moment shows a stationary point (Figure 5.24b).

Let the beam with overhanging ends of Figure 5.25a be subjected to a constant distributed load q between the two supports and to two concentrated forces F at the ends. Each of the two constraint reactions is therefore equal to $F + (ql/2)$. The moment diagram (Figure 5.25b) can be drawn by summing up graphically the trapezoidal contribution due to the forces F with the parabolic one of the load q. The procedure will then be to consider the constant central diagram equal to $Fl/2$ as reference axis for the arc of parabola. The sagitta of the arc of parabola is equal to $ql^2/8$, a notable value, being one already met with more than once. In the case where $F < ql/4$, a part of the parabola falls beneath the axis of the beam (Figure 5.25b), and at the points where the bending moment vanishes, two inflections are produced in the elastic deformed configuration; these separate the central portion, concave upwards, from the outermost portions, concave downwards (Figure 5.25c). For $F \geq ql/4$, the inflections disappear and the longitudinal fibres in tension are found only in the upper edge of the beam.

The shear diagram (Figure 5.25d) is constant at the overhanging ends and linear between the two supports. It undergoes two positive jumps at the points corresponding to the supports, equal to the reactions of the supports themselves. The infinitesimal element of beam straddling the support A, e.g. is in equilibrium with regard to vertical translation under the action of the left-hand shear F and the right-hand shear $ql/2$, both directed downwards, and of the reaction $F + (ql/2)$, directed upwards (Figure 5.25d). The moment diagram shows cusps just where the shear diagram is discontinuous and a stationary point where the shear vanishes.

Using the graphical method, we have so far examined only rectilinear beams. Now consider the beam with broken axis of Figure 5.26a, made up of three rectilinear portions and loaded by a concentrated moment m at the centre of the horizontal beam. The constraint reactions are vertical and form a couple equal and opposite to the one applied. In the portion AB, the bending moment is zero, since the reaction at A has no arm with respect to its points. In the portion BC, the moment grows in linear manner up to the value $m/3$. It then undergoes a discontinuity equal to the moment applied and, in the portion CD, it decreases linearly in absolute value until it vanishes virtually at point E'. At D, the moment is $-m/3$ and the representative segment on CD can be turned through $45°$ in a clockwise direction, so that it becomes the representative segment on DE. The diagram is, of course, the same but of opposite sign, going along the beam from E to A. In all cases, the usual procedure is to draw the

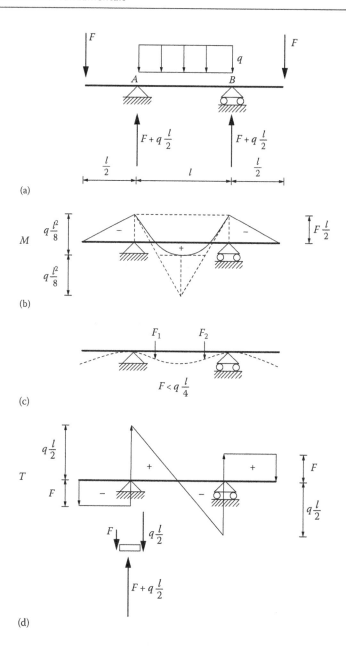

Figure 5.25

moment diagram on the side of the fibres in tension. As regards shearing force and axial force, these are represented by constant diagrams, as no distributed loads are present. The diagram of shearing force is shown in Figure 5.26b, while that of axial force is shown in Figure 5.26c.

Finally, consider the three-hinged arch of Figure 5.27a, which presents the same polygonal line of axis as the previous beam and is loaded by a concentrated force F on the left-hand portion. The triangle of forces gives the internal and external constraint reactions (Figure 5.27a). The pressure line consists of the lines of action of the reactions R_A and R_C. It is important to define where the pressure line intersects the axis of the structure, since at these points, the bending moment vanishes; in fact, the arm of the partial resultant vanishes with respect

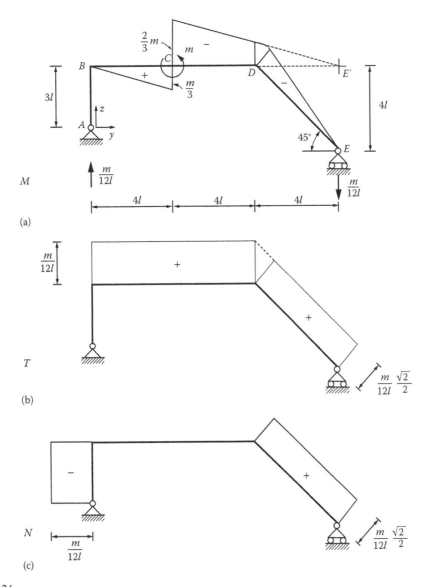

Figure 5.26

to the section. In the structure under consideration, the pressure line encounters the axis at point E, as well as at the hinges A, B and C. At these points, the moment diagram will vanish (Figure 5.27b). In the portion AD, the absolute value of moment grows linearly up to the value of $3H_Al$. At D, the diagram is turned over, remaining linear between D and F and vanishes at E. At F, there will be a cusp since a concentrated force is applied there. The linear diagram between F and G is obtained simply by joining the end of the segment representative of the moment at F with the hinge at B. The moment at F is thus H_Al, while at G it is $2H_Al$. The diagram is then turned through $45°$ and finally it is joined with the hinge C. The same diagram would, of course, have been obtained by going around the structure from C to A. The absolute value of the moment at G, calculated by means of the reaction R_C, is equal to

$$M_G = (H_C - V_C)4l \tag{5.47}$$

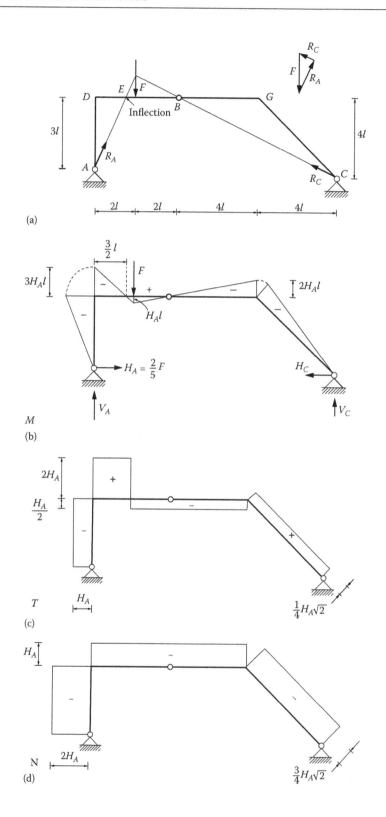

Figure 5.27

The triangle of forces (Figure 5.27a) shows, on the other hand, that $H_C = H_A$ and $V_C = H_C/2 = H_A/2$, from which we obtain $M_G = 2H_A l$, the same value as before.

The shear diagram and the axial force diagram are constant in all sections and are shown in Figure 5.27c and d.

5.4 DETERMINATION OF CHARACTERISTICS OF INTERNAL REACTION VIA THE PRINCIPLE OF VIRTUAL WORK

Just as external reactions can be calculated using the principle of virtual work, by suitably reducing the corresponding external constraints, so can the characteristics of internal reaction be obtained by transforming the internal built-in constraint into a double constraint: the hinge to obtain the bending moment and the double rod to obtain the shearing force or the axial force.

In the case, for instance, of the three-hinged portal frame of Figure 4.6a, if we wish to determine the bending moment in the midpoint of the taller upright, it is necessary to introduce a hinge at that point and to apply the corresponding unknown moments (Figure 5.28). Using the diagrams of horizontal and vertical displacements, we have

$$-ql\left(\frac{l}{2}\varphi\right) - ql\left(\frac{l}{2}\varphi\right) - M\varphi - M(2\varphi) = 0 \tag{5.48}$$

whence we obtain

$$M = -\frac{1}{3}ql^2 \tag{5.49}$$

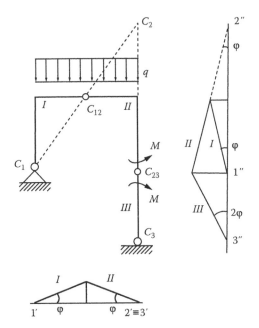

Figure 5.28

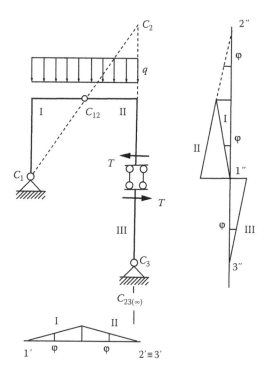

Figure 5.29

In the case where we wish to determine the shear at the same point, we must introduce a double rod parallel to the axis of the upright and apply the two unknown forces T (Figure 5.29):

$$-ql\left(\frac{l}{2}\varphi\right)-ql\left(\frac{l}{2}\varphi\right)+T(2h\varphi)+T(h\varphi)=0 \tag{5.50}$$

whence we obtain

$$T=q\frac{l^2}{3h} \tag{5.51}$$

Finally, in the case where we wish to determine the axial force in the upright, we must introduce a double rod transversely to the axis and apply the two unknown forces N (Figure 5.30)

$$-ql\left(\frac{l}{2}\varphi\right)-ql\left(\frac{5}{4}l\varphi\right)-N\left(\frac{3}{2}l\varphi\right)=0 \tag{5.52}$$

whence we obtain

$$N=-\frac{7}{6}ql \tag{5.53}$$

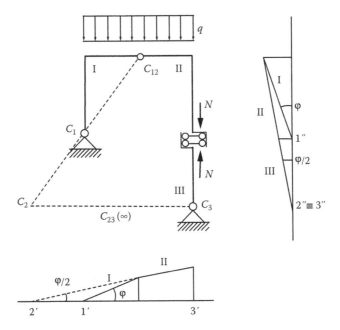

Figure 5.30

Chapter 6

Statically determinate beam systems

6.1 INTRODUCTION

Isostatic, or statically determinate, structures made up of beams are widely used in civil and industrial constructions. As will emerge more clearly later, the internal stresses induced by mechanical loads are greater in the isostatically constrained structure than in the structure that is redundantly constrained and consequently rendered hyperstatic by further constraints. On the other hand, the internal stresses induced by the so-called thermal loads (variations of temperature, uniform or otherwise, through the depth of the beam) are zero in isostatic structures, whereas in certain cases they are considerable in hyperstatic, or statically indeterminate, structures. From this, it is possible to deduce the importance, also from a practical point of view, of considering isostatic beam systems. In fact, often the mechanical loads that the structure will have to support are known to the designer, at least with a certain degree of approximation, while the thermal variations and the constraint settlements that the structure will undergo are not reasonably foreseeable, even as regards the order of magnitude. In these cases, an isostatic structure shows ample possibilities of settling with the intervention of rigid movements only (translations and global rotations), whereas a hyperstatic structure, in view of its redundant degree of constraint, will undergo deformations also of a mechanical nature, and thus internal stresses different from zero.

These subjects will be taken up again and analysed in greater depth in the chapters on statically indeterminate structures, where the consequences induced by thermal and mechanical distortions will be studied. On the other hand, the earlier discussion serves to understand why, for structures having long spans, and thus subjected to notable dilations and rotations of a thermal origin, an isostatic scheme is preferred to a hyperstatic one. The two schemes most widely used for realizing structures that have long spans and are devoid of vertical encumbrance are as follows:

1. **Gerber beams**: These consist of a rectilinear beam with a number of supports and an adequate number of disconnections and are used, for example, in the construction of motorway bridges.
2. **Trusses**: These are made up of elements whose finer structure consists of mutually hinged connecting rods and are traditionally used in railway bridges.

   For constructions with smaller spans, or ones that present also vertical encumbrance, arched structures are traditionally used, where, as we have already seen in Chapter 4, compressive stress prevails, while bending stress tends to be reduced.

For industrial sheds and, more in general, for all sorts of roofing (stations, gymnasia, football grounds, etc.), the following structural schemes are mainly used:

3. **Three-hinged arches**: These have been widely examined in the foregoing chapters in order to introduce the fundamental statical concepts.

4. **Closed-frame structures**: These are made up of chains of structural elements that close in on themselves and may in some cases also present internal statical indeterminacy.

Very often, the four structural types mentioned earlier are found combined in the global structural scheme. A typical example is represented by arched bridges, where a Gerber beam can rest on a three-hinged arch, or by suspension bridges, where a Gerber beam can be hung on a truss system of tie rods and cables. In other cases, more complex schemes, also closed ones, can be reduced to simpler three-hinged arch schemes. In this chapter, we shall look at some examples of these.

In many technically important cases, the mechanical loads and the structural geometry are such as to induce the designer to choose hyperstatic schemes. These cases, on the other hand, can be reduced to similar isostatic schemes, where, in addition to external loads, there act also hyperstatic loads exerted by redundant constraints. The calculation of these structures, which have few degrees of statical indeterminacy, is made by eliminating ideally the redundant constraints and replacing them with their respective constraint reactions. These reactions are obtained from considerations of congruence that regard the respect of the kinematic conditions imposed by the suppressed constraints. We shall return, however, to these aspects in Chapter 11, which is devoted to hyperstatic structures and their solution using the method of forces.

6.2 GERBER BEAMS

Gerber beams are rectilinear beams with $(2 + s)$ supports, in which the line of axis presents s single disconnections, so as to render the structure isostatic (Figure 6.1a). To obtain statical determinacy also with respect to horizontal forces, the supports must be all roller supports, except for one hinged to the foundation. The s simple disconnections (which may be hinges or double rods) must be well arranged, so as not to create hypostatic and/or hyperstatic portions (Figure 6.1b). Generally speaking, that is, three hinges must never be arranged consecutively (hypostatic portion) or three supports consecutively (hyperstatic portion).

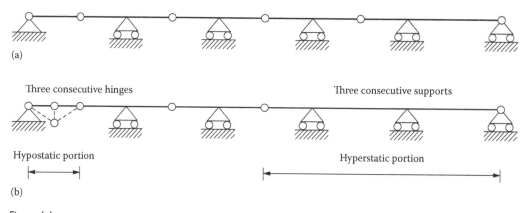

(a)

Three consecutive hinges   Three consecutive supports

Hypostatic portion   Hyperstatic portion

(b)

Figure 6.1

This having been said, Gerber beams can be resolved using analytical and graphical methods, valid for all isostatic structures, which have been presented in the foregoing chapters. Indeed, for Gerber beams, it will not usually be necessary to consider equilibrium with regard to horizontal translation, since, with only vertical loads, it is identically satisfied once the only horizontal reaction potentially present is removed. To determine the $(2 + s)$ external reactions, it is expedient to resort to the method of auxiliary equations, already introduced in Chapter 4. The global equilibrium equations are those of equilibrium with regard to vertical translation and to rotation about a suitable point of the plane. The auxiliary equations are the s equations of partial equilibrium corresponding to the portions into which the disconnections subdivide the Gerber beam. The diagrams of the characteristics of internal reaction are then drawn by isolating the individual portions ideally and by applying to them the external forces, the external constraint reactions and the internal constraint reactions, transmitted by the adjacent portions.

As an example, let us consider the Gerber beam of Figure 6.2a, subjected to the distributed load q on all three spans of length l. In this case, there is only one disconnection, and the two equations of global equilibrium are accompanied by the auxiliary equation of partial equilibrium with regard to rotation of the portion AB about the hinge B:

$$V_A + V_C + V_D = 3ql \tag{6.1a}$$

$$V_A \frac{3}{2}l = V_C \frac{l}{2} + V_D \frac{3}{2}l \tag{6.1b}$$

$$V_A l = q \frac{l^2}{2} \tag{6.1c}$$

To express equilibrium with regard to global rotation, the midpoint of the Gerber beam has been used as pole, with respect to which the moment of external load vanishes. The system (6.1) of three equations in three unknowns possesses the following solution:

$$V_A = \frac{1}{2}ql, \quad V_C = 3ql, \quad V_D = -\frac{1}{2}ql \tag{6.2}$$

The reaction V_D of the support turns out to be negative and thus acts in the opposite direction to the one assumed. On the other hand, the internal reaction transmitted by the hinge B is given by the equation of equilibrium with regard to vertical translation of the portion AB:

$$V_A + V_B = ql \tag{6.3}$$

from which we have $V_B = (1/2)ql$.

Applying the direct method, it is possible to identify the analytical functions $M(z)$ and $T(z)$. As regards bending moment, we have

$$M(z) = \frac{1}{2}qlz - \frac{1}{2}qz^2, \quad \text{for } 0 \le z \le 2l \tag{6.4a}$$

$$M(z) = \frac{1}{2}qlz - \frac{1}{2}qz^2 + 3ql(z - 2l)$$

$$= -\frac{1}{2}qz^2 + \frac{7}{2}qlz - 6ql^2, \quad \text{for } 2l \le z \le 3l \tag{6.4b}$$

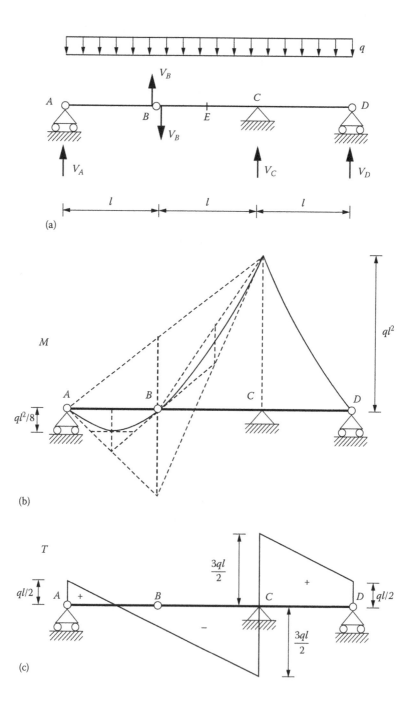

Figure 6.2

The shear function can then be obtained from the foregoing by derivation or, as was done for the moment, by summing up, one at a time, the contributions that are encountered going along the beam:

$$T(z) = \frac{1}{2}ql - qz, \quad \text{for } 0 \le z \le 2l \tag{6.5a}$$

$$T(z) = \frac{7}{2}ql - qz, \quad \text{for } 2l \le z \le 3l \tag{6.5b}$$

The moment diagram is drawn in Figure 6.2b. It is formed by two arcs of parabola with vertical axes. The portion corresponding to the span AB is the diagram, already calculated in the last chapter, corresponding to a simply supported and uniformly loaded beam (Figure 5.20b). Since no concentrated forces act on the hinge B, the shear at that point will not undergo discontinuity and thus the moment will not present cusps. The third point for identifying univocally the arc of parabola between A and C is the one representing the moment on the support C: $M(2l) = -ql^2$. Note that this moment in absolute value is as much as eight times that which loads the centre of the first span. In order to construct this first arc of parabola graphically, it is possible to proceed by considering the spans AB and BC separately and then the corresponding portions of the arc, or it is possible to proceed by considering the entire portion AC at once. The earlier graphical constructions are given in detail in Figure 6.2b. The parabolic diagram for the span CD may then be drawn immediately, if we note that the forces acting on the portion BD are symmetrical with respect to the vertical straight line passing through C. This arc of parabola will thus be specularly symmetrical with respect to the one for the span BC. It thus emerges that there is no point in studying the functions (6.4a and 6.4b) analytically. As we have already suggested, it is far more advantageous to proceed graphically and synthetically.

The shear diagram is given in Figure 6.2c. It is formed by two rectilinear segments of equal slope, since the distributed load has a constant value over the entire beam. The function $T(z)$ vanishes where the moment $M(z)$ presents a stationary point, while it undergoes a positive jump equal to the reaction $V_C = 3ql$ at the support C. Also, the extreme values of $T(z)$, for $z = 0$ and $z = 3l$, can be interpreted as jumps, positive and negative, respectively, of the function.

As a second example, let us now examine the Gerber beam of Figure 6.3a, consisting of three spans, two of which are uniformly loaded. The vertical portion AA' is constrained in the hinge A, while the single disconnection in C is, in this case, a double rod. Since only vertical loads are present, the upright is subjected only to the axial force V_A, while bending moment and shearing force are zero.

The equations of global equilibrium to vertical translation and to rotation about point P are

$$V_A + V_B + V_D = 3ql \tag{6.6a}$$

$$V_A \frac{5}{2}l + V_B \frac{3}{2}l = V_D \frac{3}{2}l \tag{6.6b}$$

while the auxiliary equation is that of partial equilibrium to the vertical translation of the portion CD,

$$V_D = 2ql \tag{6.6c}$$

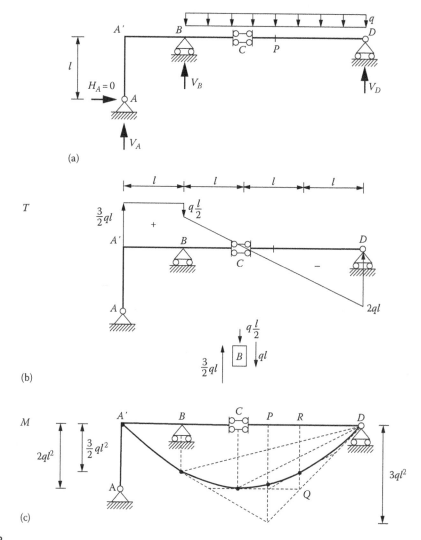

Figure 6.3

The external constraint reactions are then the following:

$$V_A = \frac{3}{2}ql, \quad V_B = -\frac{1}{2}ql, \quad V_D = 2ql \tag{6.7}$$

The reactive moment transmitted by the double rod is then obtained by considering the equilibrium with regard to rotation of the portion CD: $M_C = 2ql^2$.

The shear diagram is constant in the portion $A'B$ and linear in the portion BD (Figure 6.3b). In B, the shear undergoes a discontinuity equal to the value of the reaction but does not change its algebraic sign. At the right of B, the function $T(z)$ equals ql whereas it vanishes in C, because the double rod does not transmit the shear. Hence, two points are known of the linear function contained between B and D, which is thus defined. It is possible to verify the diagram by going along the structure from right to left. The shear at the end D, $T(4l) = -2ql$, is in absolute value equal to the reaction V_D.

The moment diagram (Figure 6.3c) is linear in the portion $A'B$ and parabolic between B and D:

$$M(z) = \frac{3}{2}qlz, \quad \text{for } 0 \le z \le l \tag{6.8a}$$

$$M(z) = \frac{3}{2}qlz - \frac{1}{2}ql(z-l) - \frac{1}{2}q(z-l)^2$$

$$= -\frac{1}{2}qz^2 + 2qzl, \quad \text{for } l \le z \le 4l \tag{6.8b}$$

Having set the scale and drawn the linear part, we can identify at once three values of the parabolic part: $M(l) = (3/2)ql^2$, $M(2l) = 2ql^2$, and $M(4l) = 0$. The pattern of the moment diagram between B and D is thus clear, all the more so, because in correspondence with the double rod C, there is a stationary point (zero shear) with a horizontal tangent. The tangent at the end D may be identified by joining D with point Q, intersection of the horizontal tangent with the vertical line through R. Finally, the graph of Figure 6.3c clearly indicates four points of the arc of parabola together with their respective tangents.

An alternative graphical construction can be made by joining the extreme points of the arc and drawing a vertical segment that starts from the midpoint of this line and drops by $\frac{1}{4}q(3l)^2 = \frac{9}{4}ql^2$. In this way again, we find the intersection of the extreme tangents at the height $\frac{3}{4}ql^2 + \frac{9}{4}ql^2 = 3ql^2$, and the weak angular point produced by the discontinuity of the shear in B emerges.

The elastic deformed configuration must present the lower longitudinal fibres in tension. On the other hand, the deformed configuration of Figure 6.4a would create an axial tensile force on the upright AA'. The axial force is, instead, compressive, and a shortening of the

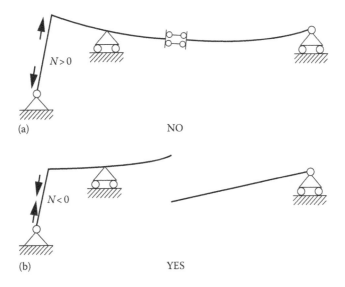

Figure 6.4

upright is compatible only with a deformed configuration that presents a discontinuity of the vertical displacement in correspondence with the double rod (Figure 6.4b). In Chapter 12, we shall re-examine this example and calculate the elastic displacements rigorously.

6.3 TRUSSES

Trusses are systems of bars connected by hinges. These hinges are referred to as **nodes** and are considered loaded by external forces and by the reactions of the bars (Figure 6.5a). This is to say that the hinges, which are normally considered only as constraints and hence as boundary conditions, in the case of trusses are considered as material points in equilibrium under the action of the forces involved. On the other hand, the bars, if they are not loaded directly from outside, are considered as connecting rods and thus as constraints.

In truss schemes, only axial force will therefore be present as a characteristic of internal reaction. In actual situations, however, bending moment and shearing force are also present, albeit frequently not in significant measure. There are substantially two reasons for this. The first is that the external loads do not always concentrate their action on the nodes, but rather often appear as distributed along the bars or concentrated at points that are other than the nodes. The second reason is that the real connection between the bars does present a certain rotational stiffness. It would therefore be closer to the true situation to represent the connections with **semi-fixed joints**, that is, with elastic hinges. The latter will be introduced further on, when we come to deal with elastic constraints. Hence, when the loads are not all concentrated on the nodes and, at the same time, the bars are fixed into one another (welded or bolted joints), trusses will work, from the static standpoint, in a way similar to that in which the so-called framed structures work. These structures will be

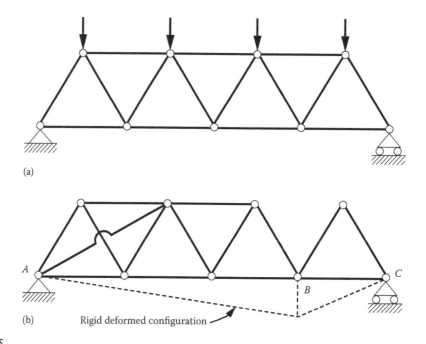

(a)

(b)   Rigid deformed configuration

Figure 6.5

dealt with in the sequel, since they have many degrees of redundancy and reveal a notable presence of bending moment.

From a static viewpoint, a truss presents $(a + 3)$ unknowns, if a is the number of bars and if the external constraint condition is statically determinate.

These unknowns are the internal reactions, and thus the axial forces of the bars, and the external reactions. On the other hand, the number of resolving equations is $2n$, if n is the number of hinge nodes. They are in fact the pairs of equations of equilibrium with regard to translation corresponding to each node. A necessary, but not sufficient, condition for the truss to be isostatic is that the following relation should hold:

$$a + 3 = 2n \tag{6.9}$$

That this relation is not of itself sufficient is shown by the truss of Figure 6.5b, which presents, as that of Figure 6.5a, 15 bars and 9 nodes. In this structure, however, the bars are ill arranged, so that globally the system is kinematically free or hypostatic. In fact, two portions are created that can turn about the two supports, and of which one is internally statically indeterminate and the other internally statically determinate.

A simpler and unequivocal way of judging the internal statical determinacy of a truss is that of checking whether it is made up of triangles of bars with adjacent sides, without intersections or joints through a single vertex (Figure 6.5b). It is therefore easy to verify the statical determinacy of the metal trusses traditionally most widely used for bridges and roofings (Figure 6.6): (a) Polonceau truss, (b) English truss, (c) Mohnié truss, (d) Howe truss, (e) Pratt truss, (f) Neville truss, (g) Nielsen parabolic truss, (h) inverted parabolic truss, (i) Fink truss and (j) K truss.

The single elements making up a structure then often consist of substructures of a truss type. Take, for example, the supported arch of Figure 6.7a or the three-hinged arch of Figure 6.7b, which can be realized by eliminating the central bar of the lower chord and adding the horizontal chain of Figure 6.7c. This latter structural scheme is known as the **tied arch**, where the horizontal action of the arch on the two supports is eliminated. It presents problems of encumbrance due to the tie bar linking the two supports.

Another example of truss elements within a primary structural scheme is provided by the supported beam of Figure 6.8a, and also by the Gerber beam of Figure 6.8b, obtained ideally from the previous one by eliminating two members of the upper chord and by adding two intermediate supports.

An example of methods of solution of trusses is given hereafter in relation to the simple structure of Figure 6.9a. This truss is statically determinate both internally and externally. The external reactions may be determined using the triangle of forces, where the horizontal external force F, the vertical reaction V_A and the reaction R_B follow one another around the triangle (Figure 6.9b), their magnitudes being

$$V_A = \frac{F}{2}, \quad R_B = \left(F^2 + \frac{1}{4} F^2 \right)^{1/2} = \frac{\sqrt{5}}{2} F$$

To determine the axial forces of the individual bars, it is generally possible to write $2n$ equations in $(a + 3)$ unknowns and so to check the values of the external reactions, already obtained by imposing global equilibrium. This mode of proceeding is known as the **method of equilibrium of nodes**, and it is possible to give a highly significant graphical version of it by taking each node as being in equilibrium and considering its polygon of forces.

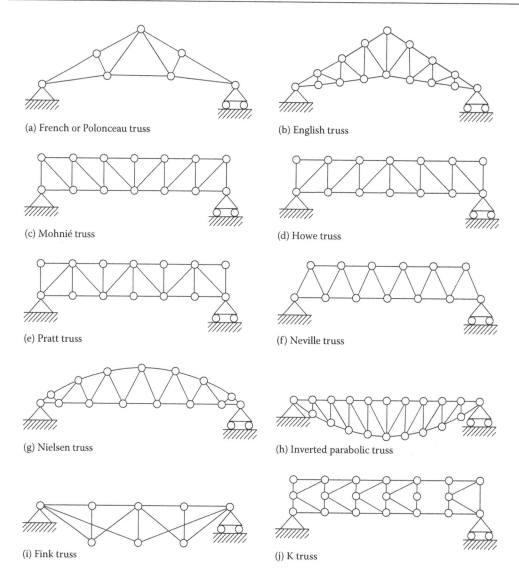

(a) French or Polonceau truss

(b) English truss

(c) Mohnié truss

(d) Howe truss

(e) Pratt truss

(f) Neville truss

(g) Nielsen truss

(h) Inverted parabolic truss

(i) Fink truss

(j) K truss

Figure 6.6

The node A is in equilibrium under the action of the external force, the reaction of the roller support V_A, the action N_{DA} of the bar DA and the action N_{EA} of the bar EA. The trapezium of forces of Figure 6.9c gives the axial forces involved. Noting that the actions of the bars on the hinges are equal and opposite to the actions of the hinges on the bars, it is possible to find that $N_{DA} = -F\sqrt{2}/2$ and $N_{EA} = -F/2$, that is, the bars DA and EA are both struts. On the other hand, the forces N_{CA}, N_{CD}, N_{GE}, N_{GB} are all zero, by virtue of the equilibrium of the nodes C and G.

The node D is in equilibrium under the action of the forces N_{AD}, N_{ED}, N_{BD}. The first is known from the previously considered force polygon, so that also the other two are determined with the triangle of forces of Figure 6.9d: $N_{AD} = N_{DA} = -F\sqrt{2}/2$; $N_{ED} = F/2$; and $N_{BD} = -F/2$.

The node E is in equilibrium under the action of the forces $N_{AE} = N_{EA} = -F/2$; $N_{DE} = N_{ED} = F/2$; and $N_{BE} = -F\sqrt{2}/2$ (Figure 6.9e). Finally, the node B proves to be in

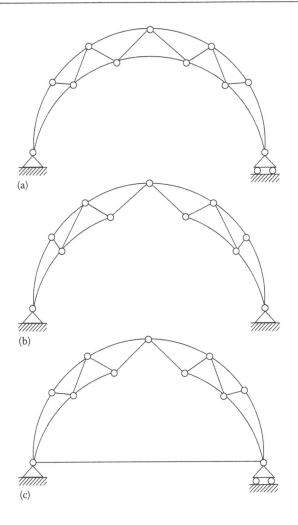

Figure 6.7

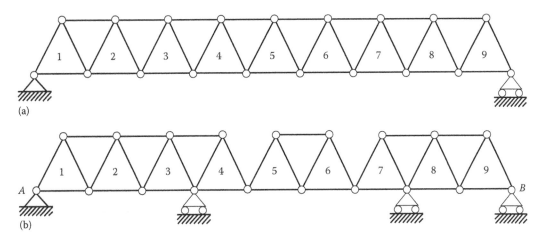

Figure 6.8

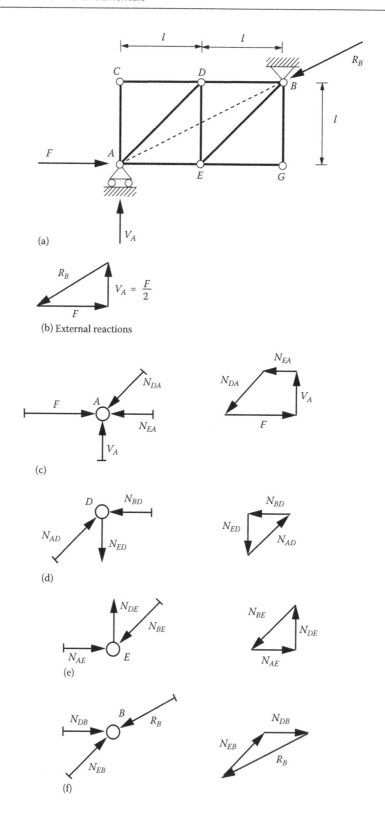

Figure 6.9

equilibrium under the action of the forces $N_{EB} = N_{BE} = -F\sqrt{2}/2$ and $N_{DB} = N_{BD} = -F/2$ and of the constraint reaction R_B, already obtained from considerations of global equilibrium (Figure 6.9f). The axial forces in the individual bars may be summarized as follows:

AC	Unloaded
CD	Unloaded
BG	Unloaded
EG	Unloaded
AE	$-F/2$
AD	$-F\sqrt{2}/2$
DE	$+F/2$
DB	$-F/2$
EB	$-F\sqrt{2}/2$

It may thus be noted that the sides of the parallelogram $ADBE$ are all struts, while the diagonal DE is the only tie bar present in the structure. Taking into account also the external forces and reactions, there is a polar–symmetrical distribution of forces with respect to the centre of the tie bar DE.

A verification of the solution just described may be carried out using the **method of sections** introduced by Ritter. A section of the truss is said to be a Ritter section in relation to a bar, if this section cuts, in addition to the bar under examination, other bars that are concurrent at a real point or at a point at infinity. The additional sectioned bars must therefore intersect in a single pole or be parallel. In the former case, it will suffice to consider the equation of partial equilibrium with regard to rotation about the pole and, in the latter case, the equation of partial equilibrium with regard to translation orthogonally to the parallel direction, to find at once the force in the bar under consideration.

The section of Figure 6.10 is a Ritter section in relation to the bar DE, as the two remaining sectioned bars DB and AE are parallel and horizontal. The equation of equilibrium with regard to vertical translation of the portion that remains to the left of the section in fact gives

$$N_{DE} = +\frac{F}{2} \tag{6.10}$$

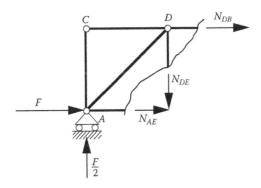

Figure 6.10

The same section is then a Ritter section also in relation to the bar AE, as the remaining two sectioned bars DE and DB are concurrent in D. The equation of equilibrium with regard to rotation about point D of the same portion of truss considered previously is written as

$$(F + N_{AE})l = \frac{F}{2}l \tag{6.11}$$

from which we obtain $N_{AE} = -F/2$.

Finally, the section is a Ritter section in relation also to the bar DB, as the two remaining sectioned bars AE and DE are concurrent in E. The equation of equilibrium with regard to rotation about E assumes the following form:

$$N_{DB}l + \frac{F}{2}l = 0 \tag{6.12}$$

from which we find $N_{DB} = -F/2$.

There follow three examples of trusses, resolved using the method of equilibrium of nodes. The first regards a cantilever truss with variable cross section (Figure 6.11), while the remaining two refer to trusses made up of diagonal struts (Figure 6.12) and of diagonal tie rods (Figure 6.13).

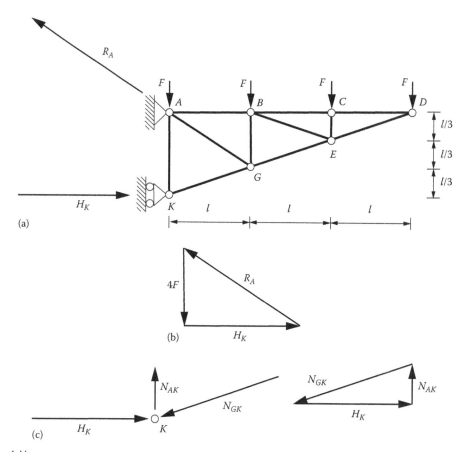

Figure 6.11

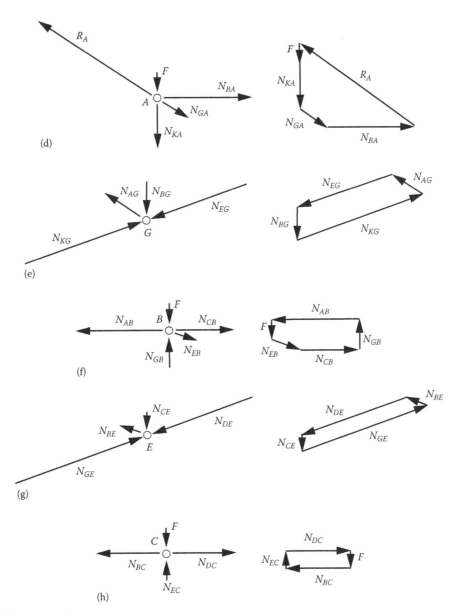

(d)

(e)

(f)

(g)

(h)

Figure 6.11 (continued)

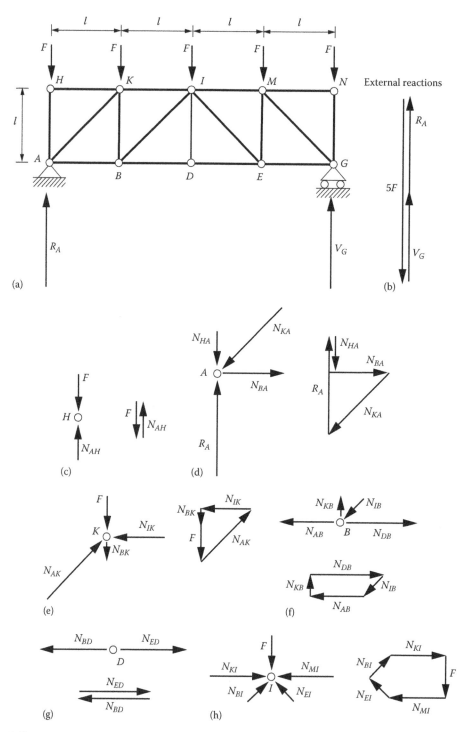

Figure 6.12

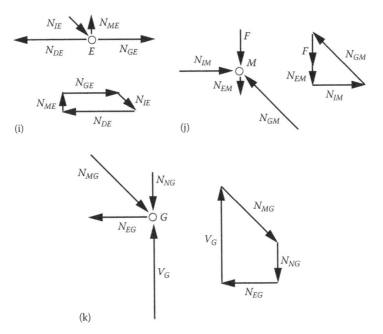

(i)   (j)

(k)

Figure 6.12 (continued)

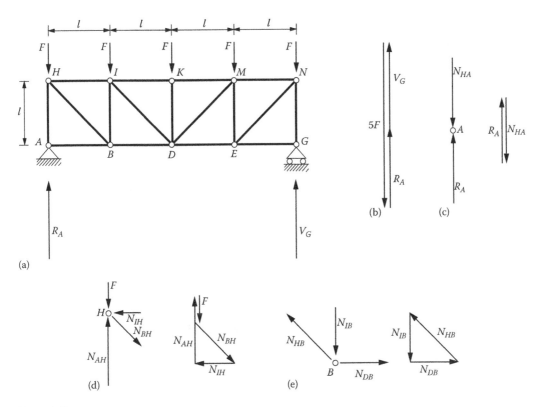

(a)

(b)   (c)

(d)   (e)

Figure 6.13

(continued)

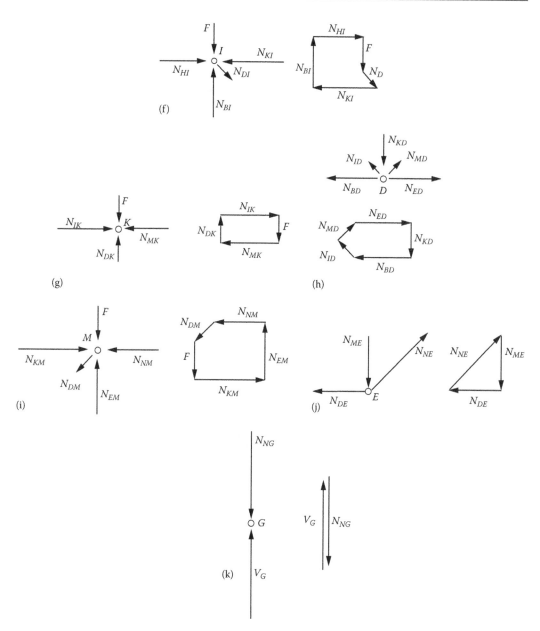

Figure 6.13 (continued)

6.4 THREE-HINGED ARCHES AND CLOSED-FRAME STRUCTURES

The schemes of three-hinged arches (where the hinges are real, ideal or at infinity) have already been extensively discussed in the previous chapters. In the present section, we shall highlight the existence of these schemes within more complex structures, thus bringing out more clearly how the entire structure works from a static viewpoint and, at the same time, providing an interesting graphical and synthetic approach to resolving such schemes.

Statically determinate closed structures are made up of internally isostatic closed frames, externally constrained in a nonredundant way. Each closed frame must thus present three single disconnections and be constrained to the foundation by three single constraints. In the cases where there are external forces concentrated on the internal hinges or external hinges coinciding with internal ones, and thus external reactions acting on the internal hinges, it will be convenient also to consider these hinges as bodies in equilibrium. We shall proceed by looking at three examples: in the first closed system, the external and internal constraints are all separate (Figure 6.14); in the second, an external constraint coincides

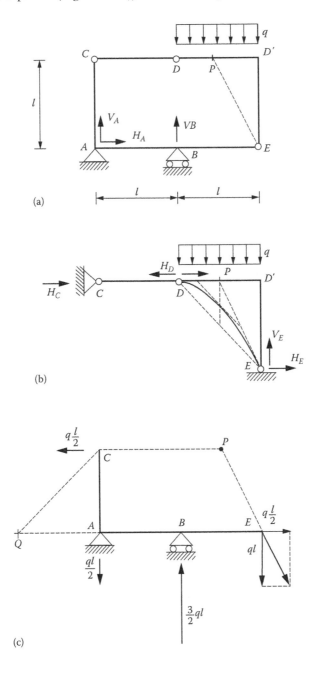

(a)

(b)

(c)

Figure 6.14

with an internal one (Figure 6.16), that is, two beams converge at one external hinge, so that the hinge is, simultaneously, both external and internal; in the third system, finally, the two external constraints both coincide with internal constraints (Figure 6.18).

The closed structure of Figure 6.14a consists of the L-shaped beam, CAE, on which the three-hinged arch CDE rests. It is possible, in the first place, to determine the external reactions, imposing equilibrium with regard to vertical translation and rotation about point P of the entire structure:

$$H_A = 0 \tag{6.13a}$$

$$V_A + V_B = ql \tag{6.13b}$$

$$V_A \frac{3}{2}l + V_B \frac{l}{2} = 0 \tag{6.13c}$$

from which we obtain

$$V_A = -\frac{1}{2}ql, \quad V_B = \frac{3}{2}ql$$

It is then expedient to resolve the three-hinged arch CDE (Figure 6.14b), determining the internal reactions H_C and R_E, and to verify the equilibrium of the beam CAE, once this is subjected to the internal reactions opposite to the previous ones and to the external reactions (Figure 6.14c). Note how the forces transmitted by the three-hinged arch to the beam constitute a system equivalent to the distribution of external forces and, at the same time, a system that balances the external constraint reactions.

The diagrams of the characteristics of internal reaction are drawn, considering each portion as isolated and subjected to all the forces involved, both active and reactive. As regards axial force (Figure 6.15a), in the portion CA it is zero, since the total force does not have an axial component. In the portion AE, the axial force is tensile and equals $\frac{1}{2}ql$, this being the component of the resultant of all the forces that precede and follow any one of its cross sections. In the portion ED', the axial force is given by the vertical component of the reaction R_E and hence is compressive and equal to $-ql$ (Figure 6.14b). Finally, in the portion $D'C$, the axial force is that of the strut CD and equals $-\frac{1}{2}ql$.

The shearing force (Figure 6.15b) is equal to the reaction of the strut CD in the portion CA and to the vertical reaction V_A in the portion AB. The diagram then undergoes a positive jump in B, where the vertical reaction V_B is applied. The shear is equal to the horizontal component of the reaction R_E on the upright ED' and varies linearly in the portion $D'D$, where the distributed load is applied. Its value at the end D' is given by the vertical component of the reaction R_E, $T(D') = -ql$, while it is zero at the end D and also in the strut CD.

The diagram of bending moment (Figure 6.15c) rises linearly in absolute value between C and A, as also between A and B. The moment $M(A)$ is obtained as the product of the reaction of the strut CD and the arm l, and its value is the same to the left as to the right of A, as opposed to what occurs in the case of the other two characteristics. The representative segment is therefore turned by 90° in a counterclockwise direction, and we thus obtain a point of the following linear diagram. A second point can then be represented by the pole Q, where the horizontal axis intersects the line of action of the resultant of the forces acting in

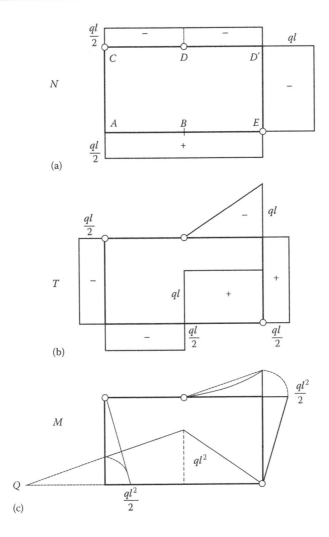

Figure 6.15

C and A (Figure 6.14c). In the portion BE, the moment diagram decreases in absolute value until it vanishes in the hinge E. The cusp that corresponds to point B reflects the discontinuity in the shear diagram (Figure 6.15b). The diagram then is again linear in the portion ED' and parabolic in the portion $D'D$, where the distributed load is applied. The arc of parabola is the same as that determined in the case of the cantilever beam (Figure 5.12b), because the strut CD transmits only an axial force.

The pressure line consists of a set of three straight lines plus the infinite number of straight lines that envelop the arc of parabola of Figure 6.14b. In particular, for the portions DC and CA, the pressure line is the straight line DC; for the portion AB, it is the straight line CQ (Figure 6.14c); for the portions BE and ED', it is the straight line EP; and, finally, for the portion $D'D$, it is the arc of parabola already indicated.

The closed structure of Figure 6.16a consists of a closed rectangular frame with three internal hinges that ensure its internal isostaticity. The roller support A constrains the portion CAD to the foundation, while the hinge B, in addition to connecting the portions CB and DB, further constrains the structure to the foundation. To resolve the structure algebraically, we

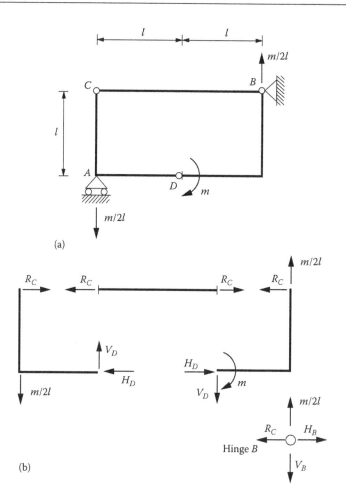

Figure 6.16

proceed by isolating each single portion of it, including the hinge B, and we replace the constraints with the actions exerted by them (Figure 6.16b). The connecting rod CB is subjected to two equal and opposite axial forces R_C. The portion CAD is subjected to the horizontal reaction R_C, to the external reaction of the roller support $m/2l$ and to the axial force H_D and the shearing force V_D, transmitted by the internal hinge D. The hinge B is subjected to the reaction of the connecting rod R_C, to the vertical external reaction $m/2l$, to the horizontal internal reaction H_B and to the vertical internal reaction V_B. For equilibrium with regard to translation of the hinge B, we must have $H_B = R_C$ and $V_B = m/2l$. It follows that the portion DB is subjected to the horizontal force R_C and the vertical force $m/2l$ at the end B, while at the end D, it is subjected to the axial force H_D and the shearing force V_D, transmitted by the hinge D, as well as to the external concentrated moment m, assumed as acting on that point. There are three unknowns, R_C, H_D and V_D. Imposing equilibrium on one of the two portions, CAD or DB, we obtain the internal unknowns. Taking the portion CAD, we have, for example,

$$H_D = R_C \tag{6.14a}$$

$$V_D = \frac{m}{2l} \tag{6.14b}$$

$$R_C l = \frac{m}{2l} l \qquad (6.14c)$$

whence it follows that $R_C = H_D = V_D = m/2l$. As a verification of this, we can note that, according to this solution, the portion DB is in equilibrium with regard to rotation and is loaded by the external moment m and two equal and mutually concordant couples, each of moment $m/2$.

A shorter way to resolve the same closed structure of Figure 6.16a is to recognize in it a three-hinged arch scheme. Note that the static and kinematic function of the connecting rod CB does not vary according to the variation of its length and thus of the position B' of the hinge to the foundation (Figure 6.17a). We thus come to identify an ideal hinge at point C and hence the simplified scheme of Figure 6.17b. In this way, the static working of the structure emerges clearly, whereas in the original scheme, it was not evident.

We now find graphically the reactions previously obtained proceeding algebraically. The pressure line of the fictitious structure consists of two parallel straight lines inclined at 45° to the horizontal. The moment diagram presents a discontinuity where the concentrated moment is applied (Figure 6.17b). The shear diagram (Figure 6.17c) and the axial force diagram (Figure 6.17d), on the other hand, do not present discontinuity in D, as no concentrated forces are applied to the structure in that point. The pressure line of the original structure (Figure 6.16a) is hence defined as follows:

Portion CB: straight line CB
Portion CA: straight line CB
Portion AD: straight line CD
Portion DB: straight line BQ

The closed structure of Figure 6.18a consists of a square frame with three internal hinges and is constrained to the foundation by two of these hinges. The only external reaction that can oppose the horizontal load ql is the horizontal one in C. There, on the other hand, a clockwise couple of moment $\frac{1}{2}ql^2$ is formed, which will be balanced by an equal and opposite couple created by the vertical external reactions. Another way to identify these reactions is that of considering the pole D, where the line of action of the external resultant and the straight line perpendicular to the plane of moving of the roller support intersect. The three elements CA, AB and BC and the two hinges A and C having been isolated, the procedure will be to impose equilibrium on each of these (Figure 6.18b). The connecting rod AC is in equilibrium under the action of two equal and opposite forces R_C. The hinge A is in equilibrium under the action of the reaction R_C of the connecting rod, the vertical external reaction $\frac{1}{2}ql$ and the reactions H_A and V_A transmitted by the portion AB. The portion AB will thus be subjected to the reactions $H_A = R_C$ and $V_A = \frac{1}{2}ql$, the external force ql and the internal reactions H_B and V_B. The three unknowns R_C, H_B and V_B may immediately be determined, considering the equilibrium of the portion AB:

$$H_B + R_C = ql \qquad (6.15a)$$

$$V_B = \frac{1}{2}ql \qquad (6.15b)$$

$$\frac{1}{2}ql^2 = V_B \frac{l}{2} + H_B l \qquad (6.15c)$$

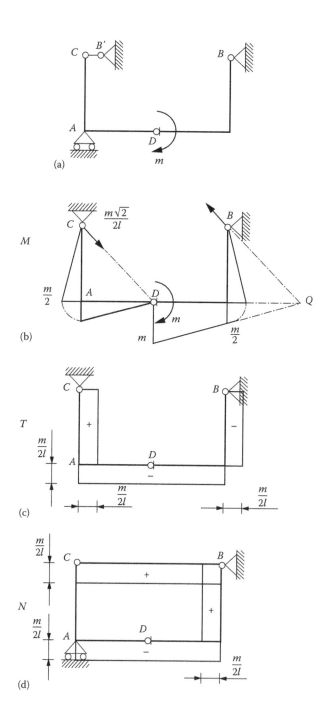

Figure 6.17

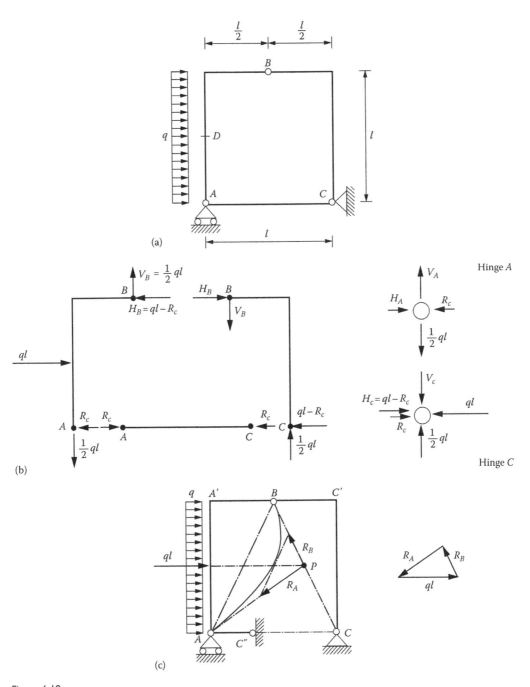

Figure 6.18

(continued)

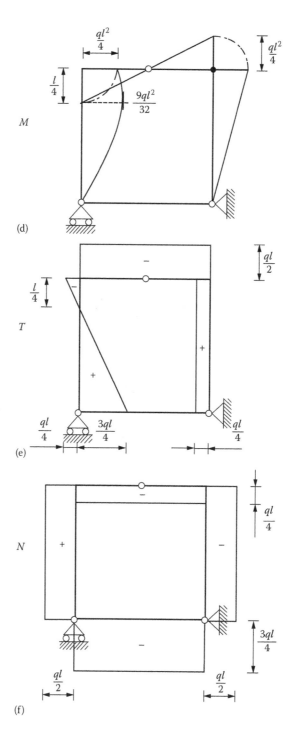

Figure 6.18 (continued)

from which we obtain

$$R_C = \frac{3}{4}ql, \quad H_B = \frac{1}{4}ql, \quad V_B = \frac{1}{2}ql \tag{6.16}$$

At this point, the equilibrium of the remaining elements is verified. The hinge C (Figure 6.18b) is subject to the horizontal external reaction ql, the vertical external one $\frac{1}{2}ql$, the reaction of the connecting rod R_C and the reactions H_C and V_C, transmitted by the portion BC. We thus have $H_C = \frac{1}{4}ql$ and $V_C = \frac{1}{2}ql$. The portion BC is found, finally, to be subject to two equal and opposite forces acting along the line joining B and C.

The graphical approach can be adopted, if we note that the static solution is not a function of the length of the connecting rod AC (Figure 6.18c). The fundamental scheme can, that is, be reduced to the three-hinged arch ABC. We thus identify the pole P as the intersection of the line of action of the external resultant with the line joining C and B. The force triangle thus furnishes the reactions R_A and R_B. The values of Equation 6.16 are again obtained if it is noted that the force triangle is geometrically similar to the triangle ACP. On the basis of the geometrical ratios of the latter, the horizontal component of R_A is three times the horizontal component of R_B, so as to give the values $H_A = \frac{3}{4}ql$ and $H_B = \frac{1}{4}ql$. At the same time, the vertical components of R_A and R_B prove to be equal in magnitude: $V_A = V_C = \frac{1}{2}ql$.

The pressure line for the portion CA' is the straight line CB, while for the portion AA', on which the distributed load acts, it is the arc of parabola shown in Figure 6.18c. The pressure line coincides, but for a scale factor, with the diagram of bending moment (Figure 6.18d). This diagram is in fact linear in the portions CC', $C'B$ and BA', vanishing at the hinges C and B, and parabolic in the loaded portion AA'. Its maximum value is reached at three-quarters of the height of the upright AA',

$$M\left(\frac{3}{4}l\right) = \left(\frac{3}{4}ql\right)\left(\frac{3}{4}l\right) - \left(\frac{3}{4}ql\right)\left(\frac{3}{8}l\right) = \frac{9}{32}ql^2$$

where the shear vanishes (Figure 6.18e). The axial force diagram is given, without further comments, in Figure 6.18f.

There follow three examples regarding three isostatic structures, along with the resolving diagrams of bending moment, shearing force, axial force and the pressure line.

Example 6.1

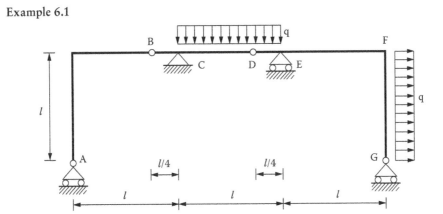

Example 6.1 (continued)

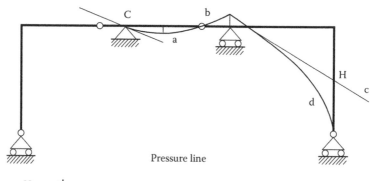

Pressure line

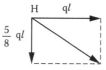

Pressure line	
Portion	Corresponding line
CD	Parabola a
DE	Parabola b
EF	Straight line c
FG	Parabola d

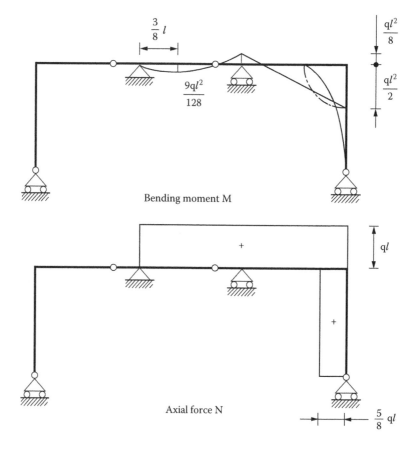

Bending moment M

Axial force N

Example 6.1 (continued)

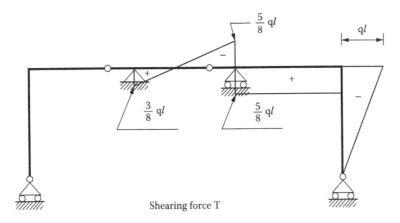

Shearing force T

Example 6.2

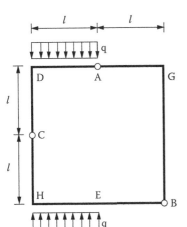

Pressure line	
Portion	Corresponding line
AB	Straight line a
DCH	Straight line b
AD	Parabola c
BE	Straight line a
HE	Parabola c

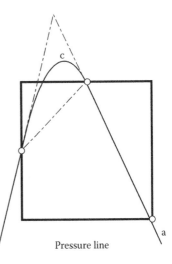

Pressure line

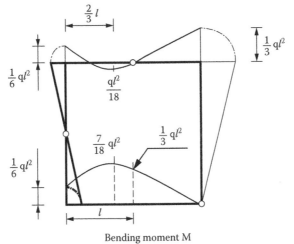

Bending moment M

Example 6.2 (continued)

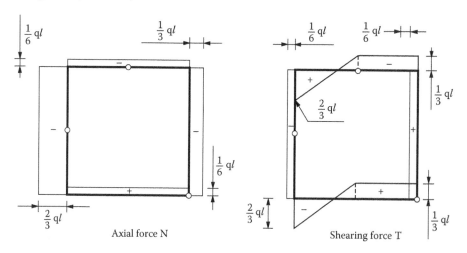

Axial force N

Shearing force T

Example 6.3

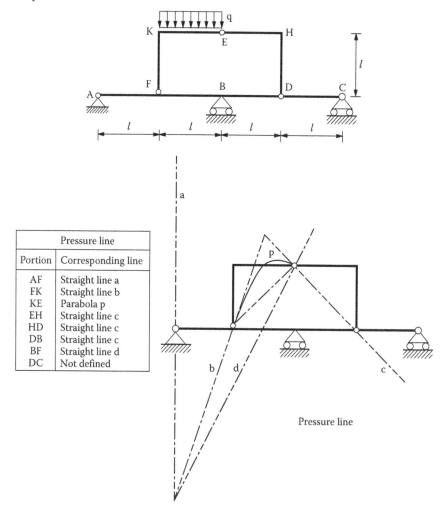

Pressure line	
Portion	Corresponding line
AF	Straight line a
FK	Straight line b
KE	Parabola p
EH	Straight line c
HD	Straight line c
DB	Straight line c
BF	Straight line d
DC	Not defined

Pressure line

Example 6.3 (continued)

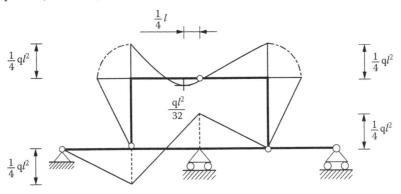

Bending moment M

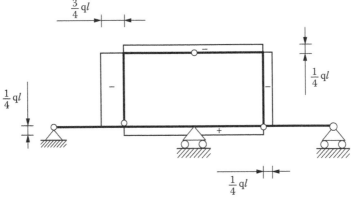

Axial force N

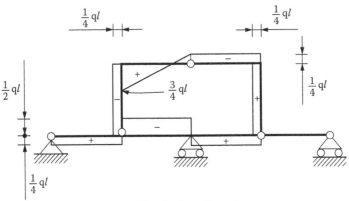

Shearing force T

Chapter 7

Analysis of strain and stress

7.1 INTRODUCTION

The analysis of strain and stress may be dealt with in a similar way, in that both these quantities, in the one case kinematic, in the other static, present a tensorial nature. In this chapter, the corresponding tensors are defined: the strain tensor, consisting of dilations and shearing strains, and the stress tensor, correspondingly made up of normal stresses and shearing stresses.

On the basis of the laws of projection of the displacement vector and of the stress vector, we arrive at the laws of transformation of the strain and stress tensors for rotations of the reference system. Once the principal directions are defined as those that diagonalize the corresponding tensor, the determination of these directions and the corresponding principal values is reduced to an eigenvalue problem. There are three principal directions, at right angles to one another. An elementary cube of the solid with the sides set parallel to the principal directions is subject to dilations only (zero shearing strains) and to normal stresses only (zero shearing stresses).

As regards the principal stress values, the graphical interpretation due to Mohr, already presented in Chapter 2 in the case of the inertia of plane sections, is reproposed.

7.2 STRAIN TENSOR

In the foregoing chapters, only rigid bodies have been considered, that is, undeformable bodies in which the distance between each pair of points does not vary, even when these bodies are loaded by external forces. We have defined the rigidity constraint and the linearization of this constraint in the assumption of small displacements. On these bases, we have recalled the kinematics of the rigid body as a study of the relationships existing between the displacements of different points belonging to one rigid body undergoing rototranslational motion. In this chapter, this study will be extrapolated to the more complex case of a deformable body; that is, we shall analyse the relationships existing between the displacements of different points (but ones sufficiently close together) belonging to one deformable body that in general undergoes rototranslational motion.

Let us thus define the **displacement function** f, as that correspondence that associates each position vector $\{r\}$ of the points of the body, in the initial position and in the undeformed configuration, to the vector $\{\eta\}$ of the displacement that these points undergo, bringing the body into the final position and into the deformed configuration (Figure 7.1). Expressed in symbols, it is

$$f: \mathcal{D} \to \mathcal{C} \tag{7.1a}$$

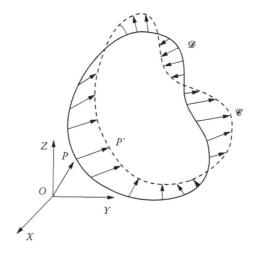

Figure 7.1

$$f: P \mapsto P' \qquad\qquad\qquad\qquad\qquad\qquad\qquad (7.1b)$$

$$f: \{r\} \mapsto \{\eta\} \qquad\qquad\qquad\qquad\qquad\qquad\qquad (7.1c)$$

where the domain $\mathscr{D}$ of the function consists of the total set of geometrical points occupied by the material points of the body in the initial state, and the codomain $\mathscr{C}$ is the volume occupied by the body in the final state. We thus have a vector field of displacements, which can be projected onto a fixed XYZ reference system (Figure 7.1):

$$\{\eta\} = u(x,y,z)\overline{i} + v(x,y,z)\overline{j} + w(x,y,z)\overline{k} \qquad\qquad\qquad (7.2)$$

Each component u, v, w is a function of the three Cartesian coordinates x, y, z. As will be shown, the function $\{\eta\}$ is the primary unknown in structural problems.

For the moment, we formulate a hypothesis of regularity of the function $f: R^3 \rightarrow R^3$, such as to rule out **fracture** and **overlapping** (Figure 7.2). In fracture mechanics, on the other hand, we consider the phenomenon of fracture explicitly and thus the discontinuity of the displacement function. The function f and its inverse function f^{-1} must be continuous and bijective (bi-univocal). These requirements can be summarized under a single term: f and f^{-1} must be **homeomorphisms**.

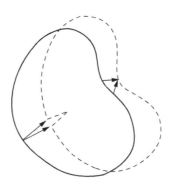

Figure 7.2

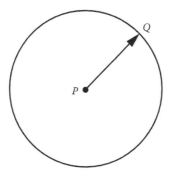

Figure 7.3

Consider an arbitrary point P, within a deformable body, and a point Q belonging to the infinitesimal volume surrounding it (Figure 7.3), such that

$$PQ = \{dr\} = (x_Q - x_P)\bar{i} + (y_Q - y_P)\bar{j} + (z_Q - z_P)\bar{k} \tag{7.3}$$

If the function f is sufficiently regular, that is, continuous together with its first partial derivatives, it is possible to expand this function applying Taylor series up to terms of the first order:

$$u_Q = u_P + \left(\frac{\partial u}{\partial x}\right)_P dx + \left(\frac{\partial u}{\partial y}\right)_P dy + \left(\frac{\partial u}{\partial z}\right)_P dz \tag{7.4a}$$

$$\upsilon_Q = \upsilon_P + \left(\frac{\partial \upsilon}{\partial x}\right)_P dx + \left(\frac{\partial \upsilon}{\partial y}\right)_P dy + \left(\frac{\partial \upsilon}{\partial z}\right)_P dz \tag{7.4b}$$

$$w_Q = w_P + \left(\frac{\partial w}{\partial x}\right)_P dx + \left(\frac{\partial w}{\partial y}\right)_P dy + \left(\frac{\partial w}{\partial z}\right)_P dz \tag{7.4c}$$

The scalar expressions (7.4) may be summarized in a single matrix expression:

$$\{\eta_Q\} = \{\eta_P\} + [J_P]\{dr\} \tag{7.5}$$

where $[J_P]$ is the Jacobian matrix of the dependent variables u, υ, w with respect to the independent variables x, y, z.

On the other hand, if the displacements were caused exclusively by a rigid motion, Equation 7.5 would be reduced to Equation 3.6, which, in compact form, can be represented thus:

$$\{\eta_Q\} = \{\eta_P\} + [\varphi_P]\{dr\} \tag{7.6}$$

In this particular case, the Jacobian matrix is represented by the rotation matrix, already defined in Chapter 3:

$$[\varphi_P] = \begin{bmatrix} 0 & -\varphi_z & \varphi_y \\ \varphi_z & 0 & -\varphi_x \\ -\varphi_y & \varphi_x & 0 \end{bmatrix} \tag{7.7}$$

Since the Jacobian matrix must be the sum of a contribution of rigid motion and a contribution of deformation, where the former consists of the skew-symmetric matrix (7.7), it may be inferred that the purely deforming contribution is equal to the difference between the Jacobian matrix and the rotation matrix. On the other hand, each square matrix consists of the sum of a symmetric matrix and a skew-symmetric matrix. The following identity in fact holds good:

$$[J_P] = \frac{1}{2}([J_P] - [J_P]^T) + \frac{1}{2}([J_P] + [J_P]^T)$$ (7.8)

Of course, the first term represents the rotational contribution (skew-symmetric matrix $[\varphi_P]$), whereas the latter represents the contribution of deformation (symmetric matrix $[\varepsilon_P]$).

Finally, Equation 7.5 expands as follows:

$$\{\eta_Q\} = \{\eta_P\} + [\varphi_P]\{dr\} + [\varepsilon_P]\{dr\}$$ (7.9)

where the first term is the translational contribution, the second one is the rotational contribution and the third is the contribution of deformation. The matrices of rotation and strain, on the basis of the identity (7.8), can be rendered explicit as functions of the first partial derivatives of the components of the displacement vector, u, v, w:

$$[\varphi_P] = \begin{bmatrix} 0 & \frac{1}{2}\left(\frac{\partial u}{\partial y} - \frac{\partial v}{\partial x}\right) & \frac{1}{2}\left(\frac{\partial u}{\partial z} - \frac{\partial w}{\partial x}\right) \\ \frac{1}{2}\left(\frac{\partial v}{\partial x} - \frac{\partial u}{\partial y}\right) & 0 & \frac{1}{2}\left(\frac{\partial v}{\partial z} - \frac{\partial w}{\partial y}\right) \\ \frac{1}{2}\left(\frac{\partial w}{\partial x} - \frac{\partial u}{\partial z}\right) & \frac{1}{2}\left(\frac{\partial w}{\partial y} - \frac{\partial v}{\partial z}\right) & 0 \end{bmatrix}_P$$ (7.10a)

$$[\varepsilon_P] = \begin{bmatrix} \frac{\partial u}{\partial x} & \frac{1}{2}\left(\frac{\partial u}{\partial y} + \frac{\partial v}{\partial x}\right) & \frac{1}{2}\left(\frac{\partial u}{\partial z} + \frac{\partial w}{\partial x}\right) \\ \frac{1}{2}\left(\frac{\partial v}{\partial x} + \frac{\partial u}{\partial y}\right) & \frac{\partial v}{\partial y} & \frac{1}{2}\left(\frac{\partial v}{\partial z} + \frac{\partial w}{\partial y}\right) \\ \frac{1}{2}\left(\frac{\partial w}{\partial x} + \frac{\partial u}{\partial z}\right) & \frac{1}{2}\left(\frac{\partial w}{\partial y} + \frac{\partial v}{\partial z}\right) & \frac{\partial w}{\partial z} \end{bmatrix}_P$$ (7.10b)

The strain matrix will henceforth be represented thus:

$$[\varepsilon] = \begin{bmatrix} \varepsilon_x & \frac{1}{2}\gamma_{yx} & \frac{1}{2}\gamma_{zx} \\ \frac{1}{2}\gamma_{xy} & \varepsilon_y & \frac{1}{2}\gamma_{zy} \\ \frac{1}{2}\gamma_{xz} & \frac{1}{2}\gamma_{yz} & \varepsilon_z \end{bmatrix}$$ (7.11)

where the ε terms represent the partial derivatives of the components of the displacement vector in the corresponding directions, while the γ terms represent the sums of the cross-partial derivatives. The elements of the strain matrix are pure numbers, to which we shall give a precise physical interpretation in the next section. In the framework of the hypotheses of small displacements and of regularity of the function f, the parameters ε and γ are small compared with unity (ε, y≪1).

7.3 DILATIONS AND SHEARING STRAINS

Consider two orthogonal segments PQ and PR of infinitesimal length within the body in the initial position and in the undeformed configuration. Choose the reference system XYZ so that the X and Y axes are parallel to the segments PQ and PR, respectively. When the body is in the final position and in the deformed configuration, the two transformed segments $P'Q'$ and $P'R'$, in addition to their having undergone a rototranslation, appear to be of different length from the initial one and no longer form a right angle (Figure 7.4). Applying Equations 7.4 to the particular case described earlier, for point Q, we have

$$u_Q = u_P + \left(\frac{\partial u}{\partial x}\right)_P dx \tag{7.12a}$$

$$v_Q = v_P + \left(\frac{\partial v}{\partial x}\right)_P dx \tag{7.12b}$$

$$w_Q = w_P + \left(\frac{\partial w}{\partial x}\right)_P dx \tag{7.12c}$$

and for point R

$$u_R = u_P + \left(\frac{\partial u}{\partial y}\right)_P dy \tag{7.13a}$$

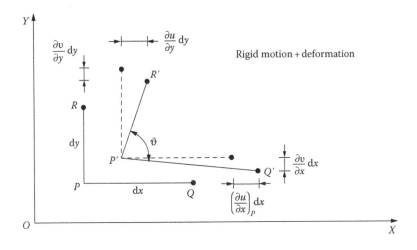

Figure 7.4

$$\upsilon_R = \upsilon_P + \left(\frac{\partial \upsilon}{\partial y}\right)_P dy \tag{7.13b}$$

$$w_R = w_P + \left(\frac{\partial w}{\partial y}\right)_P dy \tag{7.13c}$$

The projection on the XY plane of the two rototranslated and distorted segments is represented in Figure 7.4, on the basis of the relations (7.12a and 7.12b) and (7.13a and 7.13b).

Neglecting infinitesimals of a higher order, the **specific dilation** in the direction of the X axis equals

$$\frac{\left(dx + \dfrac{\partial u}{\partial x} dx\right) - dx}{dx} = \frac{\partial u}{\partial x} = \varepsilon_x \tag{7.14}$$

and thus coincides with the first diagonal term of the strain matrix. Likewise, the specific dilations in the Y and Z directions will be represented by the remaining diagonal elements ε_y and ε_z.

Next, as regards the **shearing strains** $\gamma_{xy} = \gamma_{yx}$, $\gamma_{xz} = \gamma_{zx}$, $\gamma_{yz} = \gamma_{zy}$, these represent the decreases (or negative variations) that the right angles, formed by the initial directions, undergo as a result of the deformation. For the X and Y directions, and neglecting infinitesimals of a higher order, we have (Figure 7.4)

$$\frac{\pi}{2} - \vartheta = \frac{\dfrac{\partial u}{\partial x} dy}{dy} + \frac{\dfrac{\partial \upsilon}{\partial x} dx}{dx} = \gamma_{xy} \tag{7.15}$$

where ϑ indicates the new angle formed by the earlier-mentioned axes. Note that in the diagram of Figure 7.4, the term $\partial u/\partial y$ is positive, while $\partial \upsilon/\partial x$ is negative.

We shall now check the physical meaning of the elements of the strain matrix, considering directly only the contributions of deformation (Figure 7.5). The initial segments PQ and PR in this case only undergo a variation in length and a distortion, while the contributions of rotation and translation are obliterated. On the basis of Equation 7.9 and, precisely, the third term of its second member, we can write

$$u_Q = \left(\frac{\partial u}{\partial x}\right)_P dx \tag{7.16a}$$

$$\upsilon_Q = \frac{1}{2}\left(\frac{\partial \upsilon}{\partial x} + \frac{\partial u}{\partial y}\right)_P dx \tag{7.16b}$$

$$w_Q = \frac{1}{2}\left(\frac{\partial w}{\partial x} + \frac{\partial u}{\partial z}\right)_P dx \tag{7.16c}$$

and likewise

$$u_R = \frac{1}{2}\left(\frac{\partial u}{\partial y} + \frac{\partial \upsilon}{\partial x}\right)_P dy \tag{7.17a}$$

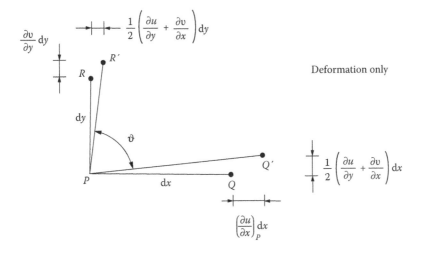

Figure 7.5

$$v_R = \left(\frac{\partial v}{\partial y}\right)_P dy \tag{7.17b}$$

$$w_R = \frac{1}{2}\left(\frac{\partial w}{\partial y}+\frac{\partial v}{\partial z}\right)_P dy \tag{7.17c}$$

The projection on the XY plane of the two purely distorted (and not rototranslated) segments is represented in Figure 7.5 on the basis of the relations (7.16a and 7.16b) and (7.17a and 7.17b). Also in this case, and neglecting infinitesimals of a higher order, the specific dilations appear to be equal to the diagonal terms of the strain matrix, just as the shearing strains coincide with the decrease in the angles formed by the straight lines passing through point P and parallel to the coordinate axes. For the X and Y directions, we have in fact (Figure 7.5)

$$\frac{\pi}{2}-\vartheta = 2\times\frac{1}{2}\left(\frac{\partial u}{\partial y}+\frac{\partial v}{\partial x}\right) = \gamma_{xy} \tag{7.18}$$

7.4 LAW OF TRANSFORMATION OF THE STRAIN TENSOR FOR ROTATIONS OF THE REFERENCE SYSTEM

Consider an infinitesimal sphere of unit radius with its centre in point P (Figure 7.6). The unit vector $\{n\}$ identifies point Q on this sphere. The displacement vector of point Q, once the components of rototranslation have been obliterated, is given by the third term on the right-hand side of Equation 7.9:

$$\{\eta_n\} = [\varepsilon]\{n\} \tag{7.19}$$

The projection of the displacement $\{\eta_n\}$ on the same direction n therefore equals

$$\eta_{nn} = \{n\}^T[\varepsilon]\{n\} \tag{7.20}$$

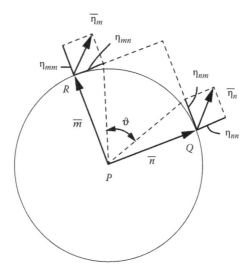

Figure 7.6

while the projection on another generic direction m, identified by a point R of the sphere, equals (Figure 7.6)

$$\eta_{mn} = \{m\}^{T}[\varepsilon]\{n\} \tag{7.21}$$

On the other hand, the displacement of the point R is

$$\{\eta_m\} = [\varepsilon]\{m\} \tag{7.22}$$

and thus its projection on the direction n is

$$\eta_{mn} = \{n\}^{T}[\varepsilon]\{m\} \tag{7.23}$$

The expressions (7.21) and (7.23) can be shown to coincide. Rendering the matrix products explicit, we obtain the following **bilinear form**:

$$\eta_{nm} = \eta_{mn} = \varepsilon_x n_x m_x + \varepsilon_y n_y m_y + \varepsilon_z n_z m_z$$

$$+ \frac{1}{2}\gamma_{xy}(n_x m_y + n_y m_x)$$

$$+ \frac{1}{2}\gamma_{xz}(n_x m_z + n_z m_x)$$

$$+ \frac{1}{2}\gamma_{yz}(n_y m_z + n_z m_y) \tag{7.24}$$

where

n_x, n_y and n_z are the direction cosines of the direction n
m_x, m_y and m_z are the direction cosines of the direction m

The equality (7.24) expresses the **law of reciprocity** for the projections of the displacement vector.

Having made the radius of the sphere of Figure 7.6 equal to unity, we can note how the projection η_{nn} also represents the specific dilation in the direction n. Hence, by Equation 7.20, we obtain

$$\varepsilon_n = \{n\}^{\mathrm{T}}[\varepsilon]\{n\} \tag{7.25}$$

If then the directions n and m are assumed to be orthogonal, we obtain the corresponding shearing strain:

$$\gamma_{nm} = \gamma_{mn} = \eta_{nm} + \eta_{mn} \tag{7.26}$$

The law of reciprocity gives us

$$\gamma_{nm} = \gamma_{mn} = 2\eta_{nm} = 2\eta_{mn} \tag{7.27}$$

and hence, from Equations 7.21 and 7.23, we obtain finally

$$\frac{1}{2}\gamma_{nm} = \frac{1}{2}\gamma_{mn} = \{m\}^{\mathrm{T}}[\varepsilon]\{n\}$$
$$= \{n\}^{\mathrm{T}}[\varepsilon]\{m\} \tag{7.28}$$

Considering three mutually orthogonal directions, n, m, l, rotated with respect to the initial reference system, X, Y, Z, it is now possible to express the law of transformation of the strain matrix for rotations of the reference system (Figure 7.7). This matrix in the rotated reference system nml is indicated by $[\varepsilon^*]$, where the asterisk implies the operation of rotation:

$$[\varepsilon^*] = \begin{bmatrix} \varepsilon_n & \dfrac{1}{2}\gamma_{mn} & \dfrac{1}{2}\gamma_{ln} \\[2mm] \dfrac{1}{2}\gamma_{nm} & \varepsilon_m & \dfrac{1}{2}\gamma_{lm} \\[2mm] \dfrac{1}{2}\gamma_{nl} & \dfrac{1}{2}\gamma_{ml} & \varepsilon_l \end{bmatrix} \tag{7.29}$$

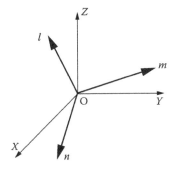

Figure 7.7

From relations (7.25) and (7.28), we obtain

$$[\varepsilon^*] = \begin{bmatrix} \{n\}^T[\varepsilon]\{n\} & \{n\}^T[\varepsilon]\{m\} & \{n\}^T[\varepsilon]\{l\} \\ \{m\}^T[\varepsilon]\{n\} & \{m\}^T[\varepsilon]\{m\} & \{m\}^T[\varepsilon]\{l\} \\ \{l\}^T[\varepsilon]\{n\} & \{l\}^T[\varepsilon]\{m\} & \{l\}^T[\varepsilon]\{l\} \end{bmatrix} = \begin{bmatrix} \{n\}^T \\ \{m\}^T \\ \{l\}^T \end{bmatrix} [\varepsilon][\{n\}\{m\}\{l\}] \tag{7.30}$$

This last matrix product only apparently and formally relates a column matrix, a scalar quantity and a row matrix. Actually, all three matrices are square (3 × 3), and the law (7.30) can be put in an even more compact form*

$$[\varepsilon^*] = [N][\varepsilon][N]^T \tag{7.31}$$

where the matrix $[N]$ is orthogonal and represents the rotation that changes the reference system from XYZ to nml (Figure 7.7):

$$[N] = \begin{bmatrix} n_x & n_y & n_z \\ m_x & m_y & m_z \\ l_x & l_y & l_z \end{bmatrix} \tag{7.32}$$

A law similar to that of Equation 7.31 has already been met with in Chapter 2, in the case of the moment of inertia tensor. It is then possible to conclude that the strain matrix in actual fact is a **tensor**, by virtue of the form that its law of transformation for rotations assumes. Also strain, like inertia, is a physical quantity that may be described solely in tensor terms.

7.5 PRINCIPAL DIRECTIONS OF STRAIN

There is now posed the problem of determining, if they exist, the points of the infinitesimal sphere of centre P that undergo only radial displacements (Figure 7.6). The task will be to define the directions along which only dilations, and not shearing strains, occur. This means that the unit vector $\{n\}$ of such a direction must be parallel to the corresponding displacement vector $\{\eta_n\}$

$$\{\eta_n\} = \varepsilon_n\{n\} \tag{7.33}$$

---

* The law of transformation (7.31) may be readily obtained by considering the relation (7.19) in the rotated reference system:

$$[N]\{\eta_n\} = [\varepsilon^*][N]\{n\}$$

Premultiplying both the sides by $[N]^T$, we obtain

$$\{\eta_n\} = [N]^T[\varepsilon^*][N]\{n\}$$

This equation coincides with Equation 7.19 if

$$[\varepsilon] = [N]^T[\varepsilon^*][N]$$

and thus in the case where Equation 7.31 holds.

In general, on the other hand, the relation (7.19) holds good, so that, by virtue of the transitive law, we obtain the characteristic equation (**eigenvalue equation**) that governs the problem:

$$([\varepsilon] - [1]\varepsilon_n)\{n\} = \{0\} \tag{7.34}$$

where [1] indicates the (3×3) identity matrix. In explicit terms, Equation 7.34 is presented thus:

$$
\begin{bmatrix}
(\varepsilon_x - \varepsilon_n) & \frac{1}{2}\gamma_{yx} & \frac{1}{2}\gamma_{zx} \\
\frac{1}{2}\gamma_{xy} & (\varepsilon_y - \varepsilon_n) & \frac{1}{2}\gamma_{zy} \\
\frac{1}{2}\gamma_{xz} & \frac{1}{2}\gamma_{yz} & (\varepsilon_z - \varepsilon_n)
\end{bmatrix}
\begin{bmatrix} n_x \\ n_y \\ n_z \end{bmatrix}
=
\begin{bmatrix} 0 \\ 0 \\ 0 \end{bmatrix}
\tag{7.35}
$$

The trivial solution of the system of linear algebraic equations (7.35) is without physical meaning, as the direction cosines must obey the relation of normality:

$$n_x^2 + n_y^2 + n_z^2 = 1 \tag{7.36}$$

The solution is different from the trivial one and represents a **principal direction**, if and only if the determinant of the matrix of coefficients vanishes. This last condition gives a third-order algebraic equation in the unknown ε_n:

$$\varepsilon_n^3 - J_I\varepsilon_n^2 - J_{II}\varepsilon_n - J_{III} = 0 \tag{7.37}$$

where the coefficients are the so-called **scalar invariants of strain**, since they remain constant as the reference system varies:

$$J_I = \varepsilon_x + \varepsilon_y + \varepsilon_z \tag{7.38a}$$

$$
J_{II} = -
\begin{vmatrix}
\varepsilon_x & \frac{1}{2}\gamma_{yx} \\
\frac{1}{2}\gamma_{xy} & \varepsilon_y
\end{vmatrix}
-
\begin{vmatrix}
\varepsilon_x & \frac{1}{2}\gamma_{zx} \\
\frac{1}{2}\gamma_{xz} & \varepsilon_z
\end{vmatrix}
-
\begin{vmatrix}
\varepsilon_y & \frac{1}{2}\gamma_{zy} \\
\frac{1}{2}\gamma_{yz} & \varepsilon_z
\end{vmatrix}
\tag{7.38b}
$$

$$J_{III} = \det[\varepsilon] \tag{7.38c}$$

The first invariant is referred to as the **trace** of the tensor and is equal to the sum of the diagonal elements. The second invariant is equal to the sum of the opposites of the determinants of the principal minors. The third invariant is equal to the determinant of the strain tensor. If these coefficients varied as the reference system varied, the solution of the physical problem would also vary, which would be absurd.

Equation 7.37 possesses three roots ε_1, ε_2, ε_3, referred to as the **eigenvalues** of the problem, so that the system (7.35) possesses three different solutions, $\{n_1\}$, $\{n_2\}$, $\{n_3\}$, called **eigenvectors** of the problem. The eigenvalues are real, since the tensor $[\varepsilon]$ is symmetric, and represent the three **principal dilations**, while the eigenvectors, if $\varepsilon_1 \neq \varepsilon_2 \neq \varepsilon_3$, are mutually orthogonal and represent the three **principal directions**. If we consider, in fact, the two principal directions i and j, the law of reciprocity gives the equality

$$\varepsilon_i \cos \vartheta_{ij} = \varepsilon_j \cos \vartheta_{ij} \tag{7.39}$$

where the angle ϑ_{ij} is that contained between the given directions. If $\varepsilon_i \neq \varepsilon_j$, we must have $\vartheta_{ij} = \pi/2$, whereas when $\varepsilon_i = \varepsilon_j$, ϑ_{ij} can assume any value. Three cases may thus present themselves:

1. $\varepsilon_1 \neq \varepsilon_2 \neq \varepsilon_3$: the three principal directions are mutually orthogonal.
2. $\varepsilon_1 = \varepsilon_2 \neq \varepsilon_3$: the direction $\{n_3\}$ is principal together with the ∞^1 directions orthogonal to it (principal plane 12).
3. $\varepsilon_1 = \varepsilon_2 = \varepsilon_3$: the ∞^2 directions are all principal.

Of course, the strain tensor in the principal reference system 123 is diagonal (Figure 7.8):

$$\begin{bmatrix} \varepsilon_1 & 0 & 0 \\ 0 & \varepsilon_2 & 0 \\ 0 & 0 & \varepsilon_3 \end{bmatrix} \tag{7.40}$$

as the shearing strains are zero. The invariants of the strain may therefore be expressed as functions of the principal dilations:

$$J_I = \varepsilon_1 + \varepsilon_2 + \varepsilon_3 \tag{7.41a}$$

$$J_{II} = -(\varepsilon_1\varepsilon_2 + \varepsilon_1\varepsilon_3 + \varepsilon_2\varepsilon_3) \tag{7.41b}$$

$$J_{III} = \varepsilon_1\varepsilon_2\varepsilon_3 \tag{7.41c}$$

The first invariant assumes the physical meaning of **cubic dilation** (or **volumetric dilation**). An elemental parallelepiped of sides $\Delta x_1, \Delta x_2, \Delta x_3$, oriented according to the principal directions, is in fact transformed, once deformation has come about, into the parallelepiped of

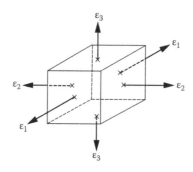

Figure 7.8

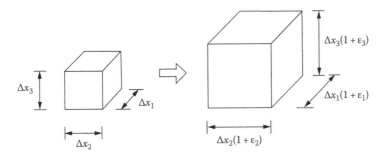

Figure 7.9

sides $\Delta x_1 (1 + \varepsilon_1)$, $\Delta x_2 (1 + \varepsilon_2)$, $\Delta x_3 (1 + \varepsilon_3)$ (Figure 7.9). The volume of the dilated element is thus

$$V' = V(1 + \varepsilon_1)(1 + \varepsilon_2)(1 + \varepsilon_3) \tag{7.42}$$

V being the volume in the undeformed configuration. Neglecting the infinitesimals of a higher order, we have

$$V' = V(1 + \varepsilon_1 + \varepsilon_2 + \varepsilon_3) \tag{7.43}$$

from which we obtain the volumetric dilation

$$\frac{\Delta V}{V} = \frac{V' - V}{V} = \varepsilon_1 + \varepsilon_2 + \varepsilon_3 = J_1 \tag{7.44}$$

7.6 EQUATIONS OF COMPATIBILITY

As we have seen, the displacement vector is a function $f: R^3 \to R^3$, which associates with each position vector the ordered triad u, v, w of the components of the corresponding displacement. On the other hand, the strain tensor, defined in the foregoing sections, can be considered as a function $g: R^3 \to R^6$, which associates with each position vector the ordered set $\varepsilon_x, \varepsilon_y, \varepsilon_z, \gamma_{xy}, \gamma_{xz}, \gamma_{yz}$ of the dilations and shearing strains. The function g may be derived from the function f, on the basis of the relations (7.10b). It thus follows that not all the continuous and derivable tensor fields produce, on integration, displacement fields that are bicontinuous functions. The six components of strain must hence be connected by three differential relations, which limit their mutual independence.

These relations, known as **equations of compatibility**, are obtained by deriving the shearing strains with respect to both the corresponding variables and noting that the third-order partial derivatives that are obtained in the displacements correspond to those of the second order in the dilations:

$$\frac{\partial^2 \gamma_{xy}}{\partial x \partial y} = \frac{\partial^3 u}{\partial x \partial y^2} + \frac{\partial^3 v}{\partial x^2 \partial y} = \frac{\partial^2 \varepsilon_x}{\partial y^2} + \frac{\partial^2 \varepsilon_y}{\partial x^2} \tag{7.45a}$$

$$\frac{\partial^2 \gamma_{xz}}{\partial x \partial z} = \frac{\partial^3 u}{\partial x \partial z^2} + \frac{\partial^3 w}{\partial x^2 \partial z} = \frac{\partial^2 \varepsilon_x}{\partial z^2} + \frac{\partial^2 \varepsilon_z}{\partial x^2} \tag{7.45b}$$

$$\frac{\partial^2 \gamma_{yz}}{\partial y \partial z} = \frac{\partial^3 v}{\partial y \partial z^2} + \frac{\partial^3 w}{\partial y^2 \partial z} = \frac{\partial^2 \varepsilon_y}{\partial z^2} + \frac{\partial^2 \varepsilon_z}{\partial y^2} \tag{7.45c}$$

7.7 STRESS TENSOR

Let us consider a body in equilibrium under the action of forces distributed over the unit external surface $\{p\}$ and in the unit volume $\{\mathscr{F}\}$ (Figure 7.10). The cardinal equations of statics impose

$$\int_S \{p\}dS + \int_V \{\mathscr{F}\}dV = \{0\} \tag{7.46a}$$

$$\int_S \{r\} \wedge \{p\}dS + \int_V \{r\} \wedge \{\mathscr{F}\}dV = \{0\} \tag{7.46b}$$

where
  S represents the boundary of the body
  V is the volume occupied by it
  $\{r\}$ is the position vector of its points

In the case where the body is deformable, let us assume that $\{r\}$ can be confused with the final position vector $\{r\} + \{\eta\}$. Now imagine sectioning the body with a plane A (Figure 7.11a), passing through one of its generic points P. Each of the two portions into which the solid has been ideally subdivided will be in equilibrium under the action of the surface forces (as well as the body forces): both those corresponding to the external partial surface S_A and

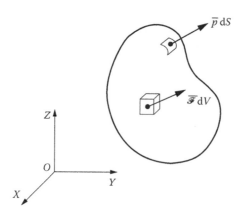

Figure 7.10

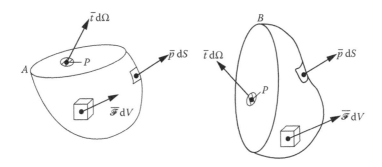

Figure 7.11

those corresponding to the surface of the section Ω_A and transmitted by the complementary portion of the body. Expressed in symbols

$$\int_{S_A} \{p\}dS + \int_{\Omega_A} \{t\}d\Omega + \int_{V_A} \{\mathscr{F}\}dV = \{0\} \tag{7.47a}$$

$$\int_{S_A} \{r\} \wedge \{p\}dS + \int_{\Omega_A} \{r\} \wedge \{t\}d\Omega + \int_{V_A} \{r\} \wedge \{\mathscr{F}\}dV = \{0\} \tag{7.47b}$$

where $\{t\}$ is the **tension vector**, that is, the force transmitted to the elementary area $d\Omega_A$, which constitutes the area surrounding point P on the plane A. This vector is not in general orthogonal to the plane A as occurs for fluids under pressure, and it is a function both of point P and of the secant plane A.

In fact, if we consider a different section of the body, obtained by a plane B passing again through point P, we find a different tension vector acting on the elementary area $d\Omega_B$, the surrounding area of point P on the plane B (Figure 7.11b). The cardinal equations of statics for this new portion of the solid will prove to be similar to Equations 7.47, if we replace the subscript A with B.

Once the position vector $\{r\}$ and the unit vector $\{n\}$ normal to the elementary area $d\Omega_B$ are known, we are able to define the tension vector:

$$\{t\} = \{t(\{r\},\{n\})\} = \{t_n\} \tag{7.48}$$

Of this vector, it is possible to consider the components with respect to the external reference system (Figure 7.10),

$$\{t_n\} = t_{nx}\overline{i} + t_{ny}\overline{j} + t_{nz}\overline{k} \tag{7.49}$$

or the components with respect to a local system, with one of the coordinate axes coinciding with the normal to the elementary area $d\Omega$ and the other two lying on the section plane (Figure 7.12)

$$\{t_n\} = \sigma_n\overline{n} + \tau_{np}\overline{p} + \tau_{nq}\overline{q} \tag{7.50}$$

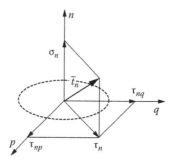

Figure 7.12

where

$\bar{n}, \bar{p}, \bar{q}$ are the unit vectors of the previously mentioned axes

σ_n is the **normal component**, while τ_{np} and τ_{nq} are the shearing stress components on the axes p and q.

The resultant of τ_{np} and τ_{nq} on the section plane is referred to as the **total shearing stress component**:

$$\tau_n = \left(\tau_{np}^2 + \tau_{nq}^2\right)^{1/2} \tag{7.51}$$

Having established the point P inside the body, we now propose to determine the law of variation of the tension vector, as the plane of the elementary area $d\Omega$ varies. For this purpose, consider a volume surrounding point P having the form of a tetrahedron with three sides parallel to the coordinate planes and the oblique face with normal unit vector $\{n\}$ (Figure 7.13a). Let this infinitesimal tetrahedron be subjected to the action of the tension vectors $-\{t_x\}, -\{t_y\}, -\{t_z\}, \{t_n\}$, and at the same time let the body force be negligible. By virtue of equilibrium with regard to translation, we have

$$\{t_n\}d\Omega_n - \{t_x\}d\Omega_x - \{t_y\}d\Omega_y - \{t_z\}d\Omega_z = \{0\} \tag{7.52}$$

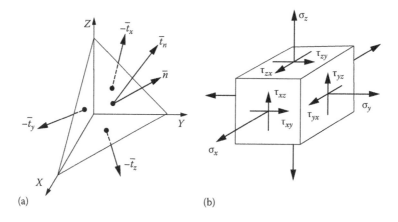

(a)   (b)

Figure 7.13

where the areas of the projections of the triangular surface $d\Omega_n$ on the coordinate planes are equal to

$$d\Omega_x = n_x d\Omega_n \tag{7.53a}$$

$$d\Omega_y = n_y d\Omega_n \tag{7.53b}$$

$$d\Omega_z = n_z d\Omega_n \tag{7.53c}$$

Dividing the expression (7.52) by $d\Omega_n$, we then obtain

$$\{t_n\} = \{t_x\}n_x + \{t_y\}n_y + \{t_z\}n_z \tag{7.54}$$

The foregoing expression, due to Cauchy, can be put in the matrix form considering the so-called **special stress components**, that is, the components of the vectors $\{t_x\}$, $\{t_y\}$, $\{t_z\}$ on the X, Y, Z axes

$$\begin{bmatrix} t_{nx} \\ t_{ny} \\ t_{nz} \end{bmatrix} = \begin{bmatrix} t_{xx} & t_{yx} & t_{zx} \\ t_{xy} & t_{yy} & t_{zy} \\ t_{xz} & t_{yz} & t_{zz} \end{bmatrix} \begin{bmatrix} n_x \\ n_y \\ n_z \end{bmatrix} \tag{7.55}$$

or using the traditional notation

$$\begin{bmatrix} t_{nx} \\ t_{ny} \\ t_{nz} \end{bmatrix} = \begin{bmatrix} \sigma_x & \tau_{yx} & \tau_{zx} \\ \tau_{xy} & \sigma_y & \tau_{zy} \\ \tau_{xz} & \tau_{yz} & \sigma_z \end{bmatrix} \begin{bmatrix} n_x \\ n_y \\ n_z \end{bmatrix} \tag{7.56}$$

where the σ terms are normal stress components and the τ terms are shearing stress components (Figure 7.13b).

The matrix relation (7.56) may be represented in compact form as follows:

$$\{t_n\} = [\sigma]\{n\} \tag{7.57}$$

which interprets the **stress matrix** $[\sigma]$ as a matrix of transformation of the normal unit vector $\{n\}$ into the corresponding tension vector $\{t_n\}$. The analogy with relation (7.19) is evident. In the latter, the strain matrix $[\varepsilon]$ may be interpreted as the matrix of transformation of the normal unit vector $\{n\}$ into the corresponding displacement vector $\{\eta_n\}$.

Considering then the equilibrium with regard to rotation of the tetrahedron of Figure 7.14, the matrix $[\sigma]$ is shown to be symmetric. The centroid G of the triangular elementary area $d\Omega_n$ has the centroids of the elementary areas $d\Omega_x$, $d\Omega_y$, $d\Omega_z$ as projections on the coordinate planes: Let the tension vectors be applied to these centroids G_x, G_y, G_z, and let the conditions of equilibrium with regard to rotation of the tetrahedron with respect to the axes GG_x, GG_y, GG_z be expressed. In the case, for instance, of the axis GG_x, the five special components $\sigma_x, \tau_{xy}, \tau_{xz}, \sigma_y, \sigma_z$ present a zero arm, while τ_{yx} and τ_{zx} are parallel to the axis. The only two components that contribute to the moment with respect to the axis GG_x are τ_{zy} and τ_{yz}:

$$\tau_{zy}\, d\Omega_z \frac{dz}{3} - \tau_{yz}\, d\Omega_y \frac{dy}{3} = 0 \tag{7.58}$$

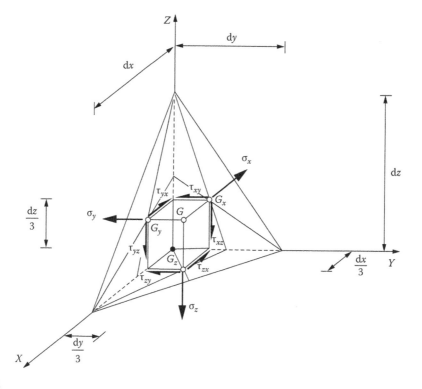

Figure 7.14

where the product of a stress component by the corresponding elementary area represents the infinitesimal resultant force, while $dz/3$ and $dy/3$ represent the arms of these two forces. Noting that

$$\frac{1}{3}d\Omega_y dy = \frac{1}{3}d\Omega_z dz = dV \tag{7.59}$$

where dV is the volume of the elementary tetrahedron, we obtain that $\tau_{yz} = \tau_{zy}$, just as $\tau_{xz} = \tau_{zx}$ and $\tau_{xy} = \tau_{yx}$. The stress matrix $[\sigma]$ is thus symmetric, with only six significant components.

7.8 LAW OF TRANSFORMATION OF THE STRESS TENSOR FOR ROTATIONS OF THE REFERENCE SYSTEM

As we have already done in the case of the displacement vector in Section 7.4, we shall now express the projection of the tension vector on a generic axis. Recalling the law of transformation expressed by Equation 7.57, the component normal to the elementary area of normal unit vector $\{n\}$ equals

$$\sigma_n = t_{nn} = \{n\}^T\{t_n\} = \{n\}^T[\sigma]\{n\} \tag{7.60}$$

More generally, the projection of the tension vector $\{t_n\}$ on a generic direction of unit vector $\{m\}$ equals

$$t_{nm} = \{m\}^{\mathrm{T}}[\sigma]\{n\} \tag{7.61}$$

On the other hand, the projection of the tension vector $\{t_m\}$ on the direction of unit vector $\{n\}$ equals

$$t_{mn} = \{n\}^{\mathrm{T}}[\sigma]\{m\} \tag{7.62}$$

Expanding the expressions (7.61) and (7.62), we obtain the **law of reciprocity** for the projections of the tension vector:

$$\begin{aligned}
t_{nm} = t_{mn} &= \sigma_x n_x m_x + \sigma_y n_y m_y + \sigma_z n_z m_z \\
&+ \tau_{xy}(n_x m_y + n_y m_x) \\
&+ \tau_{xz}(n_x m_z + n_z m_x) \\
&+ \tau_{yz}(n_y m_z + n_z m_y)
\end{aligned} \tag{7.63}$$

where n_x, n_y, n_z and m_x, m_y, m_z are the components of the unit vectors $\{n\}$ and $\{m\}$, respectively.

If then the unit vectors $\{n\}$ and $\{m\}$ are assumed to be orthogonal, the projections t_{nm} and t_{mn} become special shearing stress components:

$$\begin{aligned}
\tau_{nm} = \tau_{mn} &= \{m\}^{\mathrm{T}}[\sigma]\{n\} \\
&= \{n\}^{\mathrm{T}}[\sigma]\{m\}
\end{aligned} \tag{7.64}$$

The foregoing equality expresses the **law of reciprocity of shearing stresses**, of which the symmetry of the matrix $[\sigma]$ is a clear example (Figure 7.13b).

If n, m, l are three mutually orthogonal directions, rotated with respect to the initial reference directions X, Y, Z, on the basis of the foregoing laws of projection, it is possible to express the law of transformation of the stress matrix for rotations of the reference system (Figure 7.7). The transformed matrix is marked with an asterisk:

$$[\sigma^*] = \begin{bmatrix} \sigma_n & \tau_{mn} & \tau_{ln} \\ \tau_{nm} & \sigma_m & \tau_{lm} \\ \tau_{nl} & \tau_{ml} & \sigma_l \end{bmatrix} \tag{7.65}$$

From relations (7.60) and (7.64), we obtain

$$[\sigma^*] = \begin{bmatrix} \{n\}^{\mathrm{T}}[\sigma]\{n\} & \{n\}^{\mathrm{T}}[\sigma]\{m\} & \{n\}^{\mathrm{T}}[\sigma]\{l\} \\ \{m\}^{\mathrm{T}}[\sigma]\{n\} & \{m\}^{\mathrm{T}}[\sigma]\{m\} & \{m\}^{\mathrm{T}}[\sigma]\{l\} \\ \{l\}^{\mathrm{T}}[\sigma]\{n\} & \{l\}^{\mathrm{T}}[\sigma]\{m\} & \{l\}^{\mathrm{T}}[\sigma]\{l\} \end{bmatrix} = \begin{bmatrix} \{n\}^{\mathrm{T}} \\ \{m\}^{\mathrm{T}} \\ \{l\}^{\mathrm{T}} \end{bmatrix}[\sigma][\{n\}\{m\}\{l\}] \tag{7.66}$$

As has already been noted in the analysis of strain, the three matrices highlighted in the foregoing product are square (3 × 3), and the law of transformation sought can be put in the form

$$[\sigma^*] = [N][\sigma][N]^{\mathrm{T}} \tag{7.67}$$

where $[N]$ is the orthogonal matrix (7.32). The form of the law (7.67) makes it possible to recognize a tensor entity in the matrix $[\sigma]$, referred to as **stress tensor**. Thus, in addition to inertia (Chapter 2) and strain (Section 7.4), stress also proves to be a physical quantity of a tensor nature.

7.9 PRINCIPAL DIRECTIONS OF STRESS

The problem now is to determine, if they exist, planes with respect to which only normal stresses are present (Figure 7.15). This means that the unit vector $\{n\}$, normal to such a plane, must be parallel to the corresponding tension vector $\{t_n\}$:

$$\{t_n\} = \sigma_n\{n\} \tag{7.68}$$

In general, the relation (7.57) holds, from which, by the transitive law, we obtain the **characteristic equation** that governs the problem

$$([\sigma] - [1]\sigma_n)\{n\} = \{0\} \tag{7.69}$$

where [1] indicates the (3 × 3) identity matrix. Note the perfect formal identity of Equations 7.34 and 7.69. In explicit terms, Equation 7.69 can be presented as follows:

$$\begin{bmatrix} (\sigma_x - \sigma_n) & \tau_{yx} & \tau_{zx} \\ \tau_{xy} & (\sigma_y - \sigma_n) & \tau_{zy} \\ \tau_{xz} & \tau_{yz} & (\sigma_z - \sigma_n) \end{bmatrix} \begin{bmatrix} n_x \\ n_y \\ n_z \end{bmatrix} = \begin{bmatrix} 0 \\ 0 \\ 0 \end{bmatrix} \tag{7.70}$$

The trivial solution of the system (7.70) is without physical meaning, as the direction cosines must obey the relation of normality (7.36). The solution is different from the trivial one and represents a **principal direction**, if and only if the determinant of the matrix of

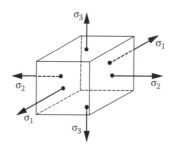

Figure 7.15

coefficients vanishes. This condition gives a third-order algebraic equation in the unknown σ_n, formally identical to Equation 7.37:

$$\sigma_n^3 - J_{\mathrm{I}}\sigma_n^2 - J_{\mathrm{II}}\sigma_n - J_{\mathrm{III}} = 0 \tag{7.71}$$

where the coefficients are the **scalar invariants of stress**

$$J_{\mathrm{I}} = \sigma_x + \sigma_y + \sigma_z \tag{7.72a}$$

$$J_{\mathrm{II}} = -\begin{bmatrix} \sigma_x & \tau_{yx} \\ \tau_{xy} & \sigma_y \end{bmatrix} - \begin{bmatrix} \sigma_x & \tau_{zx} \\ \tau_{xz} & \sigma_z \end{bmatrix} - \begin{bmatrix} \sigma_y & \tau_{zy} \\ \tau_{yz} & \sigma_z \end{bmatrix} \tag{7.72b}$$

$$J_{\mathrm{III}} = \det[\sigma] \tag{7.72c}$$

The first invariant, or trace of the tensor, is equal to the sum of the diagonal elements. The second invariant is equal to the sum of the opposites of the determinants of the principal minors. The third invariant is equal to the determinant of the stress tensor. If these coefficients varied as the reference system varied, the solution of the physical problem would also vary, and this would be absurd.

Equation 7.71 possesses three roots $\sigma_1, \sigma_2, \sigma_3$, referred to as **eigenvalues** of the problem. Consequently, the system (7.70) admits of three different solutions, $\{n_1\}, \{n_2\}, \{n_3\}$, called **eigenvectors** of the problem. The eigenvalues are real, since the tensor $[\sigma]$ is symmetric, and represent the three principal stresses, while the eigenvectors are mutually orthogonal and represent the three principal directions, if the eigenvalues are all distinct (Figure 7.15). This follows from the law of reciprocity, as has already been illustrated in the case of strain. Also in the case of stress, if two eigenvalues coincide, then there exist one principal direction and one principal plane that are mutually orthogonal. Thus, when all three eigenvalues are equal, the directions are all principal.

The stress tensor in the principal reference system 123 is of course diagonal (Figure 7.15):

$$\begin{bmatrix} \sigma_1 & 0 & 0 \\ 0 & \sigma_2 & 0 \\ 0 & 0 & \sigma_3 \end{bmatrix} \tag{7.73}$$

since the shearing stress components are all zero by definition. The invariants of stress may be expressed therefore as functions of the principal stresses:

$$J_{\mathrm{I}} = \sigma_1 + \sigma_2 + \sigma_3 \tag{7.74a}$$

$$J_{\mathrm{II}} = -(\sigma_1\sigma_2 + \sigma_1\sigma_3 + \sigma_2\sigma_3) \tag{7.74b}$$

$$J_{\mathrm{III}} = \sigma_1\sigma_2\sigma_3 \tag{7.74c}$$

The first invariant assumes the physical meaning of **mean normal stress** but for the factor 3:

$$J_{\mathrm{I}} = 3\frac{\sigma_1 + \sigma_2 + \sigma_3}{3}$$

$$= 3\frac{\sigma_x + \sigma_y + \sigma_z}{3} = 3\bar{\sigma} \tag{7.75}$$

Each stress tensor, corresponding to a generic reference system, may thus be represented as the sum of two components,

$$[\sigma] = [\sigma^i] + [\sigma^d] \tag{7.76}$$

where the first is referred to as the **hydrostatic tensor**,

$$[\sigma^i] = \begin{bmatrix} \bar{\sigma} & 0 & 0 \\ 0 & \bar{\sigma} & 0 \\ 0 & 0 & \bar{\sigma} \end{bmatrix} \tag{7.77a}$$

and the second is called the **deviatoric tensor**,

$$[\sigma^d] = \begin{bmatrix} \sigma_x - \bar{\sigma} & \tau_{yx} & \tau_{zx} \\ \tau_{xy} & \sigma_y - \bar{\sigma} & \tau_{zy} \\ \tau_{xz} & \tau_{yz} & \sigma_z - \bar{\sigma} \end{bmatrix} \tag{7.77b}$$

Whereas the hydrostatic tensor does not depend on the reference system, since it is a function only of the trace, the deviatoric tensor varies as the orientation of the reference system varies.

The component (7.77a) is given the name hydrostatic, because liquids under pressure exchange internal stresses of this sort. The principal stresses are all three equal and thus all the directions are principal (see gyroscopic areas, Section 2.5). In fluids under pressure, the stress vector is always normal to any elementary area $d\Omega$; it is compressive and its magnitude is equal to the pressure of the fluid. Perfect fluids, in fact, do not transmit shearing stresses internally, just as ropes do not transmit shearing force (and bending moment) but only axial force.

The problem of principal stresses will now be given a graphical interpretation based on the **method of Mohr's circles**. This method has already been introduced in Chapter 2, in the framework of the geometry of areas, in order to seek the principal directions of inertia.

In the principal reference system 123, Equations 7.57 and 7.60 are particularized in the following way:

$$\{t_n\} = \begin{bmatrix} \sigma_1 & 0 & 0 \\ 0 & \sigma_2 & 0 \\ 0 & 0 & \sigma_3 \end{bmatrix} \begin{bmatrix} n_1 \\ n_2 \\ n_3 \end{bmatrix} \tag{7.78}$$

$$\sigma_n = \begin{bmatrix} n_1 & n_2 & n_3 \end{bmatrix} \begin{bmatrix} \sigma_1 & 0 & 0 \\ 0 & \sigma_2 & 0 \\ 0 & 0 & \sigma_3 \end{bmatrix} \begin{bmatrix} n_1 \\ n_2 \\ n_3 \end{bmatrix} \tag{7.79}$$

where n_1, n_2, n_3 are the direction cosines of the generic direction n in the principal system.
Equation 7.79 may be developed by working out the matrix products:

$$\sigma_n = \sigma_1 n_1^2 + \sigma_2 n_2^2 + \sigma_3 n_3^2 \tag{7.80a}$$

On the other hand, the magnitude squared of the tension vector (7.78) is

$$\sigma_n^2 + \tau_n^2 = \sigma_1^2 n_1^2 + \sigma_2^2 n_2^2 + \sigma_3^2 n_3^2 \tag{7.80b}$$

while for the direction cosines, we have the condition of normality

$$n_1^2 + n_2^2 + n_3^2 = 1 \tag{7.80c}$$

Equations 7.80 constitute a system of three linear algebraic equations in the three unknowns n_1, n_2, n_3. The solution of the system is the following:

$$n_1^2 = \frac{\tau_n^2 + (\sigma_n - \sigma_2)(\sigma_n - \sigma_3)}{(\sigma_1 - \sigma_2)(\sigma_1 - \sigma_3)} \tag{7.81a}$$

$$n_2^2 = \frac{\tau_n^2 + (\sigma_n - \sigma_1)(\sigma_n - \sigma_3)}{(\sigma_2 - \sigma_1)(\sigma_2 - \sigma_3)} \tag{7.81b}$$

$$n_3^2 = \frac{\tau_n^2 + (\sigma_n - \sigma_1)(\sigma_n - \sigma_2)}{(\sigma_3 - \sigma_1)(\sigma_3 - \sigma_2)} \tag{7.81c}$$

Let us assume that between the principal stresses, there exists the order relation $\sigma_1 \geq \sigma_2 \geq \sigma_3$. As the expressions on the right-hand sides of relations (7.81) should be positive, the following inequalities are obtained:

$$\tau_n^2 + (\sigma_n - \sigma_2)(\sigma_n - \sigma_3) \geq 0 \tag{7.82a}$$

$$\tau_n^2 + (\sigma_n - \sigma_1)(\sigma_n - \sigma_3) \leq 0 \tag{7.82b}$$

$$\tau_n^2 + (\sigma_n - \sigma_1)(\sigma_n - \sigma_2) \geq 0 \tag{7.82c}$$

It is easy to verify that these inequalities are equivalent to the following ones:

$$\tau_n^2 + \left(\sigma_n - \frac{\sigma_2 + \sigma_3}{2}\right)^2 \geq \left(\frac{\sigma_2 - \sigma_3}{2}\right)^2 \tag{7.83a}$$

$$\tau_n^2 + \left(\sigma_n - \frac{\sigma_1 + \sigma_3}{2}\right)^2 \leq \left(\frac{\sigma_1 - \sigma_3}{2}\right)^2 \tag{7.83b}$$

$$\tau_n^2 + \left(\sigma_n - \frac{\sigma_1 + \sigma_2}{2}\right)^2 \geq \left(\frac{\sigma_1 - \sigma_2}{2}\right)^2 \tag{7.83c}$$

On **Mohr's plane** (Figure 7.16), all the pairs of components, consisting of the normal stress σ_n and the shearing stress τ_n, which are obtained as the unit vector $\{n\}$ varies, are represented

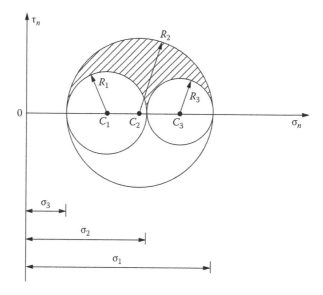

Figure 7.16

by the intersection of the three domains (7.83). The first domain is the one external to the circumference that has its centre on the axis σ_n in the point $C_1\left[\frac{1}{2}(\sigma_2+\sigma_3),0\right]$ and radius $R_1=\frac{1}{2}(\sigma_2-\sigma_3)$. The second domain is the one internal to the circumference of centre $C_2\left[\frac{1}{2}(\sigma_1+\sigma_3),0\right]$ and radius $R_2=\frac{1}{2}(\sigma_1-\sigma_3)$. Finally, the third domain is the one external to the circumference of centre $C_3\left[\frac{1}{2}(\sigma_1+\sigma_2),0\right]$ and radius $R_3=\frac{1}{2}(\sigma_1-\sigma_2)$. Note that the possible pairs (σ_n,τ_n) are ∞^2, just as the directions n issuing from a point are ∞^2. There exists a bi-univocal relation that links each unit vector $\{n\}$ to each point of the hatched area of Figure 7.16. For reasons of brevity, we shall not enter into further details of this relation.

In the case where one, or two, of the principal stresses are zero, the graphical construction described previously will present one, or two, of the intersections of the circumferences with the axis σ_n, coincident with the origin. The five possible cases are represented in Figure 7.17.

7.10 PLANE STRESS CONDITION

The stress condition in a point is said to be **plane** if the tension vector belongs in every case to one and the same plane, independently of the section chosen. A necessary and sufficient condition for the stress state to be plane is that one of the three principal stresses should vanish. If, for example, $\sigma_1 \neq 0$, $\sigma_2 \neq 0$, $\sigma_3 = 0$ (Figure 7.17c), it can easily be shown that the tension vector $\{t_n\}$ always belongs to the **plane of stresses** $\sigma_1 - \sigma_2$, whatever the orientation of the elementary area $d\Omega_n$ (Figure 7.18). Equation 7.78 becomes

$$\begin{bmatrix} t_{n1} \\ t_{n2} \\ t_{n3} \end{bmatrix} = \begin{bmatrix} \sigma_1 & 0 & 0 \\ 0 & \sigma_2 & 0 \\ 0 & 0 & 0 \end{bmatrix} \begin{bmatrix} n_1 \\ n_2 \\ n_3 \end{bmatrix} \tag{7.84}$$

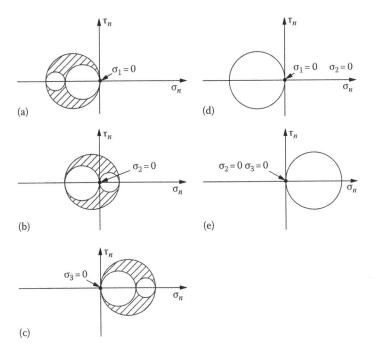

Figure 7.17

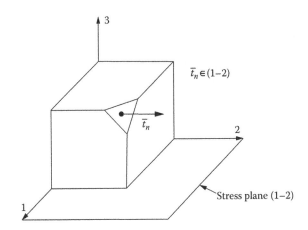

Figure 7.18

from which we obtain

$$t_{n3} = 0 \qquad (7.85)$$

Since the tension vector has always zero components in the direction 3, the stress tensor with respect to a generic system of axes $XY3$ will present the third row, and thus by symmetry the third column, identically equal to zero

$$\begin{bmatrix} \sigma_x & \tau_{yx} & 0 \\ \tau_{xy} & \sigma_y & 0 \\ 0 & 0 & 0 \end{bmatrix} \qquad (7.86)$$

The eigenvalue problem is resolved then by equating the determinant of the following matrix to zero:

$$\det\begin{bmatrix} \sigma_x - \sigma_n & \tau_{yx} & 0 \\ \tau_{xy} & \sigma_y - \sigma_n & 0 \\ 0 & 0 & -\sigma_n \end{bmatrix} = 0 \tag{7.87}$$

The three roots are obtained from the two conditions

$$\sigma_n = 0 \tag{7.88a}$$

$$\sigma_n^2 - (\sigma_x + \sigma_y)\sigma_n + (\sigma_x\sigma_y - \tau_{xy}^2) = 0 \tag{7.88b}$$

Whereas the first equation gives a result already known, because by hypothesis $\sigma_3 = 0$, the second gives the two principal stresses different from zero, σ_1 and σ_2. Note that the first coefficient $(\sigma_x + \sigma_y)$ is the trace of the significant principal minor of the stress tensor (7.86), while the second coefficient $\left(\sigma_x\sigma_y - \tau_{xy}^2\right)$ is the determinant of this minor. Resolving Equation 7.88b, we obtain the pair of roots

$$\sigma_1 = \frac{\sigma_x + \sigma_y}{2} + \frac{1}{2}\left[(\sigma_x - \sigma_y)^2 + 4\tau_{xy}^2\right]^{1/2} \tag{7.89a}$$

$$\sigma_2 = \frac{\sigma_x + \sigma_y}{2} - \frac{1}{2}\left[(\sigma_x - \sigma_y)^2 + 4\tau_{xy}^2\right]^{1/2} \tag{7.89b}$$

It is possible to obtain the same result by imposing that the significant principal minor of the tensor (7.86) should be diagonal (this approach has already been adopted in Chapter 2 for seeking the principal directions of inertia):

$$[\sigma^*] = \begin{bmatrix} \cos\vartheta & \sin\vartheta \\ -\sin\vartheta & \cos\vartheta \end{bmatrix}\begin{bmatrix} \sigma_x & \tau_{yx} \\ \tau_{xy} & \sigma_y \end{bmatrix}\begin{bmatrix} \cos\vartheta & -\sin\vartheta \\ \sin\vartheta & \cos\vartheta \end{bmatrix} \tag{7.90}$$

Equating the off-diagonal term to zero, we have

$$\tau_{xy}^* = \tau_{xy}\cos 2\vartheta - \frac{1}{2}(\sigma_x - \sigma_y)\sin 2\vartheta = 0 \tag{7.91}$$

and thus, the angle by which the XY system must turn to reach the principal system is

$$\vartheta_0 = \frac{1}{2}\arctan\left(\frac{2\tau_{xy}}{\sigma_x - \sigma_y}\right) \tag{7.92}$$

with $-\pi/4 < \vartheta_0 < \pi/4$.

The graphical construction of Mohr's circle, representing all the pairs of normal stresses σ_n and shearing stresses τ_n, which we have as the orientation of the elementary area $d\Omega_n$ varies in such a way that the unit vector $\{n\}$ should belong to the plane of stresses, is made

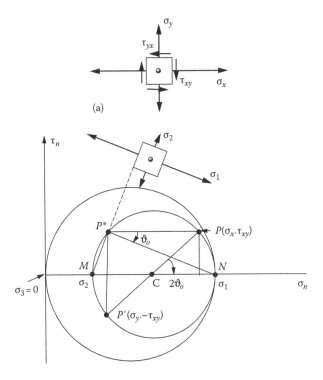

Figure 7.19

in exactly the same way as that already shown in Chapter 2 for the geometry of areas. Let us imagine that the stresses $\sigma_x, \sigma_y, \tau_{xy}$, acting on an elementary parallelepiped having sides parallel to the axes $XY3$, are known (Figure 7.19a). Let the shearing stress τ_{xy} be considered positive if it tends to rotate the element in a clockwise direction (and vice versa for τ_{yx}). Let there be fixed on Mohr's plane (Figure 7.19b) the two notable points: $P\ (\sigma_x, \tau_{xy})$, $P'(\sigma_y, -\tau_{xy})$. The intersection of the segment PP' with the axis σ_n gives the centre C of Mohr's circle, while the segments CP and CP' represent two opposite radii of this circle. The line parallel to the axis on is then drawn through P and the line parallel to the axis σ_n is drawn through P'. These two lines meet at the pole P^*. The straight lines joining P^* with the points M and N of the axis σ_n, which are the intersections of the circumference with the axis, give the directions of the two principal axes. Of course, the points M and N each have for their abscissae the value of a principal stress. In particular, in Figure 7.19b, the abscissa of M is σ_2 and the abscissa of N is σ_1, as we have assumed $\sigma_x > \sigma_y$ and the order relations are maintained:

$$\sigma_x > \sigma_y \Rightarrow \sigma_1 > \sigma_2 \tag{7.93a}$$

$$\sigma_x < \sigma_y \Rightarrow \sigma_1 < \sigma_2 \tag{7.93b}$$

When $\sigma_x = \sigma_y$ and $\tau_{xy} \neq 0$, it makes no difference whether the XY reference system is rotated $\pi/4$ clockwise or counterclockwise to obtain the principal directions. When then $\sigma_x = \sigma_y$ and $\tau_{xy} = 0$, Mohr's circle degenerates into a point. Note, on the other hand, that assuming $\sigma_3 = 0$, as the unit vector $\{n\}$ also varies outside the plane of stresses, the set of pairs

(σ_n, τ_n) is represented by the circumference with centre at the point $C\{\sigma_x/2, 0\}$ and radius $R = \sigma_x/2$. The problem, in other words, always remains three-dimensional.

It is easy to verify that the graphical construction outlined earlier reflects both the analytical solutions (7.89) and (7.92). To justify the former, consider in fact the abscissa of the centre C and the right triangle PP^*P', which has the diameter of Mohr's circle as its hypotenuse (Figure 7.19b). To justify the latter, note that $\widehat{PP^*N}$ is a circumferential angle corresponding to the central angle $\widehat{PCN}$ and that the latter has an amplitude equal to $\arctan 2\tau_{xy}/(\sigma_x - \sigma_y)$.

Chapter 8

Theory of elasticity

8.1 INTRODUCTION

In this chapter, the general problem of the non-homogeneous and anisotropic three-dimensional linear elastic body is formulated. The properties of linearity, homogeneity and isotropy concern exclusively the constitutive relations which link stresses and strains; they in any way affect neither the kinematic relations, which define dilations and shearing strains, nor the static relations provided by the indefinite equations of equilibrium. An intimate correlation exists between the static and kinematic relations, in that the two corresponding matrix operators are each the transpose of the other. The same correlation is present, at a finite level, in the case of rigid body systems, as has been seen in Chapter 3.

Static–kinematic duality leads to an extremely direct demonstration of the principle of virtual work for deformable bodies, just as it enables a representation of the elastic problem in a synthetic manner by combining the three aforementioned relations in a single operator equation which has the displacement vector as its unknown.

Having demonstrated the classical theorems of Clapeyron and Betti, which hold good also in the case of anisotropic material, we then proceed to the analysis of isotropic material, which is of particular importance for its practical engineering applications, and to the definition of the corresponding Young's modulus and Poisson's ratio. The chapter closes with the strength criteria for biaxial and triaxial stress conditions, where the modes of rupture of ductile materials are distinguished from those of brittle materials.

8.2 INDEFINITE EQUATIONS OF EQUILIBRIUM

In the last chapter, the stress tensor was defined as the matrix of transformation of the unit vector into the corresponding tension vector. It was seen at the same time how the elements of the tensor represent the special components of stress on the coordinate planes. We then studied the law of variation of the stress tensor with the variation in orientation of the reference system and identified the principal reference system, with respect to which the tensor becomes diagonal and the shearing stresses thus vanish. We shall now determine the system of differential equations that govern the variations of the stress tensor as the point under consideration varies. Having so far limited our investigation to examining what happens at point P of the body, in the present section, we shall define the differential relation that links the stresses that develop in points of the body that are very close to one another.

To this end, let us consider an elementary parallelepiped with the sides parallel to the coordinate axes, of length dx, dy, dz, respectively (Figure 8.1). On the opposite faces of the

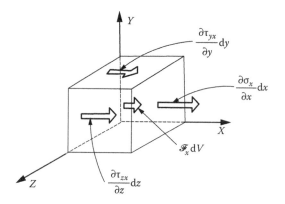

Figure 8.1

parallelepiped, there act components of stress which, but for an infinitesimal increment, are equal to one another. Equilibrium to translation in the X direction, for instance, imposes

$$\frac{\partial \sigma_x}{\partial x}dx(dy\,dz) + \frac{\partial \tau_{yx}}{\partial y}dy(dx\,dz) + \frac{\partial \tau_{zx}}{\partial z}dz(dx\,dy) + \mathcal{F}_x(dx\,dy\,dz) = 0 \tag{8.1}$$

where only the increments of stress, multiplied by the elementary areas on which they act, and the body force, multiplied by the elementary volume in which it acts, are present. Dividing Equation 8.1 by the elementary volume $dV = dx\,dy\,dz$, we obtain the first of the **indefinite equations of equilibrium**:

$$\frac{\partial \sigma_x}{\partial x} + \frac{\partial \tau_{yx}}{\partial y} + \frac{\partial \tau_{zx}}{\partial z} + \mathcal{F}_x = 0 \tag{8.2a}$$

The analogous equations of equilibrium in the Y and Z directions appear as follows:

$$\frac{\partial \tau_{xy}}{\partial x} + \frac{\partial \sigma_y}{\partial y} + \frac{\partial \tau_{zy}}{\partial z} + \mathcal{F}_y = 0 \tag{8.2b}$$

$$\frac{\partial \tau_{xz}}{\partial x} + \frac{\partial \tau_{yz}}{\partial y} + \frac{\partial \sigma_z}{\partial z} + \mathcal{F}_z = 0 \tag{8.2c}$$

Equations 8.2 may also be obtained by integration. Consider a domain V', contained in the domain V and having for boundary a closed and regular surface Ω (Figure 8.2). The equation of equilibrium to translation in vector form is written as

$$\int_\Omega \{t_n\}d\Omega + \int_{V'} \{\mathcal{F}\}dV = \{0\} \tag{8.3}$$

This vector equation is equivalent to three scalar equations, of which the first is

$$\int_\Omega t_{nx}d\Omega + \int_{V'} \mathcal{F}_x dV = 0 \tag{8.4}$$

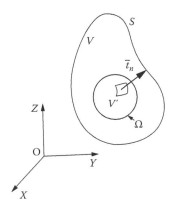

Figure 8.2

Applying the first of the Cauchy relations (7.56), the foregoing equation becomes

$$\int_{\Omega}(\sigma_x n_x + \tau_{yx} n_y + \tau_{zx} n_z)d\Omega + \int_{V'}\mathscr{F}_x dV = 0 \qquad (8.5)$$

The application of Green's theorem to the surface integral transforms it into a volume integral,

$$\int_{V'}\left(\frac{\partial \sigma_x}{\partial x} + \frac{\partial \tau_{yx}}{\partial y} + \frac{\partial \tau_{zx}}{\partial z} + \mathscr{F}_x\right)dV = 0 \qquad (8.6)$$

and this equation must hold good for any V' subdomain. The integrand must therefore be identically equal to zero, thus verifying the first of the indefinite equations of equilibrium (8.2a).

Note that Equations 8.2 constitute a system of three differential equations with partial derivatives, in the six unknown functions $\sigma_x, \sigma_y, \sigma_z, \tau_{xy}, \tau_{xz}, \tau_{yz}$. This system is hence three times indeterminate, and using the terminology introduced for beam systems, it is possible to state that the problem of a three-dimensional solid is three times hyperstatic. We shall presently see how, by adding the kinematic equations and the constitutive equations to the static equations, the problem as a whole becomes determinate.

On the boundary of the domain V (Figure 8.3), the tension vector must on the other hand coincide with the surface force $\{p\}$, externally applied:

$$t_{nx} = \sigma_x n_x + \tau_{yx} n_y + \tau_{zx} n_z = p_x \qquad (8.7a)$$

$$t_{ny} = \tau_{xy} n_x + \sigma_y n_y + \tau_{zy} n_z = p_y \qquad (8.7b)$$

$$t_{nz} = \tau_{xz} n_x + \tau_{yz} n_y + \sigma_z n_z = p_z \qquad (8.7c)$$

These relations are known as **boundary conditions of equivalence**. They represent one of the two boundary conditions for the general problem of the mechanics of elastic solids, which will be introduced in the ensuing sections.

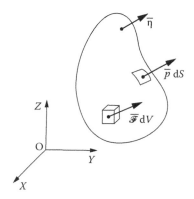

Figure 8.3

8.3 STATIC–KINEMATIC DUALITY

We are now able to express the systems of differential equations that govern, on the one hand, **congruence** and, on the other, **equilibrium** in matrix form. These two systems have intimately connected formal structures, as we shall show in this section.

As regards congruence, we recall the relations (7.10b), which define the elements of the strain tensor. The six independent components can be ordered in a **strain vector**, which can be obtained by premultiplying formally the displacement vector by a (6 × 3) matrix operator:

$$
\begin{bmatrix} \varepsilon_x \\ \varepsilon_y \\ \varepsilon_z \\ \gamma_{xy} \\ \gamma_{xz} \\ \gamma_{yz} \end{bmatrix} =
\begin{bmatrix}
\dfrac{\partial}{\partial x} & 0 & 0 \\[2mm]
0 & \dfrac{\partial}{\partial y} & 0 \\[2mm]
0 & 0 & \dfrac{\partial}{\partial z} \\[2mm]
\dfrac{\partial}{\partial y} & \dfrac{\partial}{\partial x} & 0 \\[2mm]
\dfrac{\partial}{\partial z} & 0 & \dfrac{\partial}{\partial x} \\[2mm]
0 & \dfrac{\partial}{\partial z} & \dfrac{\partial}{\partial y}
\end{bmatrix}
\begin{bmatrix} u \\ \upsilon \\ w \end{bmatrix}
\tag{8.8}
$$

The relations (8.8) can also be written in compact form

$$\{\varepsilon\} = [\partial]\{\eta\} \tag{8.9}$$

and are called **kinematic equations**.

On the other hand, as regards equilibrium, the indefinite equations of equilibrium may be reproposed, also in matrix form and with the stress components ordered in the **stress vector**:

$$
\begin{bmatrix}
\dfrac{\partial}{\partial x} & 0 & 0 & \dfrac{\partial}{\partial y} & \dfrac{\partial}{\partial z} & 0 \\[2ex]
0 & \dfrac{\partial}{\partial y} & 0 & \dfrac{\partial}{\partial x} & 0 & \dfrac{\partial}{\partial z} \\[2ex]
0 & 0 & \dfrac{\partial}{\partial z} & 0 & \dfrac{\partial}{\partial x} & \dfrac{\partial}{\partial y}
\end{bmatrix}
\begin{bmatrix}
\sigma_x \\ \sigma_y \\ \sigma_z \\ \tau_{xy} \\ \tau_{xz} \\ \tau_{yz}
\end{bmatrix}
+
\begin{bmatrix}
\mathscr{F}_x \\ \mathscr{F}_y \\ \mathscr{F}_z
\end{bmatrix}
=
\begin{bmatrix}
0 \\ 0 \\ 0
\end{bmatrix}
\tag{8.10}
$$

The **static equations** are written in compact form

$$
[\partial]^{\mathrm{T}}\{\sigma\} + \{\mathscr{F}\} = \{0\}
\tag{8.11}
$$

the static differential operator being the exact transpose of the kinematic one appearing in Equation 8.8.

In this way, just as in the mechanics of rigid bodies, the static matrix is the transpose of the kinematic one; also in the mechanics of deformable bodies, there exists the same profound interconnection between the two matrix operators. In the case of the mechanics of rigid bodies, we saw how this interconnection implies the validity of the principle of virtual work (Section 3.9). The validity of this principle will, on the other hand, be extended to the case of deformable bodies, precisely on the basis of static–kinematic duality.

To conclude, it is also possible to give an explicit matrix form to the **boundary equations of equivalence** (8.7),

$$
\begin{bmatrix}
n_x & 0 & 0 & n_y & n_z & 0 \\
0 & n_y & 0 & n_x & 0 & n_z \\
0 & 0 & n_z & 0 & n_x & n_y
\end{bmatrix}
\begin{bmatrix}
\sigma_x \\ \sigma_y \\ \sigma_z \\ \tau_{xy} \\ \tau_{xz} \\ \tau_{yz}
\end{bmatrix}
=
\begin{bmatrix}
p_x \\ p_y \\ p_z
\end{bmatrix}
\tag{8.12}
$$

or in compact form

$$
[\mathcal{N}]^{\mathrm{T}}\{\sigma\} = \{p\}
\tag{8.13}
$$

The reader's attention is drawn to the perfect correspondence that exists between the matrix operator $[\partial]^{\mathrm{T}}$ and the algebraic matrix $[\mathcal{N}]^{\mathrm{T}}$: the partial derivatives of the one are matched by the corresponding direction cosines of the other, in the spirit of Green's theorem.

8.4 PRINCIPLE OF VIRTUAL WORK

The principle of virtual work is the fundamental identity in the ambit of the mechanics of **deformable bodies**. It states the equality between external virtual work (forces multiplied by corresponding displacements) and internal virtual work (stresses multiplied by corresponding strains). More precisely, the principle of virtual work may be said to constitute the very definition of strain energy. From rational mechanics, the concept of work as scalar product of the force vector and the displacement vector is well known. However, it is not obvious that strain energy is expressible as the scalar product of the stress vector and strain vector, for the very reason that the intimate nature of these latter quantities is tensorial. On the other hand, when the body is rigid, the strains are zero and the internal virtual work vanishes, as we have already assumed in Section 3.9.

A system (a) of external forces (of volume $\{\mathcal{F}_a\}$ and surface $\{p_a\}$) and stresses $\{\sigma_a\}$ is said to be **statically admissible** when these forces satisfy the equations of statics (8.11) and the boundary conditions expressed by Equation 8.13

$$[\partial]^T\{\sigma_a\} = -\{\mathcal{F}_a\}, \quad \forall P \in V \tag{8.14a}$$

$$[\mathcal{N}]^T\{\sigma_a\} = \{p_a\}, \quad \forall P \in S \tag{8.14b}$$

where V is the three-dimensional domain occupied by the body and S is the boundary of that domain, on which the external forces $\{p_a\}$ are applied; the latter may, however, be zero over the entire boundary S or over a subset of this.

On the other hand, a system (b) of displacements $\{\eta_b\}$ and strains $\{\varepsilon_b\}$ is said to be **kinematically admissible** when the equations of kinematics (8.9) are satisfied:

$$[\partial]\{\eta_b\} = \{\varepsilon_b\}, \quad \forall P \in V \tag{8.15}$$

At this point, a digression is called for to demonstrate the rule of integration by parts on a three-dimensional domain. This rule is nothing other than an extension of Green's theorem and will be used for demonstrating the **principle of virtual work for deformable bodies**.

Consider two functions of the three Cartesian coordinates, $f(x, y, z)$ and $g(x, y, z)$, defined on a three-dimensional domain V. Perform the partial derivation of the product with respect, for instance, to the x coordinate

$$\frac{\partial}{\partial x}(fg) = \frac{\partial f}{\partial x}g + f\frac{\partial g}{\partial x} \tag{8.16}$$

and integrate both members on a generic chord parallel to the X axis and belonging to a generic section A of the domain, z = constant (Figure 8.4):

$$\int_{\alpha(y)}^{\beta(y)} \frac{\partial}{\partial x}(fg)dx = \int_{\alpha(y)}^{\beta(y)} \frac{\partial f}{\partial x}g\,dx + \int_{\alpha(y)}^{\beta(y)} f\frac{\partial g}{\partial x}dx \tag{8.17}$$

The integral of the derivative of the product is equal to the difference of the values that the product presents at the extremes of the interval of integration:

$$[fg]_{\alpha(y)}^{\beta(y)} = \int_{\alpha(y)}^{\beta(y)} \frac{\partial f}{\partial x}g\,dx + \int_{\alpha(y)}^{\beta(y)} f\frac{\partial g}{\partial x}dx \tag{8.18}$$

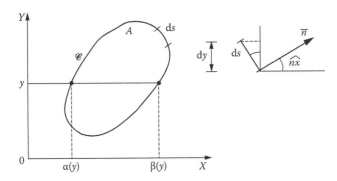

Figure 8.4

Integrating both sides of Equation 8.18 with respect to the variable y, we obtain

$$\oint_{\mathscr{C}} fg\,dy = \int_A \frac{\partial f}{\partial x}g\,dx\,dy + \int_A f\frac{\partial g}{\partial x}\,dx\,dy \tag{8.19}$$

where, on the left-hand side, there appears the counterclockwise integration of the product around a closed path performed on the boundary $\mathscr{C}$ of the two-dimensional domain A. If n_x indicates the direction cosine with respect to the X axis of the normal to the curve $\mathscr{C}$ and ds indicates the increment of the curvilinear coordinate, from the foregoing equation, we obtain

$$\oint_{\mathscr{C}} fgn_x\,dS = \int_A \frac{\partial f}{\partial x}g\,dA + \int_A f\frac{\partial g}{\partial x}\,dA \tag{8.20}$$

Finally, integrating both members with respect to the third coordinate z, we have

$$\int_S fgn_x\,dS = \int_V \frac{\partial f}{\partial x}g\,dV + \int_V f\frac{\partial g}{\partial x}\,dV \tag{8.21}$$

S being the boundary of the entire three-dimensional domain V. Equation 8.21 represents an extension of Green's theorem, as is evident if we put $f(x, y, z) = 1$, or $g(x, y, z) = 1$.

On the basis of Equation 8.21, and using the compact matrix formulation, it is not difficult to obtain the equation of virtual work. Consider the virtual work performed by the body forces $\{\mathscr{F}_a\}$ times the displacements $\{\eta_b\}$, these two fields belonging to two altogether independent systems, the first being statically admissible and the second kinematically admissible:

$$L_F = \int_V \{\mathscr{F}_a\}^T\{\eta_b\}\,dV \tag{8.22}$$

Applying the equations of statics (8.14a), we have

$$L_F = -\int_V ([\partial]^T\{\sigma_a\})^T\{\eta_b\}\,dV \tag{8.23}$$

Noting that, under the sign of transposition, there is not an algebraic matrix product, but instead a differential operator which transforms a vector function, and having recourse to Equation 8.21, we obtain

$$L_F = \int_V \{\sigma_a\}^T [\partial]\{\eta_b\} dV - \int_S \{\sigma_a\}^T [\mathcal{N}]\{\eta_b\} dS \tag{8.24}$$

Applying the equations of kinematics (8.15) and the boundary conditions of equivalence (8.14b), we have then

$$L_F = \int_V \{\sigma_a\}^T \{\varepsilon_b\} dV - \int_S \{p_a\}^T \{\eta_b\} dS \tag{8.25}$$

and thus

$$\int_V \{\sigma_a\}^T \{\varepsilon_b\} dV = \int_V \{\mathcal{F}_a\}^T \{\eta_b\} dV + \int_S \{p_a\}^T \{\eta_b\} dS \tag{8.26}$$

which constitutes the final form of the **principle of virtual work for deformable bodies**. While the right-hand side represents the external virtual work L_{ve}, the left-hand side represents and defines the internal virtual work L_{vi}, as the scalar product of the stress and strain vectors:

$$L_{vi} = L_{ve} \tag{8.27}$$

Note that up to this point, we have not framed any hypothesis on the nature of the material. This means that the principle of virtual work is of general application, whatever the constitutive law of the material.

8.5 ELASTIC CONSTITUTIVE LAW

We now introduce the concept of the **elastic body**. As will be expressed more rigorously later, a deformable body is elastic when its strain energy, i.e. the work performed from outside to bring it into a certain **strain condition** $\{\varepsilon\}$, or into a certain **stress condition** $\{\sigma\}$, does not depend on the loading process (i.e. on the previous events), but only on the final condition. It is usually said that the strain energy is in this case a **state function**.

Consider a deformable body, in equilibrium under the action of the body $\{\mathcal{F}\}$ and surface $\{p\}$ forces. There is a displacement field $\{\eta\}$ generated within it, different from zero, except on a constrained part of the external surface (Figure 8.5). Now imagine increasing the external forces by elementary quantities. Let the incremental fields be $\{d\mathcal{F}\}$ and $\{dp\}$ and let them generate an incremental displacement field $\{d\eta\}$, from which there follows an incremental strain field $\{d\varepsilon\}$.

Apply the principle of virtual work, considering as a statically admissible system that of the external forces $\{\mathcal{F}\}$ and $\{p\}$, and of the stresses produced by them $\{\sigma\}$, and as a kinematically admissible system that of the incremental displacements $\{d\eta\}$ and of the incremental strains $\{d\varepsilon\}$. Note that, in this case, the infinitesimal work is non-virtual, in the sense that

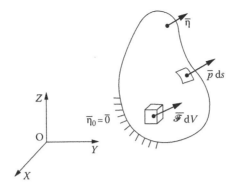

Figure 8.5

the external forces work really by the increments of displacement, while the work of the increments of the forces $\{d\mathscr{F}\}$ and $\{dp\}$, by the increments of displacement $\{d\eta\}$, is an infinitesimal of a higher order and is consequently negligible.

We have therefore

$$dL_e = \int_S \{p\}^\mathrm{T}\{d\eta\}dS + \int_V \{\mathscr{F}\}^\mathrm{T}\{d\eta\}dV \tag{8.28a}$$

$$dL_i = \int_V \{\sigma\}^\mathrm{T}\{d\varepsilon\}dV \tag{8.28b}$$

and, by the principle of virtual work, the following equality holds:

$$dL_e = dL_i \tag{8.29}$$

On the other hand, a deformable body is defined as **elastic** when the infinitesimal work expressed by Equation 8.28b is an **exact differential**. In particular, for the infinitesimal work dL, to be an exact differential, it is necessary for its integrand to be an exact differential:

$$d\Phi = \{\sigma\}^\mathrm{T}\{d\varepsilon\} \tag{8.30}$$

The function

$$\Phi = \Phi(\varepsilon_x, \varepsilon_y, \varepsilon_z, \gamma_{xy}, \gamma_{xz}, \gamma_{yz}) \tag{8.31}$$

must, i.e. be a state function and it is referred to as **elastic potential**, because it is possible to deduce from it the components of stress by means of partial derivation. The total differential of the function Φ can be expressed as

$$d\Phi = \frac{\partial \Phi}{\partial \varepsilon_x}d\varepsilon_x + \frac{\partial \Phi}{\partial \varepsilon_y}d\varepsilon_y + \frac{\partial \Phi}{\partial \varepsilon_z}d\varepsilon_z + \frac{\partial \Phi}{\partial \gamma_{xy}}d\gamma_{xy} + \frac{\partial \Phi}{\partial \gamma_{xz}}d\gamma_{xz} + \frac{\partial \Phi}{\partial \gamma_{yz}}d\gamma_{yz} \tag{8.32a}$$

while, rendering equation (8.30) explicit, we have

$$d\Phi = \sigma_x d\varepsilon_x + \sigma_y d\varepsilon_y + \sigma_z d\varepsilon_z + \tau_{xy} d\gamma_{xy} + \tau_{xz} d\gamma_{xz} + \tau_{yz} d\gamma_{yz} \tag{8.32b}$$

From Equations 8.32a and 8.32b, we obtain the components of stress

$$\sigma_x = \frac{\partial\Phi}{\partial\varepsilon_x}, \quad \sigma_y = \frac{\partial\Phi}{\partial\varepsilon_y}, \quad \sigma_z = \frac{\partial\Phi}{\partial\varepsilon_z},$$

$$\tau_{xy} = \frac{\partial\Phi}{\partial\gamma_{xy}}, \quad \tau_{xz} = \frac{\partial\Phi}{\partial\gamma_{xz}}, \quad \tau_{yz} = \frac{\partial\Phi}{\partial\gamma_{yz}} \tag{8.33}$$

In the uniaxial case, both Φ and σ_x are functions of the dilation ε_x alone,

$$\Phi = \Phi(\varepsilon_x), \quad \sigma_x = \sigma_x(\varepsilon_x) \tag{8.34}$$

so that, if we imagine loading and then unloading the one-dimensional body (e.g. a bar in tension), the paths forward and backward in the plane $\varepsilon_x - \sigma_x$ coincide (Figure 8.6a). The strain energy, represented by the potential Φ, is equal to the area under the curve $\sigma_x(\varepsilon_x)$. Hence, when the body is unloaded completely, there is no dissipation of energy and the stored elastic energy is fully recovered.

In the case of an **inelastic** one-dimensional body (Figure 8.6b), the functions Φ and σ_x no longer present the property of monodromy, i.e. to one value of ε_x, there can correspond two or more values of the work and of the force. Unloading the body, we no longer go along the curve $\sigma_x(\varepsilon_x)$ corresponding to the loading, and hence we encounter residual or permanent deformations, with dissipation of energy and only partial recovery of the strain energy. Reversing then the direction of the force and submitting the body to loading cycles, closed or spiral-shaped curves will be described in the plane $\varepsilon_x - \sigma_x$. These trajectories will be traversed in a clockwise direction, giving rise to a dissipation of energy by **hysteresis**.

Consider the infinitesimal **virtual** work

$$d\Psi = \{d\sigma\}^T\{\varepsilon\} \tag{8.35}$$

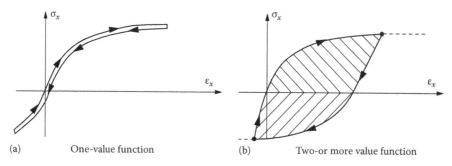

(a)   One-value function   (b)   Two-or more value function

Figure 8.6

Using the definition (8.30), we have

$$d\Phi + d\Psi = d(\{\sigma\}^{T}\{\epsilon\}) \tag{8.36}$$

The elastic potential Φ and the scalar product $\{\sigma\}^{T}\{\epsilon\}$ are both state functions, and thus $d\Phi$ and $d(\{\sigma\}^{T}\{\epsilon\})$ are exact differentials. It follows from Equation 8.36 that also $d\Psi$ is an exact differential and hence that Ψ is a state function. The function

$$\Psi = \Psi(\sigma_x, \sigma_y, \sigma_z, \tau_{xy}, \tau_{xz}, \tau_{yz}) \tag{8.37}$$

is called **complementary elastic potential,** and it is possible to deduce from it the strain components *via* partial derivation. The total differential of the function Ψ may be expressed as follows:

$$d\Psi = \frac{\partial\Psi}{\partial\sigma_x}d\sigma_x + \frac{\partial\Psi}{\partial\sigma_y}d\sigma_y + \frac{\partial\Psi}{\partial\sigma_z}d\sigma_z + \frac{\partial\Psi}{\partial\tau_{xy}}d\tau_{xy} + \frac{\partial\Psi}{\partial\tau_{xz}}d\tau_{xz} + \frac{\partial\Psi}{\partial\tau_{yz}}d\tau_{yz} \tag{8.38a}$$

whereas rendering equation (8.35) explicit, we have

$$d\Psi = \epsilon_x d\sigma_x + \epsilon_y d\sigma_y + \epsilon_z d\sigma_z + \gamma_{xy}d\tau_{xy} + \gamma_{xz}d\tau_{xz} + \gamma_{yz}d\tau_{yz} \tag{8.38b}$$

From Equations 8.38a and 8.38b, we obtain the components of strain:

$$\epsilon_x = \frac{\partial\Psi}{\partial\sigma_x}, \quad \epsilon_y = \frac{\partial\Psi}{\partial\sigma_y}, \quad \epsilon_z = \frac{\partial\Psi}{\partial\sigma_z},$$

$$\gamma_{xy} = \frac{\partial\Psi}{\partial\tau_{xy}}, \quad \gamma_{xz} = \frac{\partial\Psi}{\partial\tau_{xz}}, \quad \gamma_{yz} = \frac{\partial\Psi}{\partial\tau_{yz}} \tag{8.39}$$

In the uniaxial case, it is easy to give a graphical interpretation of Ψ. The elastic complementary energy is the area contained between the curve of loading $\epsilon_x(\sigma_x)$ and the axis σ_x (Figure 8.7a), i.e. it is the area complementary to the one representing the elastic energy Φ,

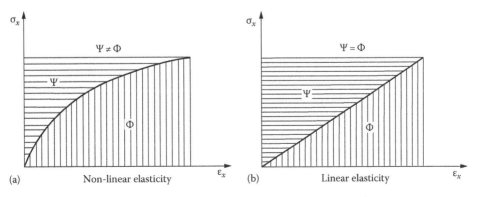

Figure 8.7

with respect to the rectangle of sides ε_x, σ_x. The latter would represent the work of deformation in the case where, during the entire deformation process, the values of stress were constant and equal to the final value. In the general case, where the relation $\sigma_x = \sigma_x(\varepsilon_x)$ is not linear (Figure 8.7a), the deformable body is said to be **non-linear elastic**. In the particular case, instead, where the relation is linear (Figure 8.7b), the body is said to be **linear elastic**. In the linear and one-dimensional case, it is obvious that $\Psi = \Phi$. We shall demonstrate, however, that equality also exists between elastic potential and complementary elastic potential in the linear and three-dimensional case.

8.6 LINEAR ELASTICITY

Let us reconsider the elastic potential function (8.31) and expand it using Maclaurin series about the origin, i.e. about the undeformed condition

$$
\begin{aligned}
\Phi(\varepsilon_x, \varepsilon_y, \ldots, \gamma_{yz}) = \Phi(0) &+ \left(\frac{\partial\Phi}{\partial\varepsilon_x}\right)_0 \varepsilon_x + \left(\frac{\partial\Phi}{\partial\varepsilon_y}\right)_0 \varepsilon_y \\
&+ \cdots + \left(\frac{\partial\Phi}{\partial\gamma_{yz}}\right)_0 \gamma_{yz} + \frac{1}{2}\left\{\left(\frac{\partial^2\Phi}{\partial\varepsilon_x^2}\right)_0 \varepsilon_x^2 + \left(\frac{\partial^2\Phi}{\partial\varepsilon_y^2}\right)_0 \varepsilon_y^2 \right. \\
&+ \cdots + \left(\frac{\partial^2\Phi}{\partial\gamma_{yz}^2}\right)_0 \gamma_{yz}^2 + 2\left(\frac{\partial^2\Phi}{\partial\varepsilon_x\partial\varepsilon_y}\right)_0 \varepsilon_x\varepsilon_y + 2\left(\frac{\partial^2\Phi}{\partial\varepsilon_x\partial\varepsilon_z}\right)_0 \varepsilon_x\varepsilon_z \\
&+ \cdots + \left. 2\left(\frac{\partial^2\Phi}{\partial\gamma_{xz}\partial\gamma_{yz}}\right)_0 \gamma_{xz}\gamma_{yz}\right\} + \cdots
\end{aligned}
\tag{8.40}
$$

If the strains are sufficiently small, a good approximation will be achieved by neglecting terms in powers above the second. On the other hand, since the stresses are obtained by derivation of Φ, the value that the function presents at the origin is an arbitrary constant, which we can take to be zero: $\Phi(0) = 0$. Also the coefficients of the first-order terms are zero, since they represent the stresses in the undeformed condition

$$
\left(\frac{\partial\Phi}{\partial\varepsilon_x}\right)_0 = \sigma_x(0) = 0
\tag{8.41a}
$$

$$
\left(\frac{\partial\Phi}{\partial\varepsilon_y}\right)_0 = \sigma_y(0) = 0
\tag{8.41b}
$$

$$
\vdots \qquad\qquad\qquad\qquad\qquad\qquad\qquad\qquad \vdots
$$

$$
\left(\frac{\partial\Phi}{\partial\gamma_{yz}}\right)_0 = \tau_{yz}(0) = 0
\tag{8.41c}
$$

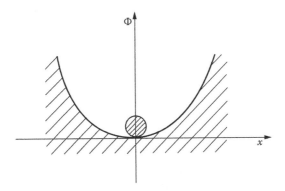

Figure 8.8

The analogy with the potential well of the harmonic oscillator is evident (Figure 8.8). Also in that case, the derivative of the potential provides the force necessary to remove the material point from the origin (or the opposite of the restoring force)

$$\Phi = \frac{1}{2}kx^2 \tag{8.42a}$$

$$\text{Force} = \frac{d\Phi}{dx} = kx \tag{8.42b}$$

where k is the stiffness of the spring. Also in this case, as is well known, the restoring force is zero in the origin.

The Maclaurin expansion (8.40) then reduces to a **quadratic form**, with 21 coefficients that can be ordered in the (6×6) Hessian matrix:

$$[H] = \begin{bmatrix} \left(\dfrac{\partial^2\Phi}{\partial\varepsilon_x^2}\right)_0 & \left(\dfrac{\partial^2\Phi}{\partial\varepsilon_x\partial\varepsilon_y}\right)_0 & \cdots & \left(\dfrac{\partial^2\Phi}{\partial\varepsilon_x\partial\gamma_{yz}}\right)_0 \\ \left(\dfrac{\partial^2\Phi}{\partial\varepsilon_y\partial\varepsilon_x}\right)_0 & \left(\dfrac{\partial^2\Phi}{\partial\varepsilon_y^2}\right)_0 & \cdots & \left(\dfrac{\partial^2\Phi}{\partial\varepsilon_y\partial\gamma_{yz}}\right)_0 \\ \vdots & & & \vdots \\ \left(\dfrac{\partial^2\Phi}{\partial\gamma_{yz}\partial\varepsilon_x}\right)_0 & \left(\dfrac{\partial^2\Phi}{\partial\gamma_{yz}\partial\varepsilon_y}\right)_0 & \cdots & \left(\dfrac{\partial^2\Phi}{\partial\gamma_{yz}^2}\right)_0 \end{bmatrix} \tag{8.43}$$

Note that the Hessian matrix is the Jacobian of the six first partial derivatives.

It is then possible to write in compact matrix form

$$\Phi = \frac{1}{2}\{\varepsilon\}^{\mathrm{T}}[H]\{\varepsilon\} \tag{8.44a}$$

This relation is analogous to Equation 8.42a, once the strain vector $\{\varepsilon\}$ is made to correspond to the elongation x and the Hessian matrix is made to correspond to the stiffness of the spring. *Via* partial derivation, it can be readily verified that the stress vector is given by

$$\{\sigma\} = [H]\{\varepsilon\} \tag{8.44b}$$

This relation is analogous to Equation 8.42b, which gives the restoring force of the spring in the case of a one-dimensional problem (harmonic oscillator). From Equations 8.44a and 8.44b, a new expression of elastic potential is derived:

$$\Phi = \frac{1}{2}\{\varepsilon\}^T\{\sigma\} \tag{8.45}$$

Since the undeformed condition must represent an absolute minimum of Φ, and not only a stationary point as Equations 8.41 ensure, the Hessian matrix must be **positive definite**, i.e. its determinant and those of its principal minors must be greater than zero. Thus, once more it is possible to discern the analogy with the one-dimensional case, in which the stiffness k of the spring, which is the second derivative of the potential, must be positive.

The positiveness of the Hessian implies the reversibility of the relation (8.44b) and hence of the Hessian matrix itself:

$$\{\varepsilon\} = [H]^{-1}\{\sigma\} \tag{8.46}$$

The relations (8.44b) and (8.46) link together linearly the stress and strain vectors and constitute the link, hitherto missing, between statics and kinematics.

From the relation (8.36), but for an arbitrary constant, we obtain

$$\Phi + \psi = \{\sigma\}^T\{\varepsilon\} = \sigma_x\varepsilon_x + \sigma_y\varepsilon_y + \sigma_z\varepsilon_z + \tau_{xy}\gamma_{xy} + \tau_{xz}\gamma_{xz} + \tau_{yz}\gamma_{yz} \tag{8.47}$$

i.e. the sum of the elastic potential and complementary elastic potential is always equal (also in non-linear cases) to the scalar product of $\{\sigma\}$ and $\{\varepsilon\}$. It has also been demonstrated that, in linear cases, Φ is half of that product (Equation 8.45). Hence, for linear elastic bodies, we have

$$\Phi = \psi = \frac{1}{2}\{\varepsilon\}^T\{\sigma\} = \frac{1}{2}\{\sigma\}^T\{\varepsilon\} \tag{8.48}$$

We can then give the following form to the complementary elastic potential:

$$\psi = \frac{1}{2}\{\sigma\}^T[H]^{-1}\sigma \tag{8.49}$$

which corresponds to Equation 8.44a and shows how Ψ is a quadratic form of the components of stress.

8.7 PROBLEM OF A LINEAR ELASTIC BODY

As has previously been noted, the three indefinite equations of equilibrium do not suffice to determine the six components of stress. On the other hand, by adding the six elastic constitutive Equations 8.44b to them, we obtain a system of nine differential equations in nine unknowns: $\sigma_x, \sigma_y, \sigma_z, \tau_{xy}, \tau_{xz}, \tau_{yz}; u, v, w$.

In matrix form, it is possible to give a very synthetic and expressive representation of the linear elastic problem, considering the displacement vector $\{\eta\}$ as the primary unknown.

If we introduce the constitutive law (8.44b), and then the kinematic equation (8.9) in the static equation (8.11), we obtain a matrix equation, called **Lame's equation in operator form**:

$$([\partial]^T[H][\partial]\{\eta\}) = -\{\mathcal{F}\} \tag{8.50}$$

The matrix and second-order differential operator in round brackets is called the **Lame operator**,

$$\underset{(3\times3)}{[\mathcal{L}]} = \underset{(3\times6)}{[\partial]^T} \underset{(6\times6)}{[H]} \underset{(6\times3)}{[\partial]} \tag{8.51}$$

It turns out to be a (3×3) matrix and, in non-homogeneous problems, where the matrix $[H]$ is a function of the point, it is a function of the point too.

Recalling the boundary equations of equivalence (8.13) and assuming that they hold good on a portion S_p of the external surface of the body and that, on the complementary portion S_η, there is imposed a congruent field of displacements $\{\eta_0\}$, the three-dimensional elastic problem can be synthesized as follows:

$$[\mathcal{L}]\{\eta\} = -\{\mathcal{F}\}, \qquad \forall P \in V \tag{8.52a}$$

$$([\mathcal{N}]^T[H][\partial])\{\eta\} = \{p\}, \quad \forall P \in S_p \tag{8.52b}$$

$$\{\eta\} = \{\eta_0\}, \qquad \forall P \in S_\eta \tag{8.52c}$$

For example, in the case of a linear elastic one-dimensional body, restrained at one end, submitted to a tensile force p at the opposite end and to a distribution $\mathcal{F}_x(x)$ of axial forces (Figure 8.9), Equations 8.52 take on a notably simplified appearance:

$$EA\frac{d^2u}{dx^2} = -\mathcal{F}_x(x), \quad \text{for } 0 \le x \le l \tag{8.53a}$$

$$EA\frac{du}{dx} = p, \qquad \text{for } x = l \tag{8.53b}$$

$$u = 0, \qquad \text{for } x = 0 \tag{8.53c}$$

where l is the length of the bar and the product EA represents the longitudinal stiffness of the bar.

Once the problem (8.52) is resolved and the displacement field $\{\eta\}$ has been identified, if we reverse the procedure and Equations 8.9 and 8.44b are reapplied, the strain field $\{\varepsilon\}$ and the

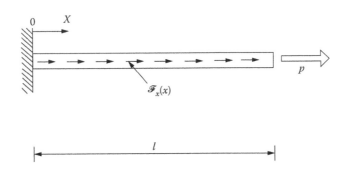

Figure 8.9

stress field $\{\sigma\}$ are respectively determined. Since Equations 8.52 are linear, the **principle of superposition** holds. It means that, if a loading system $\{\mathscr{F}_a\}$, $\{p_a\}$, $\{\eta_{0a}\}$ generates a displacement field $\{\eta_a\}$ and thus the strain and stress fields $\{\varepsilon_a\}$, $\{\sigma_a\}$, and if a different loading system $\{\mathscr{F}_b\}$, $\{p_b\}$, $\{\eta_{0b}\}$ generates the fields $\{\eta_b\}$, $\{\varepsilon_b\}$, $\{\sigma_b\}$, the loading system $\{\mathscr{F}_a\} + \{\mathscr{F}_b\}$, $\{p_a\} + \{p_b\}$, $\{\eta_{0a}\} + \{\eta_{0b}\}$ generates displacement, strain and stress fields, which are the sum of the previous ones:

$$\{\eta_a\} + \{\eta_b\}, \quad \{\varepsilon_a\} + \{\varepsilon_b\}, \quad \{\sigma_a\} + \{\sigma_b\}$$

On the basis of the principle of superposition, it is possible to demonstrate **Kirchhoff's theorem** or the **solution uniqueness theorem**: if the solution $\{\eta\}$ exists, it is the only one. The demonstration must be conducted *ab absurdo*. Imagine, i.e. one loading system, $\{\mathscr{F}\}$, $\{p\}$, $\{\eta_0\}$, can generate two different responses: $\{\eta_a\}$, $\{\varepsilon_a\}$, $\{\sigma_a\}$ or $\{\eta_b\}$, $\{\varepsilon_b\}$, $\{\sigma_b\}$. Applying the principle of virtual work to the difference system, we have

$$\int_V \{0\}^T \{\Delta\eta\} dV + \int_{S_p} \{0\}^T \{\Delta\eta\} dS + \int_{S_\eta} \{\Delta R\}^T \{0\} dS$$

$$= \int_V \{\Delta\sigma\}^T \{\Delta\varepsilon\} dV \tag{8.54}$$

$\{\Delta R\}$ being the difference in constraint reactions. The integrand on the right-hand side represents twice the elastic potential, so that we have

$$2 \int_V \Phi(\Delta\varepsilon_x, \Delta\varepsilon_y, \ldots, \Delta\gamma_{yz}) dV = 0 \tag{8.55}$$

On the other hand, it is known that Φ is a positive definite quadratic form, so that the integral (8.55) vanishes only when the integrand is zero at each point of the elastic body. This is found to be the case only when, at each point, we have

$$\Delta\varepsilon_x = \Delta\varepsilon_y = \cdots = \Delta\gamma_{yz} = 0 \tag{8.56}$$

i.e. only when solutions (*a*) and (*b*) coincide.

8.8 CLAPEYRON'S THEOREM

Consider a linear elastic body subjected to body forces $\{\mathscr{F}\}$ and to surface forces $\{p\}$. Let $\{\eta\}$ be the displacement field that is generated in the body at the end of the loading process that brings the external forces from zero to the aforesaid values. The application of the principle of virtual work gives the following equality:

$$\int_V \{\mathscr{F}\}^T \{\eta\} dV + \int_S \{p\}^T \{\eta\} dS = \int_V \{\sigma\}^T \{\varepsilon\} dV \tag{8.57}$$

if the corresponding **final** fields are considered as statically and kinematically admissible systems. Multiplying both sides of Equation 8.57 by the factor 1/2, we have

$$\frac{1}{2} \int_V \{\mathscr{F}\}^T \{\eta\} dV + \frac{1}{2} \int_S \{p\}^T \{\eta\} dS = \int_V \Phi dV \tag{8.58}$$

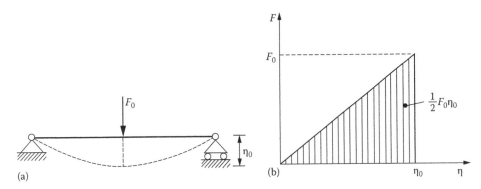

Figure 8.10

since, for a linear elastic body, the relation (8.45) holds. Equation 8.58 expresses the fact that the work of deformation performed by the external forces to bring the body from the initial undeformed condition to the final deformed condition is equal to half of the work that these forces would perform if they presented their final value during the whole deformation process. The content of **Clapeyron's theorem** was already implicitly presented, when we considered the one-dimensional case of Figure 8.7b. If, for example, a linear elastic beam is subjected to the action of a concentrated force in the centre which increases slowly (so as not to induce dynamic phenomena) from zero to the final value F_0, and at the same time the deflection at the centre increases from zero to the final value η_0, then, according to Clapeyron's theorem, the work of deformation performed on the beam is (Figure 8.10)

$$L_{\text{def.}} = \int_0^{\eta_0} F(\eta)d\eta = \frac{1}{2}F_0\eta_0 \tag{8.59}$$

8.9 BETTI'S RECIPROCAL THEOREM

Betti's reciprocal theorem shows how the principle of superposition in linear elasticity holds only for displacements, strain and stress and is not applicable, instead, to the work of deformation.

Consider a linear elastic body and submit it to a quasi-static (i.e. very slow) loading process, so that the final forces applied are $\{\mathscr{F}_a\}$, $\{p_a\}$, and the work of deformation performed is L_a. Then proceed loading with the quasi-static application of a second system of forces $\{\mathscr{F}_b\}$, $\{p_b\}$. Let the work performed by these forces be L_b; the work of the forces $\{\mathscr{F}_a\}$, $\{p_a\}$ by the displacements $\{\eta_b\}$ is called **mutual work** L_{ab}. Expressed in formulas,

$$L_{a+b} = L_a + L_b + L_{ab} \tag{8.60}$$

with

$$L_a = \frac{1}{2}\int_V \{\mathscr{F}_a\}^T\{\eta_a\}dV + \frac{1}{2}\int_S \{p_a\}^T\{\eta_a\}dS \tag{8.61a}$$

$$L_b = \frac{1}{2}\int_V \{\mathscr{F}_b\}^T \{\eta_b\}dV + \frac{1}{2}\int_S \{p_b\}^T \{\eta_b\}dS \qquad (8.61b)$$

$$L_{ab} = \int_V \{\mathscr{F}_a\}^T \{\eta_b\}dV + \int_S \{p_a\}^T \{\eta_b\}dS \qquad (8.61c)$$

Equation 8.60 clearly expresses the non-applicability of the principle of superposition to the work of deformation.

Imagine now that the process of loading described earlier is reversed, i.e. first the forces $\{\mathscr{F}_b\}$, $\{p_b\}$ are applied and then the forces $\{\mathscr{F}_a\}$, $\{p_a\}$. The total work of deformation will then be expressible as follows:

$$L_{b+a} = L_b + L_a + L_{ba} \qquad (8.62)$$

where the mutual work of the forces $\{\mathscr{F}_b\}$, $\{p_b\}$ acting through the further displacements $\{\eta_a\}$ is

$$L_{ba} = \int_V \{\mathscr{F}_b\}^T \{\eta_a\}dV + \int_S \{p_b\}^T \{\eta_a\}dS \qquad (8.63)$$

Comparing expressions (8.60) and (8.62) and noting that the total work of deformation must not depend on the loading path (i.e. on the order in which the external forces are applied), since the body was assumed as being elastic, we obtain the equality of the two mutual work expressions (8.61c) and (8.63):

$$L_{ab} = L_{ba} \qquad (8.64)$$

In general, for an elastic body subjected to two systems of surface and body forces, the work done by the first system acting through the displacements resulting from the second and that done by the second system acting through the displacements resulting from the first are equal and different from zero. When, in particular, these are both zero, the two systems of forces are said to be **energetically orthogonal** and the principle of superposition becomes valid also for the work of deformation. Consider, for instance, a supported beam made of linear elastic material, subjected to a concentrated force F in the centre (Figure 8.11a) or

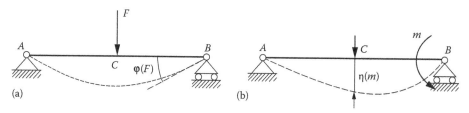

Figure 8.11

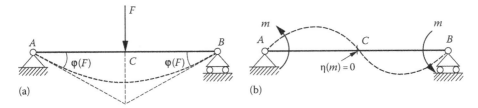

Figure 8.12

to a concentrated moment m at one end (Figure 8.11b). The two systems, F and m, are not energetically orthogonal, since their mutual work is different from zero,

$$F\eta(m) = m\varphi(F) \neq 0 \tag{8.65}$$

where
    $\eta(m)$ is the deflection in the centre caused by the moment m
    $\varphi(F)$ is the angle of elastic rotation at the ends caused by the force F

Instead, the two systems of external forces of Figure 8.12, i.e. the concentrated force F in the centre and the two concentrated moments m applied in the same direction at the ends, are energetically orthogonal. In fact, their mutual work is zero,

$$F \times 0 = m\varphi(F) - m\varphi(F) = 0 \tag{8.66}$$

the deflection in the centre being zero in the diagram of Figure 8.12b, which is skew-symmetrical, and the end elastic rotations being equal and opposite in the diagram of Figure 8.12a, which is symmetrical.

    As will be shown later, the characteristics of the internal reaction of the beams are, except for marginal cases, energetically orthogonal.

8.10 ISOTROPY

The deformable body is considered in this section also as **isotropic**, as well as linear elastic. This means that the mechanical properties are considered as identical in all directions issuing from the generic point P (the case of anisotropic material is dealt with in Appendix C). As there do not exist preferential directions, the complementary elastic potential Ψ will depend on the values of the three principal stresses, and not on the orientation of the principal reference system:

$$\Psi = \Psi(\sigma_1, \sigma_2, \sigma_3) \tag{8.67}$$

As Ψ is a quadratic form of the stress components, it may be cast in the following form:

$$\Psi = \frac{1}{2E}\left\{\left(\sigma_1^2 + \sigma_2^2 + \sigma_3^2\right) - 2\nu\left(\sigma_1\sigma_2 + \sigma_1\sigma_3 + \sigma_2\sigma_3\right)\right\} \tag{8.68}$$

where $(1/2)E$ and $-v/E$ are two coefficients which multiply the squares and the mutual products, respectively. The coefficients thus reduce from 21 to 2, on the hypothesis of isotropy alone. The constants E and v, as we shall see presently, have a precise physical meaning.

Recalling the expressions (7.74) of the invariants as functions of the principal stresses, Equation 8.68 can be written as follows:

$$\psi = \frac{1}{2E}\left\{J_I^2 + 2J_{II}(1+v)\right\} \tag{8.69}$$

Expressing then the invariants as functions of the generic components of the stress vector, as Equations 7.72 show, we have

$$\Psi = \frac{1}{2E}\left(\sigma_x^2 + \sigma_y^2 + \sigma_z^2\right) - \frac{v}{E}\left(\sigma_x\sigma_y + \sigma_x\sigma_z + \sigma_y\sigma_z\right) + \frac{1}{2G}\left(\tau_{xy}^2 + \tau_{xz}^2 + \tau_{yz}^2\right) \tag{8.70}$$

where we have set

$$G = \frac{E}{2(1+v)} \tag{8.71}$$

Equation 8.70 thus represents the complementary elastic potential in the case of linear and isotropic elasticity.

The components of strain are obtained according to Equations 8.39, by partial derivation of Ψ:

$$\varepsilon_x = \frac{\partial\psi}{\partial\sigma_x} = \frac{\sigma_x}{E} - \frac{v}{E}\sigma_y - \frac{v}{E}\sigma_z \tag{8.72a}$$

$$\varepsilon_y = \frac{\partial\psi}{\partial\sigma_y} = \frac{\sigma_y}{E} - \frac{v}{E}\sigma_x - \frac{v}{E}\sigma_z \tag{8.72b}$$

$$\varepsilon_z = \frac{\partial\psi}{\partial\sigma_z} = \frac{\sigma_z}{E} - \frac{v}{E}\sigma_x - \frac{v}{E}\sigma_y \tag{8.72c}$$

$$\gamma_{xy} = \frac{\partial\psi}{\partial\tau_{xy}} = \frac{\tau_{xy}}{G} \tag{8.72d}$$

$$\gamma_{xz} = \frac{\partial\psi}{\partial\tau_{xz}} = \frac{\tau_{xz}}{G} \tag{8.72e}$$

$$\gamma_{yz} = \frac{\partial\psi}{\partial\tau_{yz}} = \frac{\tau_{yz}}{G} \tag{8.72f}$$

Note that, whereas the shearing strains are linearly dependent, with a relation of proportionality, only on the respective shearing stresses, the dilations each depends on all three normal stresses. In explicit matrix form, we may write

$$
\begin{bmatrix} \varepsilon_x \\ \varepsilon_y \\ \varepsilon_z \\ \gamma_{xy} \\ \gamma_{xz} \\ \gamma_{yz} \end{bmatrix} = \begin{bmatrix} \dfrac{1}{E} & -\dfrac{\nu}{E} & -\dfrac{\nu}{E} & 0 & 0 & 0 \\[2mm] -\dfrac{\nu}{E} & \dfrac{1}{E} & -\dfrac{\nu}{E} & 0 & 0 & 0 \\[2mm] -\dfrac{\nu}{E} & -\dfrac{\nu}{E} & \dfrac{1}{E} & 0 & 0 & 0 \\[2mm] 0 & 0 & 0 & \dfrac{1}{G} & 0 & 0 \\[2mm] 0 & 0 & 0 & 0 & \dfrac{1}{G} & 0 \\[2mm] 0 & 0 & 0 & 0 & 0 & \dfrac{1}{G} \end{bmatrix} \begin{bmatrix} \sigma_x \\ \sigma_y \\ \sigma_z \\ \tau_{xy} \\ \tau_{xz} \\ \tau_{yz} \end{bmatrix}
\tag{8.73}
$$

The foregoing matrix relation in compact form is represented by Equation 8.46. The inverse of Equation 8.73 is as follows:

$$
\frac{1}{2G}\begin{bmatrix} \sigma_x \\ \sigma_y \\ \sigma_z \\ \tau_{xy} \\ \tau_{xz} \\ \tau_{yz} \end{bmatrix} = \begin{bmatrix} \dfrac{1-\nu}{1-2\nu} & \dfrac{\nu}{1-2\nu} & \dfrac{\nu}{1-2\nu} & 0 & 0 & 0 \\[2mm] \dfrac{\nu}{1-2\nu} & \dfrac{1-\nu}{1-2\nu} & \dfrac{\nu}{1-2\nu} & 0 & 0 & 0 \\[2mm] \dfrac{\nu}{1-2\nu} & \dfrac{\nu}{1-2\nu} & \dfrac{1-\nu}{1-2\nu} & 0 & 0 & 0 \\[2mm] 0 & 0 & 0 & \dfrac{1}{2} & 0 & 0 \\[2mm] 0 & 0 & 0 & 0 & \dfrac{1}{2} & 0 \\[2mm] 0 & 0 & 0 & 0 & 0 & \dfrac{1}{2} \end{bmatrix} \begin{bmatrix} \varepsilon_x \\ \varepsilon_y \\ \varepsilon_z \\ \gamma_{xy} \\ \gamma_{xz} \\ \gamma_{yz} \end{bmatrix}
\tag{8.74}
$$

Since the Hessian matrix has been assumed as positive definite (the undeformed condition must represent an absolute minimum of the work of deformation), also its inverse is positive definite (i.e. likewise, the unstressed condition must represent an absolute minimum of the work of deformation). All the principal minors of the matrix (8.73) must therefore be greater than zero. The following principal minor must therefore be positive definite:

$$
\begin{bmatrix} \dfrac{1}{E} & \dfrac{-\nu}{E} & \dfrac{-\nu}{E} \\[2mm] \dfrac{-\nu}{E} & \dfrac{1}{E} & \dfrac{-\nu}{E} \\[2mm] \dfrac{-\nu}{E} & \dfrac{-\nu}{E} & \dfrac{1}{E} \end{bmatrix}
\tag{8.75a}
$$

and the following condition must at the same time hold:

$$
\frac{1}{G} > 0
\tag{8.75b}
$$

From Equations 8.75, four inequalities are drawn:

$$\frac{1}{E} > 0 \tag{8.76a}$$

$$\frac{1}{E^2}(1 - v^2) > 0 \tag{8.76b}$$

$$\frac{1}{E^3}(1 + v)^2(1 - 2v) > 0 \tag{8.76c}$$

$$\frac{E}{2(1 + v)} > 0 \tag{8.76d}$$

which, when resolved, impose

$$E > 0 \tag{8.77a}$$

$$-1 < v < 1 \tag{8.77b}$$

$$v < \frac{1}{2} \tag{8.77c}$$

$$v > -1 \tag{8.77d}$$

Combining the conditions (8.77b and 8.77c) and noting that Equation 8.77d contains Equation 8.77b, we obtain finally the following bounds:

$$E > 0 \tag{8.78a}$$

$$-1 < v < \frac{1}{2} \tag{8.78b}$$

On the other hand, physically, we note that the coefficient v is never negative. In the sequel, we shall discuss the physical meaning of the parameters E, v, G, and their limitations.

Consider an elementary parallelepiped subjected to the normal tensile stress component σ_x only (Figure 8.13a). Equation 8.72 in this case takes, on a particular form:

$$\varepsilon_x = \frac{\sigma_x}{E} \tag{8.79a}$$

$$\varepsilon_y = -\frac{v}{E}\sigma_x = -v\varepsilon_x \tag{8.79b}$$

$$\varepsilon_z = -\frac{v}{E}\sigma_x = -v\varepsilon_x \tag{8.79c}$$

$$\gamma_{xy} = \gamma_{xz} = \gamma_{yz} = 0 \tag{8.79d}$$

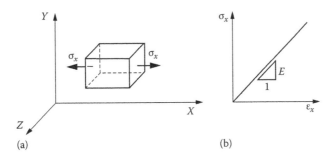

Figure 8.13

From Equation 8.79a, the physical meaning of stiffness of the material for E is derived. On the plane $\varepsilon_x - \sigma_x$, E represents in fact the positive slope of the straight line passing through the origin, which describes the process of loading (Figure 8.13b). The slope E is known as the **normal elastic modulus** or **Young's modulus**.

The coefficient ν, on the other hand, represents the ratio between the dilations induced in the directions perpendicular to that of stress and the dilation in the direction of stress:

$$\nu = \left| \frac{\varepsilon_y}{\varepsilon_x} \right| = \left| \frac{\varepsilon_z}{\varepsilon_x} \right| \tag{8.80}$$

Since the former are experimentally always of the opposite algebraic sign with respect to the latter (Figure 8.14), physically ν is positive, and thus the limitations (8.78b) corresponding to it are in actual fact more severe:

$$0 < \nu < \frac{1}{2} \tag{8.81}$$

The constant ν is called **ratio of transverse contraction** or **Poisson's ratio**. It is interesting to note that the reversibility of the Hessian matrix implies a volumetric dilation concordant with the normal stress σ_x, and *vice versa*. In fact, from relations (8.79), we have

$$\frac{\Delta V}{V} = \varepsilon_x + \varepsilon_y + \varepsilon_z = \frac{\sigma_x}{E}(1 - 2\nu) \tag{8.82}$$

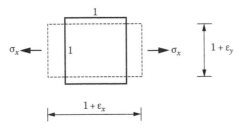

Figure 8.14

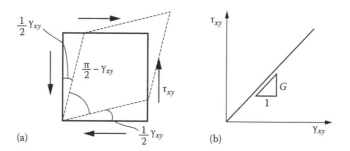

Figure 8.15

Since from Equation 8.78a we have $E > 0$, and from Equation 8.78b $\nu < 1/2$, Equation 8.82 shows how $\Delta V/V$ and σ_x are concordant, i.e. if σ_x is a tensile stress, a positive volumetric dilation is produced, whereas if σ_x is a compressive stress, a negative volumetric dilation (i.e. a volumetric contraction) is produced.

Consider once more the elementary parallelepiped, stressed in this case only by the shearing stress component τ_{xy} (Figure 8.15a). Equations 8.72 are particularized as follows:

$$\varepsilon_x = \varepsilon_y = \varepsilon_z = \gamma_{xz} = \gamma_{yz} = 0 \tag{8.83a}$$

$$\gamma_{xy} = \frac{\tau_{xy}}{G} \tag{8.83b}$$

The parameter G thus represents the stiffness that the solid opposes to shearing strain. This is called the **shear elastic modulus** and graphically is the positive angular coefficient of the line of loading in the plane $\gamma_{xy} - \tau_{xy}$ (Figure 8.15b). From the relation (8.83b) and the similar relations that show the proportionality between shearing strains and corresponding shearing stresses, we deduce how, in the linear elastic and isotropic body, the principal directions of strain and stress coincide.

A linear elastic and isotropic body is characterized by the values that the two parameters E and ν assume at each point. If the point functions E and ν prove to be constant, then the body is said to be **homogeneous.**

Two typical building materials, steel and concrete, are with good approximation considered linear elastic, isotropic and homogeneous. On the other hand, the hypothesis of linear elasticity is acceptable only in the conditions where the materials are not excessively stressed. Beyond certain threshold values, of which we shall speak presently, the behaviour of the material becomes non-linear and inelastic. When the materials are subjected to increasing stresses, critical conditions are eventually reached, on account of which the body can no longer be considered even a deformable continuum. In other words, fractures form and hence produce discontinuities in the displacement function.

In Table 8.1, indicative values of E, ν and σ_{max} for the two materials mentioned earlier are given, σ_{max} being the value of tensile normal stress which causes yielding or fracturing.

Table 8.1

	E (kg/cm^2)	ν	σ_{max} (kg/cm^2)
Steel	2,100,000	0.30	2400
Concrete	250,000	0.15	30

The ratio σ_{max}/E indicates the order of magnitude of the dilation, below which the linear elastic idealization has a physical meaning. It may be noted that this order of magnitude lies between 10^{-3} and 10^{-4}. The strains which are usually found in structural elements are in fact very small.

As regards the **anisotropic elastic constitutive law**, the reader is referred to Appendix C.

8.11 STRENGTH, DUCTILITY, FRACTURE ENERGY

Once the elastic stress field of a structural element has been calculated, on the basis of the forces applied $\{\mathscr{F}\}$ and $\{p\}$ and the displacements imposed $\{\eta_0\}$, using the equations of statics and kinematics and the constitutive equations and then resolving Lame's equation, we are faced with the problem of evaluating whether the theoretically determined stresses exceed, albeit in only one point of the body or in one of its portions, the strength of the material of which the body is made. In fact, as has already been mentioned in the foregoing section, even though the law $\sigma(\varepsilon)$ is linear and elastic in the initial portion of the curve, it loses its linearity in the next portion, giving rise to phenomena of yielding, plastic deformation and ultimately fracture. The stress field, on the other hand, is proportional to the forces applied so far as the structural behaviour is linearly elastic. Consequently, if calculation, in this conventionally elastic case, gives excessively high values of stress, it is possible to consider the external loads reduced by a suitable factor to obtain a really elastic stress field. This factor will therefore be equal to the ratio between the conventionally elastic and excessively high stresses and the really elastic and hence admissible stresses for the material. The latter are chosen by the structural designer on the basis of criteria of resistance and safety which we shall look at later on.

Structural materials are traditionally catalogued, on the basis of the characteristics of the $\sigma(\varepsilon)$ curve, into two distinct categories: **ductile materials** and **brittle materials**. Whereas the former shows large portions of the $\sigma(\varepsilon)$ diagram that are not linear, before they reach the fracture point, the latter breaks suddenly, when the response is still substantially elastic and linear. A second characteristic which distinguishes them clearly is the ratio between tensile strength and compressive strength. Whereas for ductile materials this ratio is close to unity, for brittle materials it is a good deal lower (in some cases, $10^{-1} \div 10^{-2}$). The differences in behaviour depend to a great extent upon the microscopic mechanisms of damage and fracture, which, in the various structural materials present notable differences. In metal alloys, for instance, sliding takes place between the planes of atoms and crystals which gives rise to a behaviour of a plastic and ductile kind, with considerable permanent deformations. In concrete and rocks, on the other hand, the microcracks and debondings between the granular components and the matrix can extend and combine to form a macroscopic crack which splits the structural element suddenly into two parts. This unstable fracturing process causes the material to behave in a brittle manner.

On the other hand, it is not always easy to determine the microscopic magnitude of the mechanisms of damage. It may present very different dimensions according to the nature of the mechanisms and the heterogeneity of the material. In crystals, damage occurs at an atomic level, with vacancies and dislocations; in metal alloys, cracks spread at an intergranular or transgranular level; and in concrete, the cracking occurs at the interface between the aggregates and the cement matrix. It is thus understandable how the scale of damage comes to depend upon the regularity of the solid and hence upon the size of heterogeneities present in it. Alongside the traditional building materials, more recent times have witnessed the advent of a large number of new materials which present highly heterogeneous and anisotropic features on account of their being reinforced with fibres and composed of laminas. These materials, called **composites**, may have a polymer, metal, ceramic or cement matrix.

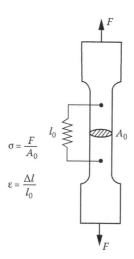

Figure 8.16

In these, there are essentially two mechanisms of damage: fibre pull-out and delamination (i.e. the debonding of the layers).

The distinction between ductile materials and brittle materials is not always so clear in practice, because the ductility of the material also depends upon the ambient temperature and, as we shall see later, upon the size of the structural element. Of the two, the latter is the factor that is harder to grasp, because in this case ductility ceases to be a property of the material and becomes a property of the structure as a whole.

Let us consider a **uniaxial tensile test** carried out on a test specimen of ductile material, for instance, steel (Figure 8.16). Let the test specimen have the usual hourglass shape, to prevent fracture occurring in the vicinity of the ends where the specimen is clamped to the testing equipment. Let A_0 be the area of the initial cross section of the tensile specimen in the middle zone, and l_0 the initial distance between two sensors glued at two distinct points of the middle zone. Let this distance be measured by an electrical device that connects the two points. Let the **nominal stress** σ be defined as the ratio between the force F transmitted by the testing equipment and the initial area A_0:

$$\sigma = \frac{F}{A_0} \tag{8.84}$$

In this way, the elastic and possible plastic transverse contractions are neglected. Then, let the **conventional dilation** ε be defined as the ratio between the variation in the distance between the two sensors, Δl, and the initial distance l_0:

$$\varepsilon = \frac{\Delta l}{l_0} \tag{8.85}$$

This dilation is the average dilation for the zone being checked. It is very likely that during the test, and especially in the non-linear regime, dilation is not uniform and consequently at a given point does not coincide with the average.

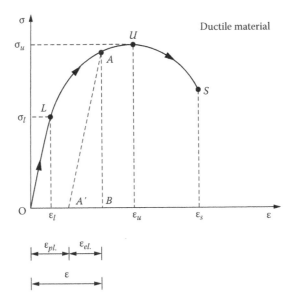

Figure 8.17

Now let all the pairs of points recorded during the loading process be plotted on the $\sigma - \varepsilon$ plane (Figure 8.17). Between points O and L, the diagram is linear and elastic. From L onwards, the response is no longer linear and the material begins to yield. When the specimen is unloaded, there is evidence of permanent deformation ε_{pl}. This means that part of the strain energy has been recovered (triangle ABA'), i.e. corresponding to the strain ε_{el}, whereas the remainder has dissipated plastically (area $OLAA'$). When the test specimen is again loaded, once more it covers elastically the path $A'A$, which is parallel to the path OL. When it arrives at A, the specimen yields again at a stress $\sigma > \sigma_l$. Virgin material, then, yields at lower levels of stress than does material that has already undergone yielding. This phenomenon is referred to as **hardening**. When the applied force F is further increased, the curve ceases to be linear (portion AU). In this phase, the increase in stress per unit increase in dilation (usually called tangential stiffness) continues to diminish, until it vanishes at point U. When the point U is reached, if the loading process is controlled by the external force F, the specimen breaks, because F cannot increase any further.

On the other hand, if the loading process is controlled by the variation in distance Δl (i.e. if a slope is electronically imposed on that quantity in time), it is possible to investigate the behaviour of the material beyond the point of ultimate strength U. Beyond the point U, in fact, the tangential stiffness becomes negative and, to positive increments of displacement Δl, there correspond negative increments of the force F. This is due to the phenomenon of **plastic transverse contraction** or **necking** (Figure 8.18), whereby the area A of the actual cross section becomes notably less than A_0, in a localized band between the two sensors. Finally, once a terminal point S is reached, the specimen gives suddenly, even though the loading process is deformation controlled.

In the case of certain metal alloys, such as low carbon steels, a sudden yield follows the proportional limit L, so that dilation increases by a finite quantity under constant loading (Figure 8.19). In these cases, it is thus easy to identify the value of **uniaxial yielding stress** σ_P, as this coincides with the proportional limit σ_l. When, instead, the proportional limit is followed by the hardening portion of the curve, it is more difficult to define σ_P. In this case, it is conventional to use the stress value of which the permanent deformation ε_{pl} at unloading is equal to 2‰.

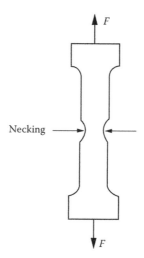

Necking

Figure 8.18

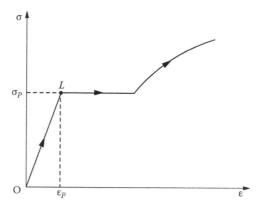

Figure 8.19

Whereas ductile materials present similar behaviours in tension and compression, brittle materials behave in considerably different ways. Concrete, for instance, is ductile in compression but brittle in tension, and presents an ultimate compressive strength that is about one order of magnitude greater than its ultimate tensile strength. A tensile test on a specimen of concrete, if conducted by applying a load or, as is usually said, under controlled loading conditions, shows an approximately linear elastic response up to a point where the load drops sharply, corresponding to the sudden formation of a crack. However, today's electronic techniques allow us to control the strain (*input* = strain ε, *output* = stress σ). By so doing, the post-peak response curve of the cement material is highlighted (Figure 8.20). Only recently has it been realized that there exists an extensive branch of **softening** and that it is possible for concrete to dissipate a considerable amount of energy per unit volume. This energy is represented by the area under the curve $\sigma(\varepsilon)$.

Even more recently, it has been possible to demonstrate that energy, in actual fact, is not dissipated uniformly in the volume, but rather is dissipated over a localized band, which subsequently becomes a crack (the same phenomenon occurs in ductile materials with necking). In other words, point dilation between the two sensors of Figure 8.16 is not a constant function. On the contrary, it presents a notable peak corresponding to the crack that is forming.

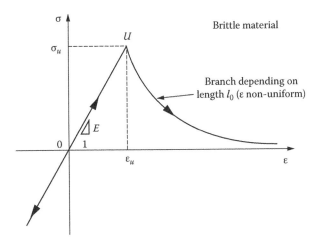

Figure 8.20

Ideally, it is possible to imagine that the dilation ε is a Dirac δ function, since the dilation is infinite where a discontinuity occurs in the **axial displacement** function.

As a consequence of the localization of the strain ε, the decreasing branch of the σ(ε) curve comes to depend on the length l_0 of the measurement base. What, instead, emerges as a true characteristic of the material is the σ(w) diagram, which represents the stress transmitted through the crack, as a function of the opening (or width) of the crack itself (Figure 8.21). This law of decay indicates, of course, a weakening of the interaction with the increase in the distance w between the faces (or free surfaces) of the crack. When w reaches the limiting value w_c, the interaction ceases totally and the crack becomes a complete disconnection which divides the specimen into two distinct parts. The area under the curve σ(w) represents the energy dissipated over the unit surface of fracture. Since the **cohesive law** σ(w) is a characteristic of the material, which depends on the intimate structure and on the mechanisms of damage of the material, the **fracture energy** $\mathcal{G}_{IC}$ is also an intrinsic property of the material:

$$\mathcal{G}_{IC} = \int_0^{w_c} \sigma(w)\mathrm{d}w \tag{8.86}$$

The energy dissipated over the surface of the crack is equal to $\mathcal{G}_{IC}A_0$ since $\mathcal{G}_{IC}$ is work per unit surface and thus force per unit length, $[F][L]^{-1}$. Since, however, we have assumed that

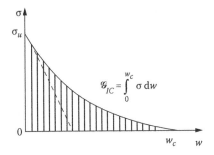

Figure 8.21

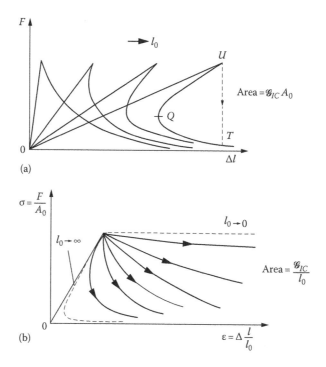

Figure 8.22

the dissipation of energy has occurred only on the fracture surface and not in the volume of the undamaged material, the energy dissipated globally in the volume $A_0 l_0$ is still equal to $\mathcal{G}_{IC} A_0$ (this is rigorously valid only in the absence of hardening). If the response curves are then plotted on the plane $F - \Delta l$, with the increase in the length l_0 of the specimen, we obtain elastic portions of curve having a decreasing stiffness and softening portions having a growing negative slope and, beyond a certain limit, having a positive slope (Figure 8.22a). The area under each curve must in fact be constant and equal to $\mathcal{G}_{IC} A_0$.

On the $\sigma - \varepsilon$ plane (Figure 8.22b), the transition just described is represented by a single linear elastic portion of curve and by a fan of softening branches, as the length l_0 varies. The area under the curve, in fact, in this case varies with l_0, as it is equal to $\mathcal{G}_{IC} l_0$. For $l_0 \to 0$, the softening branch becomes horizontal and represents a **perfectly plastic** structural response. On the other hand, for $l_0 \to \infty$, the area between the $\sigma(\varepsilon)$ curve and the axis ε must tend to zero, and thus the softening branch tends to coincide with the elastic portion (Figure 8.22b).

The positive slope of the softening branch may be justified not only, as we have seen, by considering the dissipated energy, but also by analytical derivation of the function $\varepsilon(\sigma)$. In the post-peak regime, we have (Figure 8.23)

$$\varepsilon = \frac{\Delta l}{l_0} = \frac{\varepsilon_{el} l_0 + w}{l_0} \tag{8.87}$$

where ε_{el} indicates the specific longitudinal dilation of the undamaged zone:

$$\varepsilon_{el} = \frac{\sigma}{E} \tag{8.88}$$

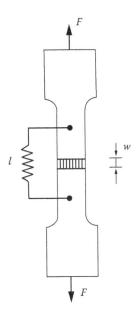

Figure 8.23

From Equation 8.87, we then draw

$$\varepsilon = \frac{\sigma}{E} + \frac{1}{l_0} w(\sigma)$$

(8.89)

and derive with respect to σ

$$\frac{d\varepsilon}{d\sigma} = \frac{1}{E} + \frac{1}{l_0} \frac{dw}{d\sigma}$$

(8.90)

This derivative, and consequently also the inverse $d\sigma/d\varepsilon$, is greater than zero for

$$l_0 > E \left| \frac{dw}{d\sigma} \right|$$

(8.91)

It follows that there are portions of softening having a positive slope for

$$l_0 > \frac{E}{\left| \dfrac{d\sigma}{dw} \right|_{max}}$$

(8.92)

i.e. when the length of the specimen, or rather the distance l_0, between the points of which the relative displacement is measured, is larger than the ratio between the elastic modulus and the maximum slope of the cohesive law. This is due to the fact that, during the softening phase, the stress σ diminishes and, while the point which represents the fracture zone drops along the curve $\sigma(w)$ (Figure 8.21), the point representing the undamaged zone drops along the straight line $\sigma(\varepsilon)$ (Figure 8.20) and describes an elastic unloading. If the length l_0 is sufficiently great, the elastic contraction prevails over the dilation of the fracture zone, giving rise to the phenomenon described before. Softening with a positive slope represents a phenomenon that falls within the scope of catastrophe theory. If, in fact, the loading process is governed by the conventional dilation ε, or by the elongation Δl, once the point U is reached (Figure 8.22a), there is a vertical drop in the load, until the lower softening portion of curve, which has a negative slope, is encountered. The portion UQT is thus ignored and becomes virtual. To record this portion experimentally, it is necessary to govern the process of loading *via* the opening w of the crack, a procedure rendered possible by modern-day electronic techniques. The instability described earlier is called **snap-back**. All relatively brittle materials (e.g. concrete, cast iron, plexiglass, glass) which possess a low **fracture energy** $\mathscr{G}_{IC}$ with the nominal lengths l_0 of the measurement base present a sharp drop in load when the global behaviour of the specimen is still linear elastic.

To conclude this section, it is expedient to note how strength and fracture energy are intrinsic properties of the material, whereas ductility depends on a structural factor, such as the length of the specimen. Among the factors that affect structural ductility, or brittleness, the size scale of the structural element must be included.

8.12 STRENGTH CRITERIA

In the case of bodies subjected to a condition of uniaxial stress, such as ropes or columns, the check on strength is immediate, once the service stress and the yielding or ultimate strength are known. In ductile materials, the material does not reach the critical point, i.e. yielding, in the case where the following relation holds:

$$-\sigma_P < \sigma < \sigma_P \tag{8.93a}$$

as the yielding stress has approximately the same absolute value in tension as it does in compression. In brittle materials, the behaviour in compression is usually different from that in tension and the critical point for the material is avoided if

$$-\sigma_c < \sigma < \sigma_u \tag{8.93b}$$

where σ_c is the ultimate compressive strength.

The relations (8.93) would provide real safety limits if all the quantities involved were known with certainty and without statistical oscillations. However, in the physical world, the quantities will always possess a degree of approximation. For instance, the strength measured in the laboratory, the dimensions of the body and the forces actually applied are not deterministic quantities. **Admissible stress**, to permit conditions of safety, will then be represented by a fraction of the nominal strength, in such a way that the so-called **safety criteria**, for ductile and brittle materials, will be presented as follows:

$$-\frac{\sigma_P}{s} < \sigma < \frac{\sigma_P}{s} \tag{8.94a}$$

$$-\frac{\sigma_c}{s} < \sigma < \frac{\sigma_u}{s} \tag{8.94b}$$

The parameter $s > 1$ is called the **safety factor**. The higher this factor, the less foreseeable is the behaviour of the material and hence the less repeatable are the laboratory results. Whereas for ductile materials it is common practice to take $s = 1.5$, for brittle materials the factor of safety is usually higher (in some cases $s = 6$), because in the case of these materials, the mechanisms of damage are more unstable, and hence, as has already been shown, there do not exist ductile or plastic reserves beyond the ultimate load.

In general, however, the service of structural elements is not restricted to only one axis. Nonetheless, the tests that are usually performed in laboratories are uniaxial, hence the need to correlate biaxial and triaxial stress states to the uniaxial ones. In other words, the procedure must be to define a function of the stress tensor, called **ideal stress** or **equivalent stress**, to compare with the stress of uniaxial yield:

$$\sigma_{eq} = \sigma_{eq}(\sigma_x, \sigma_y, \sigma_z, \tau_{xy}, \tau_{xz}, \tau_{yz}) < \sigma_P \tag{8.95}$$

In the case of isotropy of the material, the ideal stress is a function of the principal components of stress alone:

$$\sigma_{eq} = \sigma_{eq}(\sigma_1, \sigma_2, \sigma_3) < \sigma_P \tag{8.96}$$

As regards uniaxial tensile and compressive tests on metal specimens, it may be noted that the hydrostatic pressure of the environment in which the tests are conducted does not influence the yield stress value. This important experimental fact has given rise to two strength criteria, which have been widely confirmed by experience: Tresca's criterion and Von Mises' criterion.

Tresca's criterion or the **criterion of maximum shearing stress** considers shearing stress responsible for the yielding of the material when it is subjected to a triaxial stress condition. It is thus implicitly assumed that not only in the uniaxial case but also in the triaxial case the super-position of a hydrostatic condition does not affect the strength of the material. It is very simple and expressive to represent Tresca's criterion on Mohr's plane. The maximum shearing stress τ_{max} is in fact equal to half the difference between the extreme principal stresses (Figure 8.24)

$$\tau_{max} = \frac{1}{2} \max\left\{ |\sigma_1 - \sigma_2|, |\sigma_1 - \sigma_3|, |\sigma_2 - \sigma_3| \right\} \tag{8.97}$$

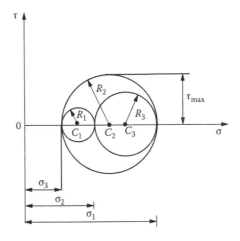

Figure 8.24

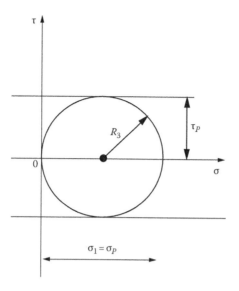

Figure 8.25

and, from the condition of uniaxial yielding, we obtain (Figure 8.25)

$$\tau_P = \frac{1}{2}\sigma_P \tag{8.98}$$

where τ_P is the yield shearing stress. Tresca's condition $\tau_{\max} < \tau_P$, applying Equations 8.97 and 8.98, is translated into the following:

$$\max\left\{|\sigma_1 - \sigma_2|, |\sigma_1 - \sigma_3|, |\sigma_2 - \sigma_3|\right\} < \sigma_P \tag{8.99}$$

where the term on the left-hand side is called **Tresca's equivalent stress**. Equation 8.99 is thus a particular case of Equation 8.96. On Mohr's plane, the original condition $\tau_{\max} < \tau_P$ is represented by an infinite strip, bounded by the two parallel straight lines $\tau_{\max} = \pm\tau_P$ (Figure 8.25).

In the case of a **plane stress condition**, one of the three principal stresses vanishes, for example $\sigma_3 = 0$, and the condition (8.99) becomes

$$\max\left\{|\sigma_1 - \sigma_2|, |\sigma_1|, |\sigma_2|\right\} < \sigma_P \tag{8.100}$$

The inequality (8.100) may be interpreted as the intersection of three different inequalities:

$$|\sigma_1| < \sigma_P \tag{8.101a}$$

$$|\sigma_2| < \sigma_P \tag{8.101b}$$

$$|\sigma_1 - \sigma_2| < \sigma_P \tag{8.101c}$$

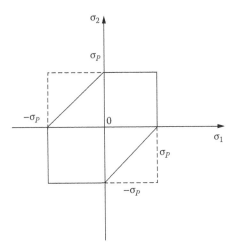

Figure 8.26

The first two, on the plane $\sigma_1 - \sigma_2$, represent **Rankine's criterion** or the **criterion of maximum normal stress**. The strength domain would in this case be a square (Figure 8.26), but, in actual fact, such a simple criterion has been proved inadequate by the experimental evidence. The inequality (8.101c) in fact further sections the square domain and finally furnishes a hexagonal domain, called **Tresca's hexagon** (Figure 8.26). Note how the four points of intersection of Tresca's hexagon with the axes σ_1 and σ_2 represent the uniaxial critical points, of tension and compression, in the two principal directions.

It is worthwhile mentioning here the extrapolation of Tresca's criterion for brittle materials, which present different values for tensile and compressive strength. Instead of considering the critical shearing stress as a constant independent of the state of stress, the **Mohr–Coulomb criterion** proposes a limit τ_p that is a function of the corresponding normal stress

$$\tau_{\max} < \tau_c - \mu\sigma \tag{8.102}$$

where τ_c represents the **cohesiveness** of the material and μ the **coefficient of internal friction**. The strength increases with the increase in normal compression, so that the strength domain on Mohr's plane is represented by a strip that broadens in the direction of the negative stresses σ (Figure 8.27). Cutting this domain with the two vertical straight lines that

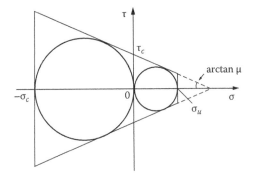

Figure 8.27

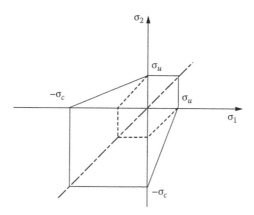

Figure 8.28

represent the uniaxial critical states of tension and compression, we obtain a trapezoidal domain which reproduces satisfactorily the critical states of brittle or incoherent materials (e.g. concrete, rocks, soil). On the plane of the principal stresses $\sigma_1 - \sigma_2$, the Mohr–Coulomb domain is represented by a symmetrical hexagon with respect to the bisector of the first and third quadrants (Figure 8.28), which reduces to Tresca's hexagon for $\mu = 0$ (zero internal friction).

Von Mises' criterion, or the **criterion of maximum energy of distortion,** considers the strain energy corresponding to the **deviatoric tensor** (7.77b) responsible for the critical condition of the material. Since, in general, the strain energy per unit volume (or strain energy density) may be expressed as a function of the first two invariants of stress, as is shown by Equation 8.69,

$$\psi = \Phi = \frac{1}{2E}\left\{J_{\mathrm{I}}^2 + 2J_{\mathrm{II}}(1+\nu)\right\} \tag{8.103}$$

this expression can be particularized to the case of the deviatoric tensor. The first invariant of the deviatoric tensor is zero by definition, so that the energy Φ_{d} associated to this tensor is

$$\Phi_{\mathrm{d}} = \frac{1+\nu}{E} J_{\mathrm{IId}} \tag{8.104}$$

where J_{IId} indicates the second deviatoric invariant. This equals

$$J_{\mathrm{IId}} = -\left[(\sigma_1 - \bar\sigma)(\sigma_2 - \bar\sigma) + (\sigma_1 - \bar\sigma)(\sigma_3 - \bar\sigma) + (\sigma_2 - \bar\sigma)(\sigma_3 - \bar\sigma)\right] \tag{8.105}$$

where $\bar\sigma$ indicates the mean stress:

$$\bar\sigma = \frac{1}{3}(\sigma_1 + \sigma_2 + \sigma_3) \tag{8.106}$$

Substituting Equation 8.106 in Equation 8.105, we obtain

$$J_{\mathrm{IId}} = -\frac{1}{9}\left[(2\sigma_1 - \sigma_2 - \sigma_3)(2\sigma_2 - \sigma_1 - \sigma_3) + \cdots\right] \tag{8.107}$$

and, then, multiplying the three pairs of trinomials and ordering the terms, we obtain

$$J_{\text{IId}} = -\frac{1}{9}\Big[\sigma_3^2 - 2\sigma_1^2 - 2\sigma_2^2 + 5\sigma_1\sigma_2 - \sigma_1\sigma_3 - \sigma_2\sigma_3$$

$$+ \sigma_2^2 - 2\sigma_1^2 - 2\sigma_3^2 + 5\sigma_1\sigma_3 - \sigma_1\sigma_2 - \sigma_3\sigma_2$$

$$+ \sigma_1^2 - 2\sigma_2^2 - 2\sigma_3^2 + 5\sigma_2\sigma_3 - \sigma_2\sigma_1 - \sigma_3\sigma_1\Big]$$

$$= \frac{1}{3}\Big[\big(\sigma_1^2 + \sigma_2^2 + \sigma_3^2\big) - \big(\sigma_1\sigma_2 + \sigma_1\sigma_3 + \sigma_2\sigma_3\big)\Big] \tag{8.108}$$

The distortion energy is then

$$\Phi_{\text{d}} = \frac{1+\nu}{3E}\Big[\big(\sigma_1^2 + \sigma_2^2 + \sigma_3^2\big) - \big(\sigma_1\sigma_2 + \sigma_1\sigma_3 + \sigma_2\sigma_3\big)\Big] \tag{8.109}$$

In the uniaxial critical condition, the distortion energy reaches its limit value:

$$\Phi_{\text{d}P} = \frac{1+\nu}{3E}\sigma_P^2 \tag{8.110}$$

The original condition of Von Mises,

$$\Phi_{\text{d}} < \Phi_{\text{d}P} \tag{8.111}$$

therefore translates into the following:

$$\Big[\big(\sigma_1^2 + \sigma_2^2 + \sigma_3^2\big) - \big(\sigma_1\sigma_2 + \sigma_1\sigma_3 + \sigma_2\sigma_3\big)\Big] < \sigma_P^2 \tag{8.112}$$

where the term on the left-hand side represents the square of **Von Mises' equivalent stress.** Equation 8.112 is thus a particular case of Equation 8.96.

We now intend to express the condition of Von Mises as a function of the special components of stress, instead of as a function of the principal stresses. Noting that the inequality (8.112) can be expressed as a function of the invariants

$$\big(J_{\text{I}}^2 + 3J_{\text{II}}\big) < \sigma_P^2 \tag{8.113}$$

and using expressions (7.72a and 7.72b), we obtain

$$\Big[\big(\sigma_x^2 + \sigma_y^2 + \sigma_z^2\big) - \big(\sigma_x\sigma_y + \sigma_x\sigma_z + \sigma_y\sigma_z\big) + 3\big(\tau_{xy}^2 + \tau_{xz}^2 + \tau_{yz}^2\big)\Big] < \sigma_P^2 \tag{8.114}$$

As regards **plane stress conditions,** for example $\sigma_3 = 0$, the condition (8.112) becomes

$$\big(\sigma_1^2 + \sigma_2^2 - \sigma_1\sigma_2\big) < \sigma_P^2 \tag{8.115}$$

The inequality (8.115) represents all the points inside an ellipse which has its major axis coincident with the bisector of the first and third quadrants (Figure 8.29). **Von Mises' ellipse** is circumscribed to Tresca's hexagon and intersects the axes σ_1 and σ_2 at the same four

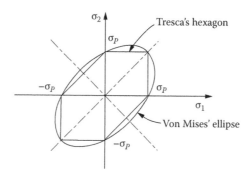

Figure 8.29

notable points. The two boundaries have in common also the two points of intersection with the bisector of the first and third quadrants.

Von Mises' criterion is less conservative than Tresca's criterion, since the elliptical domain is wider than the hexagonal one. On the other hand, Von Mises' criterion is simpler to use than is Tresca's, since it imposes only one inequality (8.115) as against the three of Equation 8.101.

Finally, it should be noted how the triaxial strength domains, both according to Tresca and according to Von Mises, are represented in the space $\sigma_1 - \sigma_2 - \sigma_3$ by a cylinder with the generators parallel to the trisector of the first octant. The inequality (8.99) represents in fact the cylinder that has the hexagon of Figure 8.26 as its directrix on the plane $\sigma_1 - \sigma_2$, while the inequality (8.112) represents the cylinder that has the ellipse of Figure 8.29 as its directrix on the plane $\sigma_1 - \sigma_2$. That the trisector of the first octant should be entirely contained in the aforementioned strength domains is a result consistent with the basic experimental observation, i.e. with the independence of the conditions of yielding from a superimposed hydrostatic state. This means that, according to these criteria, even severe hydrostatic conditions of loading do not cause in any case the failure of the material.

As regards the **strength criteria for anisotropic materials,** the reader is referred to Appendix C.

Chapter 9

Saint Venant problem

9.1 INTRODUCTION

This chapter deals with the particular case of a cylindrical, homogeneous and isotropic, linearly elastic solid loaded exclusively on its end planes. This solid, known as the **Saint Venant solid**, represents a relatively simple and highly useful model in the case of beams. All the fundamental loadings are studied, which correspond to the internal reaction characteristics already introduced in Chapter 5: axial force, shearing force, twisting moment and bending moment. For each loading, both the stress condition produced and the corresponding deformation characteristic are obtained: axial dilation, mean shearing strain, unit angle of torsion and curvature. It is shown that, in the case of symmetrical sections, the fundamental reactions are all mutually energetically orthogonal, and that, in this case, the strain energy is a diagonal quadratic form of the reactions themselves.

The chapter further deals with combined loadings consisting of eccentric axial force, with the definition of the central core of inertia, and of shear–torsion, with the definition of the corresponding centre of shear or of torsion. As regards the case of torsion of thin-walled sections, the considerable difference existing between closed sections (tubular sections) and open sections is emphasized. The latter are in fact subjected to stresses and torsional rotations that are far higher.

The chapter closes with a number of examples of strength assessments, carried out on beams having the areas considered in the closing part of Chapter 2 as sections.

9.2 FUNDAMENTAL HYPOTHESES

The **Saint Venant problem** constitutes a particular elastic problem regarding a cylindrical solid, loaded at its ends.

Consider a generic area with its central (i.e. centroidal and principal) XY reference system. Imagine translating this area perpendicularly to its own plane, so as to cause its centroid G to describe a rectilinear trajectory, normal to the XY coordinate plane and of length l (Figure 9.1). Let the oriented straight line of this trajectory constitute the third reference axis Z. The cylindrical volume thus described constitutes the domain of the Saint Venant solid.

Having thus described the geometry of the Saint Venant solid, it is necessary to specify the material of which it is made and the forces by which it is loaded. As regards the material, it is assumed to be linear elastic, isotropic and homogeneous. The last of these assumptions may be omitted, as is shown in Appendix D. As far as the external forces applied are concerned, on the other hand, these are assumed to be only surface forces, which act exclusively on the end planes (Figure 9.1). Body forces are hence excluded, as are displacements imposed on

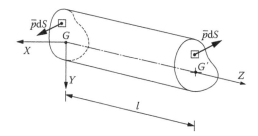

Figure 9.1

the boundary. Of course, the surface forces acting on both ends, A and A', must make up a balanced system and consequently satisfy the cardinal equations of statics:

$$\int_{A \cup A'} \{p\}\, dA = \{0\} \tag{9.1a}$$

$$\int_{A \cup A'} (\{r\} \wedge \{p\})\, dA = \{0\} \tag{9.1b}$$

The equations governing the Saint Venant problem are the same as those governing the elastic problem, the former being merely a particular case in the framework of the latter. These are those of static equations (8.2), kinematic equations (8.8) and the constitutive equations (8.73). As shall be noted in the sequel, in the case in point, some of these are identically satisfied, precisely on account of the simplifying hypotheses introduced.

On the other hand, the boundary conditions of equivalence on the lateral surface, where $n_z = 0$, are

$$\sigma_x n_x + \tau_{yx} n_y = 0 \tag{9.2a}$$

$$\tau_{xy} n_x + \sigma_y n_y = 0 \tag{9.2b}$$

$$\tau_{xz} n_x + \tau_{yz} n_y = 0 \tag{9.2c}$$

while those on the end planes, where $n_x = 0$, $n_y = 0$, $n_z = 1$, take the following form:

$$\tau_{zx} = p_x \tag{9.3a}$$

$$\tau_{zy} = p_y \tag{9.3b}$$

$$\sigma_z = p_z \tag{9.3c}$$

We may now state Saint Venant's fundamental hypothesis, one which has been amply borne out both experimentally and theoretically (Figure 9.2):

At a sufficient distance from each end plane, the strain and stress fields depend only upon the resultant $\{R\}$ of the forces acting on the end itself and upon the resultant moment $\{M\}$ of the forces with respect to the centroid of the end considered.

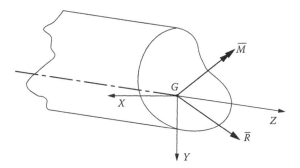

Figure 9.2

In line with this hypothesis, the conditions of equivalence at the ends can be global, and not local, as in the case of Equations 9.3:

$$\int_A \{t_z\}\, dA = \int_A \{p\}\, dA = \{R\} \tag{9.4a}$$

$$\int_A (\{r\} \wedge \{t_z\})\, dA = \int_A (\{r\} \wedge \{p\})\, dA = \{M\} \tag{9.4b}$$

Consider, for instance, a rectilinear beam having rectangular cross section, of base b and depth h, subjected on each end plane, in one case to a couple made up of two forces F with arm h (Figure 9.3a) and in the other to a couple consisting of two forces $2F$ with arm $h/2$ (Figure 9.3b). The strain and stress fields are approximately the same at distances from the ends greater than the depth h. Reversing one of the two systems of forces and applying the principle of superposition, two self-balancing systems acting on the end planes are obtained, which generate approximately zero strain and stress fields, except in the end regions. The damping of the perturbation created by the self-balancing systems of forces occurs at distances from the ends approximately greater than the maximum dimension of the cross section, when the cross section itself is compact and of a regular shape. When, instead, the cross section is thin walled, the distance of damping may prove far greater.

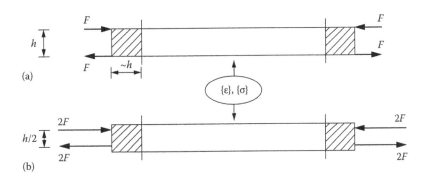

Figure 9.3

As we have already said in Chapter 5, the components of the resultant force $\{R\}$ and the resultant moment $\{M\}$ are called **elementary internal reactions:**

$R_x = T_x$ = shearing force along the X axis
$R_y = T_y$ = shearing force along the Y axis
$R_z = N$ = axial force
M_x = bending moment along the X axis
M_y = bending moment along the Y axis
M_z = twisting moment

The cases of **elementary loading** for the Saint Venant solid are thus the following:

1. Centred axial force N
2. Flexure M_x
3. Flexure M_y
4. Shear T_y (and flexure M_x)
5. Shear T_x (and flexure M_y)
6. Torsion M_z

It should be noted that, since the shear is the derivative of the moment function, its presence also presupposes that of the corresponding flexure, whereas the reverse is not true.

In the sequel, we shall see how the three elementary loadings, N, M_x, M_y, produce an axial one-dimensional stress field, where only normal stress σ_z is present. The remaining three elementary loadings, T_y, T_x, M_z, produce instead, on each cross section, a field of shearing stress with the presence of the components τ_{zx}, τ_{zy}, and thus of τ_z only. Also for this reason, it is customary to combine the elementary loadings, so as to obtain **complex loadings** which produce only normal stresses, σ_z, or only shearing stresses, τ_z:

1. Biaxial flexure: M_x, M_y
2. Eccentric axial force: N, M_x, M_y
3. Shear–torsion: T_x, T_y, M_z

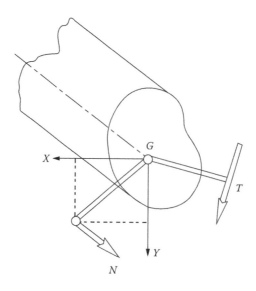

Figure 9.4

It should be noted that the eccentric axial force is equivalent to an axial force N, exerted at a point of the XY plane which does not coincide with the centroid. On the other hand, the shear–torsion is equivalent to a force T, with the line of action belonging to the XY plane and not passing necessarily through the centroid. In the last analysis, then, the resultant force $\{R\}$ and the resultant moment $\{M\}$ are a system equivalent to that of the two skew and orthogonal forces N and T (Figure 9.4).

9.3 CENTRED AXIAL FORCE

As regards centred axial force and, as we shall see, also the other elementary loadings that produce normal stress σ_z, we start by assuming a stress solution and then verify that this solution satisfies all the equations of the elastic problem, including the boundary conditions.

Let us assume then

$$\sigma_x = \sigma_y = \tau_{xy} = \tau_{xz} = \tau_{yz} = 0 \tag{9.5a}$$

$$\sigma_z = c \tag{9.5b}$$

where c is a constant that can be determined on the basis of the global boundary conditions of equivalence (9.4). These conditions are all identically satisfied, except for the projection of the first on the Z axis

$$\int_A \sigma_z \, dA = N \tag{9.6}$$

whence we obtain

$$cA = N \tag{9.7}$$

and hence the constant stress on the cross section

$$\sigma_z = \frac{N}{A} \tag{9.8}$$

On the other hand, the indefinite equations of equilibrium (8.2), as well as the boundary conditions of equivalence (9.2), are all identically satisfied by the solution given by Equations 9.5a and 9.8.

The elastic constitutive equations (8.73) then give the strain field

$$\varepsilon_x = \varepsilon_y = -\nu \frac{N}{EA} \tag{9.9a}$$

$$\varepsilon_z = \frac{N}{EA} \tag{9.9b}$$

$$\gamma_{xy} = \gamma_{xz} = \gamma_{yz} = 0 \tag{9.9c}$$

If the axial force N is tensile, there is thus a uniform dilation in the axial direction and contractions that are all equal to one another in the transverse directions, while the shearing strains are zero.

Integrating the strain field, it is possible finally to obtain the displacement field, save for components of rigid rototranslation. We have in fact

$$\frac{\partial u}{\partial x} = -v \frac{N}{EA} \tag{9.10a}$$

$$\frac{\partial v}{\partial y} = -v \frac{N}{EA} \tag{9.10b}$$

$$\frac{\partial w}{\partial z} = \frac{N}{EA} \tag{9.10c}$$

$$\frac{\partial u}{\partial y} + \frac{\partial v}{\partial x} = 0 \tag{9.10d}$$

$$\frac{\partial u}{\partial z} + \frac{\partial w}{\partial x} = 0 \tag{9.10e}$$

$$\frac{\partial v}{\partial z} + \frac{\partial w}{\partial y} = 0 \tag{9.10f}$$

Integrating the first three equations, we obtain

$$u = -v \frac{N}{EA} x + u_0(y,z) \tag{9.11a}$$

$$v = -v \frac{N}{EA} y + v_0(x,z) \tag{9.11b}$$

$$w = \frac{N}{EA} z + w_0(x,y) \tag{9.11c}$$

Substituting Equations 9.11 into Equations 9.10d through f, we have

$$\frac{\partial u_0}{\partial y} + \frac{\partial v_0}{\partial x} = 0 \tag{9.12a}$$

$$\frac{\partial u_0}{\partial z} + \frac{\partial w_0}{\partial x} = 0 \tag{9.12b}$$

$$\frac{\partial v_0}{\partial z} + \frac{\partial w_0}{\partial y} = 0 \tag{9.12c}$$

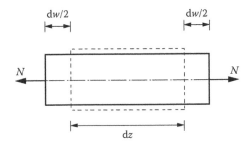

Figure 9.5

as well as, of course,

$$\frac{\partial u_0}{\partial x} = 0 \tag{9.12d}$$

$$\frac{\partial v_0}{\partial y} = 0 \tag{9.12e}$$

$$\frac{\partial w_0}{\partial z} = 0 \tag{9 12f}$$

From Equations 9.12, it follows that the field of displacements u_0, v_0, w_0 does not have strain components. It, therefore, represents a generic, rigid rototranslation.

The solution obtained consists of stresses (9.5a) and (9.8) and displacements (9.11). It is the only one possible by virtue of Kirchhoff's solution uniqueness theorem.

The elementary work of deformation, for an infinitesimal segment of length dz of the Saint Venant solid, may be obtained by applying Clapeyron's theorem (Figure 9.5):

$$dL = \frac{1}{2} N \, dw = \frac{1}{2} N \varepsilon_z \, dz \tag{9.13}$$

Substituting the expression (9.9b) in Equation 9.13, we have then

$$\frac{dL}{dz} = \frac{1}{2} \frac{N^2}{EA} \tag{9.14}$$

The factor 1/2 (characteristic of linear elasticity) is therefore multiplied by the square of the static characteristic N and divided by the product of the elastic characteristic E and the geometric characteristic A. It will be noted in the sequel how the structure of the formula (9.14) is also conserved in the case of the other elementary loadings.

9.4 FLEXURE

Also in the case of flexure, a uniaxial stress field σ_z is assumed, in this case linearly variable on the cross section

$$\sigma_x = \sigma_y = \tau_{xy} = \tau_{xz} = \tau_{yz} = 0 \tag{9.15a}$$

$$\sigma_z = ax + by + c \tag{9.15b}$$

The constants a, b, c may be determined on the basis of the boundary conditions expressed by Equations 9.4:

$$\int_A \sigma_z \, dA = N = 0 \qquad (9.16a)$$

$$\int_A \tau_{zx} \, dA = T_x = 0 \qquad (9.16b)$$

$$\int_A \tau_{zy} \, dA = T_y = 0 \qquad (9.16c)$$

$$\int_A \sigma_z y \, dA = M_x \neq 0 \qquad (9.16d)$$

$$\int_A \sigma_z x \, dA = -M_y = 0 \qquad (9.16e)$$

$$\int_A (\tau_{zy} x - \tau_{zx} y) \, dA = M_z = 0 \qquad (9.16f)$$

Whilst the conditions (9.16b,c,f) are identically satisfied, the conditions (9.16a, 9.16d, 9.16e) give the constants a, b, c. From Equation 9.16a, we have in fact

$$\int_A (ax + by + c) \, dA = aS_y + bS_x + cA = 0 \qquad (9.17)$$

and, since the static moments S_x and S_y are zero as the XY system is a centroidal reference system, we obtain

$$c = 0 \qquad (9.18)$$

From Equation 9.16e, we have then

$$\int_A (ax + by)x \, dA = aI_y + bI_{xy} = 0 \qquad (9.19)$$

and, since the product of inertia I_{xy} is zero as XY is a principal reference system, we obtain

$$a = 0 \qquad (9.20)$$

Finally, from Equation 9.16d, we have

$$\int_A by^2 \, dA = bI_x = M_x \qquad (9.21)$$

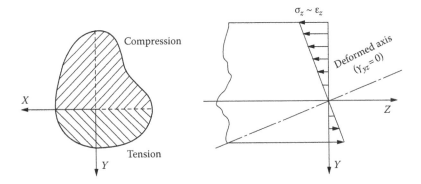

Figure 9.6

from which we obtain

$$b = \frac{M_x}{I_x} \tag{9.22}$$

and thus

$$\sigma_z = \frac{M_x}{I_x} y \tag{9.23}$$

If only the elementary load M_x is present, the stress σ_z depends upon the y coordinate alone and increases in absolute value as it moves away from the X axis (Figure 9.6). It is directly proportional to the applied moment and inversely proportional to the relative moment of inertia. The stress σ_z is zero for $y = 0$, i.e. on the X axis, which, for this reason, is called the **neutral axis**. The Y axis is called the **loading axis**, because the couple M_x belongs to the plane YZ, which in turn is called the **loading plane**. In this case, the loading axis and the neutral axis are mutually orthogonal. The case of the **heterogeneous beam in flexure** is dealt with in Appendix D.

The stress solution, represented by Equations 9.15a and 9.23, identically satisfies the indefinite equations of equilibrium (8.2) and the equations of equivalence (9.2), and, *via* the elastic constitutive equations (8.73), gives the following strain field:

$$\varepsilon_x = \varepsilon_y = -\nu \frac{M_x}{EI_x} y \tag{9.24a}$$

$$\varepsilon_z = \frac{M_x}{EI_x} y \tag{9.24b}$$

$$\gamma_{xy} = \gamma_{xz} = \gamma_{yz} = 0 \tag{9.24c}$$

Integrating the strains of Equations 9.24, it is possible to obtain the displacement field, except for the components of rigid rotation. From Equations 9.24, we have

$$\frac{\partial u}{\partial x} = -\nu \frac{M_x}{EI_x} y \tag{9.25a}$$

$$\frac{\partial \upsilon}{\partial y} = -\nu \frac{M_x}{EI_x} y \tag{9.25b}$$

$$\frac{\partial w}{\partial z} = \frac{M_x}{EI_x} y \tag{9.25c}$$

$$\frac{\partial u}{\partial y} + \frac{\partial \upsilon}{\partial x} = 0 \tag{9.25d}$$

$$\frac{\partial u}{\partial z} + \frac{\partial w}{\partial x} = 0 \tag{9.25e}$$

$$\frac{\partial \upsilon}{\partial z} + \frac{\partial w}{\partial y} = 0 \tag{9.25f}$$

Integrating Equations 9.25a through 9.25c, we obtain

$$u = -\nu \frac{M_x}{EI_x} xy + u_0(y,z) \tag{9.26a}$$

$$\upsilon = -\nu \frac{M_x}{EI_x} \frac{y^2}{2} + \upsilon_0(x,z) \tag{9.26b}$$

$$w = \frac{M_x}{EI_x} yz + w_0(x,y) \tag{9.26c}$$

Substituting Equations 9.26 in Equations 9.25d through 9.25f, we have

$$\frac{\partial u_0}{\partial y} + \frac{\partial \upsilon_0}{\partial x} = \nu \frac{M_x}{EI_x} x \tag{9.27a}$$

$$\frac{\partial u_0}{\partial z} + \frac{\partial w_0}{\partial x} = 0 \tag{9.27b}$$

$$\frac{\partial \upsilon_0}{\partial z} + \frac{\partial w_0}{\partial y} = -\frac{M_x}{EI_x} z \tag{9.27c}$$

Integrating Equation 9.27a with respect to x, we obtain

$$\upsilon_0(x,z) = \nu \frac{M_x}{EI_x} \frac{x^2}{2} + \upsilon_1(x,z) \tag{9.28}$$

while from Equation 9.27c, we draw

$$\frac{\partial \upsilon_1}{\partial z} + \frac{\partial w_0}{\partial y} = -\frac{M_x}{EI_x} z \tag{9.29}$$

and hence

$$v_1(x,z) = -\frac{M_x}{EI_x}\frac{z^2}{2} + v_2(x,z) \tag{9.30}$$

In conclusion, we can express the displacement field as follows:

$$u = -v\frac{M_x}{EI_x}xy + u_0(y,z) \tag{9.31a}$$

$$v = -\frac{M_x}{2EI_x}[z^2 + v(y^2 - x^2)] + v_2(x,z) \tag{9.31b}$$

$$w = \frac{M_x}{EI_x}yz + w_0(x,y) \tag{9.31c}$$

Substituting the foregoing equations (9.31) in Equations 9.25d through 9.25f, we have

$$\frac{\partial u_0}{\partial y} + \frac{\partial v_2}{\partial x} = 0 \tag{9.32a}$$

$$\frac{\partial u_0}{\partial z} + \frac{\partial w_0}{\partial x} = 0 \tag{9.32b}$$

$$\frac{\partial v_2}{\partial z} + \frac{\partial w_0}{\partial y} = 0 \tag{9.32c}$$

as well as, of course,

$$\frac{\partial u_0}{\partial x} = 0 \tag{9.32d}$$

$$\frac{\partial v_2}{\partial y} = 0 \tag{9.32e}$$

$$\frac{\partial w_0}{\partial z} = 0 \tag{9.32f}$$

From Equations 9.32, it follows that the field of displacements u_0, v_2, w_0 does not have any strain components. It, therefore, represents a generic, rigid rototranslation, which shall henceforth be neglected.

The points of the Z axis, of coordinates $P(0, 0, \bar{z})$, are transformed, once deformation has occurred, into the points $P'(0, v_P, \bar{z})$, since by Equation 9.31b, we have

$$v_P = v(0,0,\bar{z}) = -\frac{M_x}{2EI_x}\bar{z}^2 \tag{9.33}$$

In the foregoing equation, the term of rigid rototranslation v_2 has been omitted (Figure 9.7).

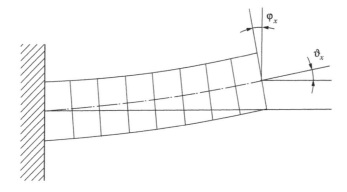

Figure 9.7

On the other hand, the points of a generic cross section, of coordinates $Q(x, y, \bar{z})$, are transformed, as a consequence of the deformation, into the points $Q'(x + u_Q, y + v_Q, \bar{z} + w_Q)$, Equations 9.31 yielding

$$u_Q = u(x, y, \bar{z}) = -v \frac{M_x}{EI_x} xy \tag{9.34a}$$

$$v_Q = v(x, y, \bar{z}) = -\frac{M_x}{2EI_x} \left[\bar{z}^2 + v\left(y^2 - x^2\right) \right] \tag{9.34b}$$

$$w_Q = w(x, y, \bar{z}) = \frac{M_x}{EI_x} y\bar{z} \tag{9.34c}$$

Hence the transformation $Q \to Q'$ will be, in general, non-linear:

$$x' = x - v \frac{M_x}{EI_x} xy \tag{9.35a}$$

$$y' = y - \frac{M_x}{2EI_x} \left[\bar{z}^2 + v\left(y^2 - x^2\right) \right] \tag{9.35b}$$

$$z' = \bar{z} + \frac{M_x}{EI_x} y\bar{z} \tag{9.35c}$$

If, on the other hand, we consider the Saint Venant solid sufficiently slender to allow us to consider the coordinates x and y as always being infinitesimal compared to the $\bar{z}$ coordinate, Equations 9.35 may be rewritten as follows:

$$x' = x \tag{9.36a}$$

$$y' = y - \frac{M_x}{2EI_x} \bar{z}^2 \tag{9.36b}$$

$$z' = \bar{z}\left(1 + \frac{M_x}{EI_x}y\right)$$

(9.36c)

In Equations 9.36a and 9.36b, the second-order infinitesimal terms, which represent the phenomenon of transverse contraction, have been omitted.

In the framework of the aforementioned hypotheses, the **principle of conservation of plane sections** holds, according to which each individual cross section rotates rigidly by the angle φ_x about the X axis and translates by the quantity v_P in the Y direction (Figure 9.7). Furthermore, each individual cross section, after deformation, remains perpendicular to the deformed axis, since $\gamma_{yz} = 0$, whilst the latter undergoes bending but not variations in length, in that ε_z $(x = 0, y = 0) = 0$.

A further confirmation of what is stated earlier comes from the relation (9.36c):

$$\varphi_x = \frac{z' - \bar{z}}{y} = \frac{M_x}{EI_x}\bar{z}$$

(9.37)

In fact, the angle ϑ_x, which the geometrical tangent to the deformed configuration of the axis expressed by Equation 9.33 forms with the Z axis, is approximately equal to (Figure 9.7)

$$\theta_x \simeq -\frac{dv_P}{d\bar{z}} = \frac{M_x}{EI_x}\bar{z}$$

(9.38)

Comparing expressions (9.37) and (9.38), which represent the rotation of the section and the rotation of the axis, respectively, we find what has been previously stated, *viz.* that $\varphi_x = \vartheta_x$.

Referring to Equation 9.37, the differential of the angle φ_x equals

$$d\varphi_x = \frac{M_x}{EI_x}dz$$

(9.39)

On the other hand, if we denote the radius of curvature of the deformed axis by R_x and the curvature due to flexure by $\chi_x = 1/R_x$, we have (Figure 9.8)

$$d\varphi_x = \frac{dz}{R_x} = \chi_x\,dz$$

(9.40)

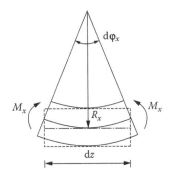

Figure 9.8

From the comparison of the two equations (9.39) and (9.40), we obtain the **curvature**

$$\chi_x = \frac{M_x}{EI_x} \tag{9.41}$$

Likewise, the curvature produced by the flexure M_y equals

$$\chi_y = \frac{M_y}{EI_y} \tag{9.42}$$

The foregoing formula presents the same structure as Equation 9.9b, which gives the axial dilation produced by the centred axial force. In both cases, we have the static characteristic divided by the product of the modulus of elasticity and the geometrical characteristic.

A much easier way to obtain Equation 9.41, avoiding the explicit calculation of the displacement field, is to consider the bent beam segment in Figure 9.8. The geometrical similitude between the two curvilinear triangles of radii R_x and $(R_x + y)$ provides

$$d\varphi_x = \frac{dz}{R_x} = \frac{dz(1 + \varepsilon_x)}{R_x}$$

and then $\varepsilon_z = y/R_x$. From Equation 9.23, we can obtain Equation 9.41, considering $\chi_x = 1/R_x$.

By then applying Clapeyron's theorem to the infinitesimal beam segment of Figure 9.8, we obtain the elementary work of deformation,

$$dL = \frac{1}{2} M_x \, d\varphi_x = \frac{1}{2} M_x \chi_x \, dz \tag{9.43}$$

Substituting relation (9.41) into Equation 9.43, we have finally

$$\frac{dL}{dz} = \frac{1}{2} \frac{M_x^2}{EI_x} \tag{9.44}$$

9.5 ECCENTRIC AXIAL FORCE AND BIAXIAL FLEXURE

Consider the combined loading of **eccentric axial force** consisting of an axial force N, having eccentricity e_y with respect to the X axis and e_x with respect to the Y axis, and thus eccentricity $e = \left(e_x^2 + e_y^2\right)^{1/2}$ with respect to the centroidal axis (Figure 9.9). This force is equivalent, on the other hand, to the system of elementary loadings made up of the centred axial force N and the flexures

$$M_x = Ne_y \tag{9.45a}$$

$$M_y = -Ne_x \tag{9.45b}$$

Summing up the uniaxial stress fields that correspond to the aforesaid elementary loadings, we obtain

$$\sigma_z = \frac{N}{A} + \frac{M_x}{I_x} y - \frac{M_y}{I_y} x \tag{9.46}$$

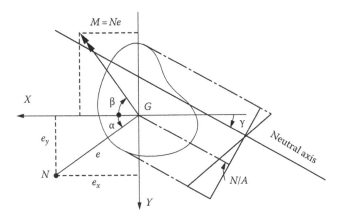

Figure 9.9

The negative sign of the last term is due to the fact that the moment M_y, if positive, stretches the longitudinal fibres of the half-plane $x < 0$.

Substituting Equations 9.45 in Equation 9.46, we deduce the following relation:

$$\sigma_z = \frac{N}{A} + \frac{Ne_y}{I_x}y + \frac{Ne_x}{I_y}x \tag{9.47}$$

which, expressing the moments of inertia as functions of the area A and of the respective central radii of gyration, becomes

$$\sigma_z = \frac{N}{A}\left(1 + \frac{e_y}{\rho_x^2}y + \frac{e_x}{\rho_y^2}x\right) \tag{9.48}$$

Defining the **neutral axis**, also in the case of combined loading, as the straight line on which the stress σ_z vanishes, we obtain, by simply equating the expression (9.48) to zero,

$$1 + \frac{e_y}{\rho_x^2}y + \frac{e_x}{\rho_y^2}x = 0 \tag{9.49}$$

The neutral axis may cut the cross section or not, according to the eccentricity e and the angle α (Figure 9.9). For small eccentricities, the neutral axis does not intersect the cross section, and the stresses σ_z all have the same sign, whereas, for large eccentricities, the neutral axis intersects the cross section, and the stresses σ_z change sign through it. The term **central core of inertia** is given to the area within which the eccentric axial force falls, in such a way that the neutral axis does not intersect the cross section of the beam. This concept is of particular importance in reference to compressed beams made of brittle or non-traction-bearing materials (concrete, masonry, etc.). We shall shortly see two particularly significant examples.

The linear variation of the stresses σ_z on the cross section can be represented by drawing a reference line perpendicular to the neutral axis (Figure 9.9) and by marking in the two values corresponding to the neutral axis, $\sigma_z = 0$, and to the centroid, $\sigma_z = N/A$. The diagram of stresses σ_z will therefore be butterfly shaped in the case where the neutral axis intersects the cross section (Figure 9.9) or trapezoidal in the case where the neutral axis is external to the cross section (Figure 9.10).

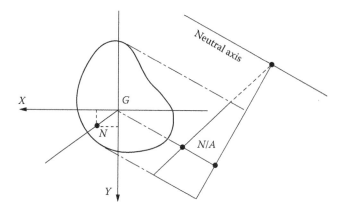

Figure 9.10

As the eccentricities e_x, e_y vary, three particular cases may arise.

1. $e_x = e_y = 0$: centred axial force. The neutral axis degenerates into the straight line at infinity.

2. (a) $e_x = 0$: centred axial force plus flexure. The neutral axis is parallel to the X axis:

$$y = -\frac{\rho_x^2}{e_y} \tag{9.50a}$$

(b) $e_y = 0$: centred axial force plus flexure. The neutral axis is parallel to the Y axis:

$$x = -\frac{\rho_y^2}{e_x} \tag{9.50b}$$

Note that the centre of pressure and the centroid always belong to the same one of the two half-planes into which the neutral axis divides the XY plane. In this regard, see also the general equation (9.49).

3. $e_y/e_x = \tan\alpha$; $e_x \to \infty$, $e_y \to \infty$. The latter is the case of **biaxial flexure**. The general equation (9.49) then becomes

$$\frac{y}{\rho_x^2}\tan\alpha + \frac{x}{\rho_y^2} = 0 \tag{9.51}$$

which represents a centroidal straight line. That is, when only the flexures M_x, M_y are present, the neutral axis is centroidal; however, in general, this is not perpendicular to the loading axis.

The **loading axis** is the straight line NG (Figure 9.9) represented by the equation

$$y = x\tan\alpha \tag{9.52}$$

On the other hand, the **neutral axis** can be represented by the equation

$$y = -\frac{1}{\tan\alpha}\left(\frac{\rho_x}{\rho_y}\right)^2 x - \frac{\rho_x^2}{e_y} \tag{9.53}$$

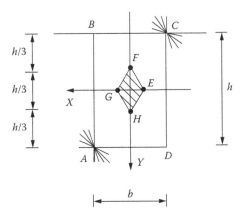

Figure 9.11

It is readily recognizable that the lines (9.52) and (9.53) are orthogonal, not only for $e_x = 0$ or for $e_y = 0$, but also when the central radii of gyration are equal: $\rho_x = \rho_y$. In this case, we have $I_x = I_y$ and thus the section is of a gyroscopic nature (Chapter 2).

We shall now determine the central cores of inertia of two cross sections that are very frequently encountered in building practice: the rectangle and the circle.

In the case of the rectangular cross section of Figure 9.11, the neutral axis can be assumed to coincide with the straight line to which the upper side BC belongs. Using Equation 9.50a, we obtain

$$-\frac{h}{2} = \frac{(-h^2/12)}{e_y} \tag{9.54}$$

from which

$$e_y = \frac{h}{6} \tag{9.55}$$

The eccentric axial force corresponding to the neutral axis BC is applied to the point H, belonging to the Y axis at a distance $h/6$ from the X axis. Similarly, the centres of pressure corresponding to the straight lines CD, DA, AB are G, F, E, respectively (Figure 9.11).

It is easy to show that the central core of inertia is represented by the rhombus $EFGH$. The neutral axis corresponding to each point of the segment GH

$$\frac{e_x}{(b/6)} + \frac{e_y}{(h/6)} = 1 \tag{9.56}$$

is given by the Equation 9.53, once we insert the ratio

$$\tan\alpha = \frac{e_y}{e_x} = \frac{h(b-6e_x)}{6be_x} \tag{9.57}$$

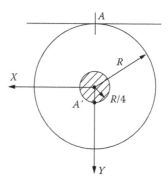

Figure 9.12

We thus have

$$y = \frac{6e_x b}{h(6e_x - b)} \left(\frac{h}{b}\right)^2 x - \frac{h^2}{12} \frac{1}{e_y} \tag{9.58}$$

Equation 9.58 is satisfied by the coordinates of the point $C(-b/2, -h/2)$ for any pair of values (e_x, e_y) which satisfies the relation (9.56).

The cross section is traditionally said to be entirely in compression if the centre of pressure is within the **middle third**. In fact, in the case of compressive axial force and flexure, when the force N acts between F and H ($\overline{FH} = h/3$), the neutral axis is external to the cross section, and we have $\sigma_z < 0$ in each point (Figure 9.11).

A similar line of reasoning, though simpler on account of polar symmetry, leads us to conclude that the central core of inertia of a circular cross section of radius R is the concentric circle of radius $R/4$. Equation 9.50a, in the case of the neutral axis being tangent in point A (Figure 9.12), gives in fact

$$-R = -\frac{(R^2/4)}{e_y} \tag{9.59}$$

from which there follows

$$e_y = \frac{R}{4} \tag{9.60}$$

As regards deformation, eccentric axial force produces a rigid rotation of each individual cross section about the neutral axis, as well as an axial translation of the section. Referring to Figure 9.9 and to the relation (9.37), we can see that the moment $M_x = Ne_y$ produces the rotation

$$\varphi_x = \frac{M_x}{EI_x} z = \frac{Nz}{EA} \frac{e_y}{\rho_x^2} \tag{9.61a}$$

just as the moment $M_y = -Ne_x$ produces the rotation

$$\varphi_y = \frac{M_y}{EI_y} z = -\frac{Nz}{EA} \frac{e_x}{\rho_y^2} \tag{9.61b}$$

Since the rotations φ_x and φ_y are infinitesimal, the vector summation of them may be made and will give the total rotation vector

$$\{\varphi\} = \varphi_x \overline{i} + \varphi_y \overline{j} \tag{9.62}$$

where $\overline{i}$ and $\overline{j}$ are the unit vectors of the X axis and the Y axis, respectively. The angular coefficient of the axis of rotation is thus equal to

$$\frac{\varphi_y}{\varphi_x} = -\frac{e_x}{e_y}\left(\frac{\rho_x}{\rho_y}\right)^2 = -\frac{1}{\tan\alpha}\left(\frac{\rho_x}{\rho_y}\right)^2 \tag{9.63}$$

and coincides with that of the neutral axis (9.53).

Finally, it may be shown that the three elementary loadings, N, M_x, M_y, are energetically orthogonal, and the work associated with the eccentric axial force is therefore equal to the sum of the work of each of the individual characteristics. The elementary work, corresponding to an infinitesimal segment of beam, is equal to

$$dL = \int_A \Psi \, dA \, dz \tag{9.64}$$

where Ψ represents the work per unit volume. Recalling expression (8.70), which furnishes the complementary elastic potential as a function of the stress condition, we have

$$dL = \int_A \frac{\sigma_z^2}{2E} \, dA \, dz \tag{9.65}$$

If we use the solution (9.46), we obtain

$$\frac{dL}{dz} = \frac{1}{2E} \int_A \left(\frac{N}{A} + \frac{M_x}{I_x}y - \frac{M_y}{I_y}x\right)^2 dA \tag{9.66}$$

Expanding the square of the trinomial under the integral sign, we split the elementary work into two parts; in the first part there appear the squares of the characteristics, whilst in the second there appear the mutual products:

$$\frac{dL}{dz} = \frac{1}{2}\left[\frac{N^2}{EA} + \frac{M_x^2}{EI_x} + \frac{M_y^2}{EI_y}\right] + \frac{1}{E}\left[\frac{NM_x}{AI_x}S_x - \frac{NM_y}{AI_y}S_y - \frac{M_xM_y}{I_xI_y}I_{xy}\right] \tag{9.67}$$

Since the X and Y axes are central, the corresponding static moments vanish, as does the product of inertia, and hence the second part of expression (9.67) vanishes:

$$\frac{dL}{dz} = \frac{1}{2}\left[\frac{N^2}{EA} + \frac{M_x^2}{EI_x} + \frac{M_y^2}{EI_y}\right] \tag{9.68}$$

The relation (9.68) proves that the three characteristics so far considered are energetically orthogonal.

9.6 TORSION IN BEAMS OF CIRCULAR CROSS SECTION

Let us consider a beam of circular cross section, subjected at its ends to the action of two twisting moments M_z of equal magnitude and opposite direction (Figure 9.13). Let l denote the length of the beam and R the radius of the cross section.

Whereas in the case of axial force and flexure we have assumed the stresses σ_z (static hypothesis), in the case of torsion the hypothesis regards the displacement field (kinematic hypothesis). We assume in fact that each cross section rotates rigidly about the longitudinal axis, remaining at the same time plane. Expressed in formulas (Figure 9.14),

$$u = -\varphi_z y \tag{9.69a}$$

$$\upsilon = \varphi_z x \tag{9.69b}$$

$$w = 0 \tag{9.69c}$$

where φ_z is the angle of infinitesimal rotation. For this purpose, we have had recourse to Equations 3.8. Since, if we exclude the end regions, the relative rotation per unit length of the beam must be constant and equal to Θ, we shall have, for example,

$$\varphi_z = \Theta z \tag{9.70}$$

in the case where the beam is built-in in the cross section $z = 0$. The characteristic of deformation Θ is called the **unit angle of torsion**. Equations 9.69 then become

$$u = -\Theta y z \tag{9.71a}$$

$$\upsilon = \Theta x z \tag{9.71b}$$

$$w = 0 \tag{9.71c}$$

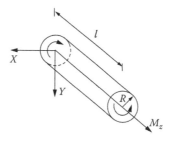

Figure 9.13

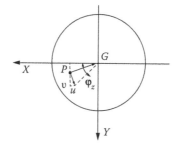

Figure 9.14

From the displacement field expressed by Equations 9.71, the strain field may be immediately derived *via* the **kinematic equations**

$$\varepsilon_x = \varepsilon_y = \varepsilon_z = \gamma_{xy} = 0 \tag{9.72a}$$

$$\gamma_{zx} = -\Theta y \tag{9.72b}$$

$$\gamma_{zy} = \Theta x \tag{9.72c}$$

Applying then the **elastic constitutive equations** (8.74), from the strain field (9.72), it is possible to obtain the corresponding stress field:

$$\sigma_x = \sigma_y = \sigma_z = \tau_{xy} = 0 \tag{9.73a}$$

$$\tau_{zx} = -G\Theta y \tag{9.73b}$$

$$\tau_{zy} = G\Theta x \tag{9.73c}$$

From the shearing stress components (9.73b and 9.73c), the magnitude of the shearing stress vector is obtained:

$$\tau_z = G\Theta r \tag{9.74}$$

where r is the radial distance of the generic point from the centre of the circular cross section.

The **indefinite equations of equilibrium** (8.2) are identically satisfied by the stress field (9.73), just as are the **equations of equivalence on the lateral surface** (9.2). Equation 9.2c in fact becomes

$$G\Theta(-yn_x + xn_y) = 0 \tag{9.75}$$

from which relation we obtain the proportion

$$\frac{x}{y} = \frac{n_x}{n_y} \tag{9.76}$$

which is identically satisfied on the basis of the similarity of the right-angled triangles ABC and $A'B'B$ in Figure 9.15. Note that Equation 9.2c is equivalent to equating to zero the scalar product of the stress vector and the unit vector normal to the lateral surface

$$\{\tau_z\}^T \{n\} = 0 \tag{9.77}$$

whereby the stress vector $\{\tau_z\}$ is always tangential to the lateral surface, and thus, in the case of a circular cross section, it is also perpendicular to the radius vector. On the other hand,

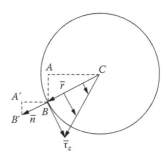

Figure 9.15

not only on the lateral surface, but indeed at each point of the circular cross section, the vector $\{\tau_z\}$ is perpendicular to the radius vector (Figure 9.15). From Equations 9.73b and 9.73c, we have in fact

$$\{\tau_z\}^T\{r\} = -G\Theta yx + G\Theta xy = 0 \tag{9.78}$$

Since $\sigma_z = 0$, the **conditions of equivalence on the end planes** $N = M_x = M_y = 0$ are identically satisfied. On the other hand, the shear along the X axis vanishes,

$$T_x = \int_A \tau_{zx} dA = -G\Theta \int_A y\, dA = -G\Theta S_x = 0 \tag{9.79}$$

since the static moment of the cross section with respect to the centroidal axis X vanishes. The same applies for shear along the Y axis:

$$T_y = \int_A \tau_{zy}\, dA = G\Theta \int_A x\, dA = G\Theta S_y = 0 \tag{9.80}$$

The only significant condition of equivalence remains the one corresponding to the twisting moment M_z:

$$M_z = \int_A \{r\} \wedge \{\tau_z\} dA = \int_A (x\tau_{zy} - y\tau_{zx}) dA \tag{9.81}$$

which, on the basis of Equations 9.73b and 9.73c, gives

$$M_z = G\Theta \int_A r^2\, dA = G\Theta I_p \tag{9.82}$$

where I_p is the polar moment of inertia of the circular cross section. From the foregoing relation (9.82), we obtain the **unit angle of torsion**

$$\Theta = \frac{M_z}{GI_p} \tag{9.83}$$

which is the characteristic of deformation corresponding to the twisting moment. As in the case of longitudinal dilation (9.9b) produced by the centred axial force and in that of the curvature (9.41) produced by flexure, the angle Θ is directly proportional to the static characteristic M_z and inversely proportional to the elastic characteristic G and the geometric characteristic I_p.

Reconsidering Equation 9.74, it is now possible to formulate the definitive expression of **global shearing stress:**

$$\tau_z = \frac{M_z}{I_p} r \tag{9.84}$$

This expression has a structure similar to Equation 9.23 for flexure. If the twisting moment is counterclockwise (positive), the vector $\{\tau_z\}$ will always give positive moment with respect to the centre and will follow concentric circular lines of flux. It will increase linearly from zero, at the centre, to its maximum value, at the boundary (Figure 9.16a):

$$\tau_{max} = \frac{M_z}{I_p} R \tag{9.85}$$

Since the polar moment of inertia is equal to $\pi R^4/2$, the expression (9.85) may be rewritten as a function of the radius R alone, as well as of the static characteristic M_z:

$$\tau_{max} = \frac{2M_z}{\pi R^3} \tag{9.86}$$

In the case of a cross section with the form of an annulus (Figure 9.16b), it is possible to repeat the entire line of reasoning so far developed for the case of a circular cross section. It will be sufficient to take into account the different polar moment of inertia

$$I_p = \frac{\pi}{2}\left(R_2^4 - R_1^4\right) \tag{9.87}$$

in the expression of stress (9.84):

$$\tau_z = \frac{2M_z}{\pi\left(R_2^4 - R_1^4\right)} r \tag{9.88}$$

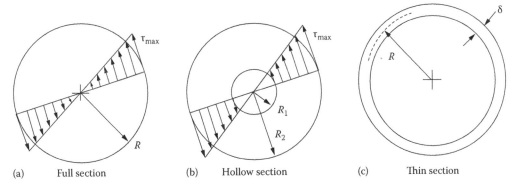

(a)   Full section   (b)   Hollow section   (c)   Thin section

Figure 9.16

from which we obtain the maximum shearing stress

$$\tau_{max} = \frac{2M_z R_2}{\pi\left(R_2^4 - R_1^4\right)} \tag{9.89}$$

Hence, given an equal maximum radius R_2, the hollow cross section is subjected to a higher stress.

Finally, let us consider a thin-walled circular cross section of thickness δ (Figure 9.16c), which may be understood as a limit case of the preceding example:

$$R_1 \simeq R_2 = R \tag{9.90a}$$

$$R_2 - R_1 = \delta \tag{9.90b}$$

Expanding the difference of the fourth powers, the relation (9.89) becomes

$$\tau_{max} = \frac{2M_z R_2}{\pi\left(R_2^2 + R_1^2\right)(R_2 + R_1)(R_2 - R_1)} \tag{9.91}$$

and applying the approximations (9.90)

$$\tau_{max} = \frac{M_z}{2\pi R^2 \delta} \tag{9.92}$$

The foregoing formula may also be deduced from Equation 9.85, setting $I_p = (2\pi R\delta)R^2$.

On Mohr's plane, the stress condition is represented by a circumference with its centre in the origin. This stress condition is referred to as the state of **pure shear** and implies two principal directions, one of tension and the other of compression, rotated by 45° with respect to the Z axis (Figure 9.17a). It is thus understandable why cylindrical elements of concrete subjected to torsion are reinforced with spiral-shaped bars, inclined at an angle of 45° with respect to the axis and disposed in such a way as to stand up to the principal tensile stress (Figure 9.17b).

The work of deformation of an element of the beam (Figure 9.13) is obtained once more by applying Clapeyron's theorem

$$dL = \frac{1}{2}M_z \Theta\, dz \tag{9.93}$$

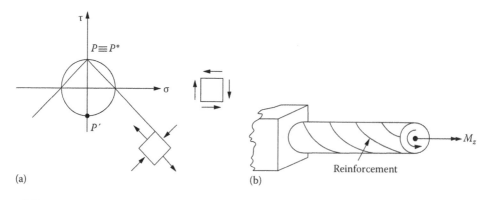

(a)

(b)

Reinforcement

M_z

Figure 9.17

and using the expression (9.83) of the unit angle of torsion:

$$\frac{dL}{dz} = \frac{1}{2}\frac{M_z^2}{GI_p} \tag{9.94}$$

The quadratic expression (9.94) is altogether analogous to Equations 9.14 and 9.44, obtained for the centred axial force and for the flexure, respectively.

9.7 TORSION IN BEAMS OF GENERIC CROSS SECTION

In the case of beams of generic cross section, each cross section is assumed to rotate about a longitudinal axis (which may not coincide with the centroidal one), called the **axis of torsion,** at the same time not remaining plane. Expressed in formulas and analogously to Equations 9.71, we have (Figure 9.18)

$$u = -\Theta z(y - y_C) \tag{9.95a}$$

$$v = \Theta z(x - x_C) \tag{9.95b}$$

$$w = \Theta\omega(x, y) \tag{9.95c}$$

where x_C and y_C are the coordinates of the **centre of torsion** C, while $\omega(x, y)$, called the **warping function,** represents the axial displacements of the points of the cross section.

From the kinematic equations (8.8), we obtain the strain field

$$\varepsilon_x = \varepsilon_y = \varepsilon_z = \gamma_{xy} = 0 \tag{9.96a}$$

$$\gamma_{zx} = -\Theta(y - y_C) + \Theta\frac{\partial\omega}{\partial x} \tag{9.96b}$$

$$\gamma_{zy} = -\Theta(x - x_C) + \Theta\frac{\partial\omega}{\partial y} \tag{9.96c}$$

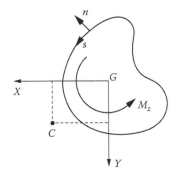

Figure 9.18

and, consequently, applying the elastic constitutive equations (8.74), we obtain the stress field

$$\sigma_x = \sigma_y = \sigma_z = \tau_{xy} = 0 \tag{9.97a}$$

$$\tau_{zx} = G\Theta\left[\frac{\partial\omega}{\partial x} - (y - y_C)\right] \tag{9.97b}$$

$$\tau_{zy} = G\Theta\left[\frac{\partial\omega}{\partial y} + (x - x_C)\right] \tag{9.97c}$$

In this case, the third of the indefinite equations of equilibrium (8.2c) is satisfied if and only if

$$\frac{\partial^2\omega}{\partial x^2} + \frac{\partial^2\omega}{\partial y^2} = 0 \tag{9.98}$$

i.e. if and only if ω is a harmonic function. On the other hand, the third of the equations of equivalence on the lateral surface (9.2c) is satisfied if and only if the following boundary condition is satisfied:

$$\left[\frac{\partial\omega}{\partial x} - (y - y_C)\right]n_x + \left[\frac{\partial\omega}{\partial y} + (x - x_C)\right]n_y = 0 \tag{9.99}$$

Laplace's equation (9.98) associated with the boundary condition (9.99) constitutes a problem for which the solution exists and is the unique one, but for an arbitrary additional constant (Neumann problem). It should be noted, however, that, up to this point, the coordinates x_C, y_C of the centre of torsion are still unknown.

The conditions of equivalence on the end planes

$$N = M_x = M_y = 0 \tag{9.100}$$

are identically satisfied, since $\sigma_z = 0$, whilst from the annihilation of the shear, we obtain the coordinates of the centre of torsion C. We have in fact

$$T_x = \int_A \tau_{zx}\,dA = G\Theta\int_A\left[\frac{\partial\omega}{\partial x} - (y - y_C)\right]dA = 0 \tag{9.101a}$$

$$T_y = \int_A \tau_{zy}\,dA = G\Theta\int_A\left[\frac{\partial\omega}{\partial y} + (x - x_C)\right]dA \tag{9.101b}$$

whence we obtain

$$\int_A \frac{\partial\omega}{\partial x}\,dA - S_x + y_C A = 0 \tag{9.102a}$$

$$\int_A \frac{\partial\omega}{\partial y}\,dA - S_y - x_C A = 0 \tag{9.102b}$$

and, thus, since the centroidal static moments are zero,

$$x_C = \frac{1}{A}\int_A \frac{\partial\omega}{\partial y}\,dA \tag{9.103a}$$

$$y_C = -\frac{1}{A}\int_A \frac{\partial\omega}{\partial x}\,dA \tag{9.103b}$$

Applying Green's theorem, the coordinates of the centre of torsion finally appear as follows:

$$x_C = \frac{1}{A}\oint_\mathscr{C} \omega n_y\,ds \tag{9.104a}$$

$$y_C = -\frac{1}{A}\oint_\mathscr{C} \omega n_x\,ds \tag{9.104b}$$

where $\mathscr{C}$ indicates the boundary of the cross section or, in the alternative form,

$$x_C = -\frac{1}{A}\oint_\mathscr{C} \omega\,dx \tag{9.105a}$$

$$y_C = -\frac{1}{A}\oint_\mathscr{C} \omega\,dy \tag{9.105b}$$

The boundary condition expressed by Equation 9.99 for the warping function then transforms into the following integrodifferential equation:

$$\left[\frac{\partial\omega}{\partial x} - y - \frac{1}{A}\oint_\mathscr{C} \omega\,dy\right]n_x + \left[\frac{\partial\omega}{\partial y} + x + \frac{1}{A}\oint_\mathscr{C} \omega\,dx\right]n_y = 0 \tag{9.106}$$

Equations 9.98 and 9.106 give the warping function, and consequently, from Equations 9.105, we obtain the coordinates of the centre of torsion. For a numerical solution to this problem, the reader is referred to Appendix F.

There remains to be imposed the last condition of equivalence on the end planes, the one corresponding to twisting moment

$$M_z = \int_A \{r\}\wedge\{\tau_z\}\,dA = \int_A (x\tau_{zy} - y\tau_{zx})\,dA \tag{9.107}$$

which, on the basis of Equations 9.97b and 9.97c, gives

$$M_z = G\Theta\int_A \left(x^2 + y^2 + x\frac{\partial\omega}{\partial y} - y\frac{\partial\omega}{\partial x}\right)\,dA \tag{9.108}$$

The integral represents the so-called **factor of torsional rigidity** I_t, which is a quantity that is always less than or, at the most, equal to the polar moment of inertia I_p. The **unit angle of torsion** can thus be expressed in general as follows:

$$\Theta = \frac{M_z}{GI_t} \tag{9.109}$$

with

$$I_t = \int_A \left(x^2 + y^2 + x\frac{\partial \omega}{\partial y} - y\frac{\partial \omega}{\partial x} \right) dA \tag{9.110}$$

Note how I_t is an exclusively geometrical characteristic of the cross section.

From Equations 9.97b and 9.97c and 9.109, we obtain the shearing stress vector

$$\tau_{zx} = \frac{M_z}{I_t}\left[\frac{\partial \omega}{\partial x} - (y - y_C) \right] \tag{9.111a}$$

$$\tau_{zy} = \frac{M_z}{I_t}\left[\frac{\partial \omega}{\partial y} + (x - x_C) \right] \tag{9.111b}$$

Finally, Clapeyron's theorem gives the elementary work of deformation:

$$\frac{dL}{dz} = \frac{1}{2}\frac{M_z^2}{GI_t} \tag{9.112}$$

9.8 TORSION IN OPEN THIN-WALLED SECTIONS

The solution previously outlined regarding a cross section of generic shape can be applied to the case of the **rectangular** cross section. It may thus be deduced how the maximum stress is reached in the intermediate point of the larger side a (Figure 9.19):

$$\tau_{max} = \alpha\frac{M_z}{ab^2} \tag{9.113}$$

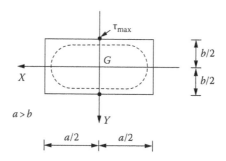

Figure 9.19

Table 9.1

a/b	1	1.5	2	3	10	∞
α	4.80	4.33	4.06	3.74	3.20	3
β	0.141	0.196	0.229	0.263	0.312	1/3

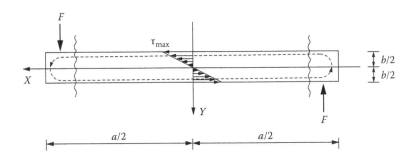

Figure 9.20

where α is a dimensionless coefficient which depends only on the ratio a/b between the sides of the cross section. At the same time, it is possible to obtain the factor of torsional rigidity to be inserted into equation

$$I_t = \beta a b^3 \tag{9.114}$$

where β is another dimensionless coefficient which depends only on the ratio a/b between the sides of the cross section. The two coefficients, α and β, are given in Table 9.1 as functions of a/b.

In the case of a **thin rectangular cross section** ($a/b \to \infty$), the lines of flux of the vector $\{\tau_z\}$ are closed curves with two portions practically parallel to the larger sides (Figure 9.20). These portions reverse their course only in the end regions, so that the shearing stress vector $\{\tau_z\}$ presents the sole component τ_{zx} for the most part of the cross section. The shearing stress component τ_{zx} is shown to have a linear distribution over the thickness b,

$$\tau_{zx} = -\frac{6M_z}{ab^3}y \tag{9.115}$$

with a maximum absolute value at the boundary,

$$\tau_{max} = \frac{3M_z}{ab^2} \tag{9.116}$$

Since on the basis of Equation 9.114 and of the tabulated value, we have

$$I_t = \frac{1}{3}ab^3 \tag{9.117}$$

we also obtain

$$\tau_{max} = \frac{M_z}{I_t}b \tag{9.118}$$

Note that the stresses parallel to the larger sides are not, however, sufficient of themselves to ensure equivalence with the applied twisting moment M_z. In fact, on the basis of Equation 9.16f, we obtain

$$M_z(\tau_{zx}) = -\int_A y\tau_{zx}\,\mathrm{d}A \tag{9.119}$$

Substituting the expression (9.115) in Equation 9.119, we obtain

$$M_z(\tau_{zx}) = \frac{6M_z}{ab^3}\int_{-a/2}^{a/2}\mathrm{d}x\int_{-b/2}^{b/2}y^2\mathrm{d}y \tag{9.120}$$

from which we have

$$M_z(\tau_{zx}) = \frac{M_z}{2} \tag{9.121}$$

The other half of the moment M_z is furnished by the components τ_{zy}, which are important only in the end regions of the cross section, where they present an arm with respect to the centroid (or centre of torsion) which is much greater than that of the components τ_{zx} $(a \gg b)$.

Let us consider the case of an **open thin-walled** cross section made up of a number of thin rectangles welded together so as not to create any closed path (Figure 9.21), and let us allow the applied moment M_z to be distributed in such a way that the ith section takes up the amount M_z^i. The maximum stress on the ith section will be given by Equation 9.118,

$$\tau_{max}^i = \frac{M_z^i}{I_t^i}b_i \tag{9.122}$$

where b_i is the thickness of the ith section and

$$I_t^i = \frac{1}{3}a_i b_i^3 \tag{9.123}$$

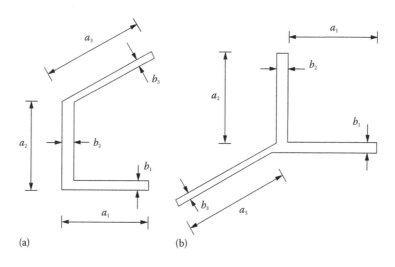

(a)   (b)

Figure 9.21

from Equation 9.117. On the other hand, for reasons of congruence, since each section must rotate by the same angle Θ, we have

$$\Theta^i = \frac{M_z^i}{GI_t^i} = \Theta = \frac{M_z}{GI_t} \tag{9.124}$$

where I_t is the factor of torsional rigidity of the entire section. From Equations 9.122 and 9.124, we obtain

$$\tau_{max}^i = \frac{M_z}{I_t} b_i \tag{9.125}$$

The factor I_t is obtained from considerations of equilibrium or rather of equivalence. Since the sum of the partial moments M_z^i must equal the applied moment M_z, we have

$$M_z = G\Theta I_t = \sum_i M_z^i = G\Theta \sum_i I_t^i \tag{9.126}$$

from which we obtain

$$I_t = \sum_i I_t^i \tag{9.127}$$

The global factor of torsional rigidity is thus equal to the sum of the partial factors, as is the case when the elements are in parallel. From Equation 9.124, we then have

$$M_z^i = M_z \frac{I_t^i}{\sum_i I_t^i} \tag{9.128}$$

where the ratio $I_t^i / \sum_i I_t^i$ is called **coefficient of distribution**. Recalling Equation 9.127, the global factor may be expressed as follows:

$$I_t = \frac{1}{3} \sum_i a_i b_i^3 \tag{9.129}$$

If the open thin-walled section has as its midline a regular curve (Figure 9.22), and not a broken line as in the previous case, Equations 9.125 and 9.129 will be replaced by the following:

$$\tau_{zs}(max) = \frac{M_z}{I_t} b(s) \tag{9.130}$$

$$I_t = \frac{1}{3} \int_{\mathscr{C}} b^3(s)\, ds \tag{9.131}$$

where s is the curvilinear coordinate on the midline $\mathscr{C}$.

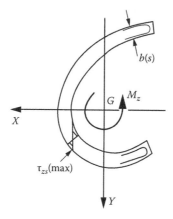

Figure 9.22

The distribution of the stress τ_{zs} is linear on the thickness and vanishes on the midline. The maximum absolute value (9.130), which is recorded on the boundary, is in turn maximum in the points in which the thickness $b(s)$ is maximum. The flux lines of the vector $\{\tau_z\}$ are thus parallel to the midline and reverse their course only in the end regions (Figure 9.22).

9.9 TORSION IN CLOSED THIN-WALLED SECTIONS

Closed thin-walled sections are also called **tubular sections**. These are frequently used in building applications, because, as we shall see shortly, they present notable strength and stiffness. Their disparity of behaviour in regard to open thin-walled sections is due substantially to the flux of the vector $\{\tau_z\}$, which in the case of closed sections manages to develop more conveniently, i.e. with the shearing stresses τ_{zs} that are approximately constant on each chord that is perpendicular to the midline $\mathscr{C}$ (Figure 9.23a). In addition, even though the thickness $b(s)$ is not constant, the product of shearing stress τ_{zs} and thickness b must be constant:

$$\tau_{zs}(s)b(s) = \text{constant} \tag{9.132}$$

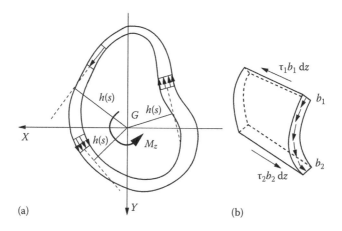

Figure 9.23

This follows from the simple application of the law of reciprocity of shearing stresses to an element of the thin wall, obtained by sectioning the beam with two planes perpendicular to the axis of the beam, located at a distance dz apart, and with two planes perpendicular to the midline of the cross section (Figure 9.23b). If we denote the chords of intersection of the longitudinal and the transverse sections as b_1, b_2, and their respective shearing stresses as τ_1, τ_2, by virtue of the axial equilibrium of the element, we must have

$$\tau_1 b_1 \, dz = \tau_2 b_2 \, dz \tag{9.133}$$

from which there follows the constancy of the product (9.132), since Equation 9.133 holds good for any pair of planes which cut the midline at right angles to it.

From equivalence, we must find

$$M_z = \oint_{\mathscr{C}} \tau_{zs}(s) b(s) h(s) \, ds \tag{9.134}$$

where $h(s)$ represents the arm of the elementary force ($\tau_{zs} \, bds$) with respect to the centroid (Figure 9.23a).

From Equations 9.132 and 9.134, we deduce

$$M_z = \tau_{zs}(s) b(s) \oint_{\mathscr{C}} h(s) \, ds \tag{9.135}$$

and, thus, noting that the integral given earlier represents twice the area Ω enclosed by the midline $\mathscr{C}$, we obtain

$$\tau_{zs} = \frac{M_z}{2\Omega b(s)} \tag{9.136}$$

The expression (9.136), called **Bredt's formula**, underlines how the maximum stress is produced where the thickness $b(s)$ is minimum, in sharp contrast with what occurs in the case of open thin-walled sections, as described by Equation 9.130.

An application of Clapeyron's theorem then furnishes the **factor of torsional rigidity**. The elementary work of deformation is equal to

$$dL = \frac{1}{2} \frac{M_z^2}{GI_t} \, dz = \int_A \Psi \, dV \tag{9.137}$$

where Ψ is the complementary elastic potential expressed by Equation 8.70. Since the stress component τ_{zs} alone is different from zero, we have

$$dL = \frac{dz}{2G} \int_A \tau_{zs}^2 \, dA \tag{9.138}$$

and, thus, substituting Equation 9.136,

$$dL = \frac{dz}{2G} \frac{M_z^2}{4\Omega^2} \oint_{\mathscr{C}} \frac{1}{b^2(s)} (b(s) \, ds) \tag{9.139}$$

From Equations 9.137 and 9.139, we obtain the factor of torsional rigidity:

$$I_t = \frac{4\Omega^2}{\oint_\mathscr{C} ds/b(s)} \tag{9.140}$$

This formula simplifies when the thickness of the section is constant:

$$I_t = \frac{4\Omega^2 b}{s} \tag{9.141}$$

where s is the length of the midline. It should be noted that, whereas the thickness b appears raised to the first power in Equation 9.141, it appears raised to the third power in Equation 9.117, thus jeopardizing the stiffness of the open sections.

In the case of a thin circular section of thickness δ, we find again Equation 9.92, applying Bredt's formula

$$\tau_{zs} = \frac{M_z}{2\pi R^2 \delta} \tag{9.142}$$

while Equation 9.141 gives

$$I_t = \frac{4(\pi R^2)^2 \delta}{2\pi R} \tag{9.143}$$

which proves to be the polar moment of inertia of the cross section

$$I_t = 2\pi R^3 \delta = I_p \tag{9.144}$$

In certain cases, it may happen that a thin section is made up of tubular parts and open parts (flanges). It may readily be shown that practically the whole of the applied moment M_z is sustained by the tubular part, while the remaining open thin parts are subjected to much lower amounts.

Consider, for instance, the **box section** of Figure 9.24. Cross sections of this sort are frequently used for beams of road bridges. It consists of a thin rectangular cross section (1) and of two flanges (2).

Angular congruence gives

$$\Theta_1 = \frac{M_1}{GI_1} = \Theta_2 = \frac{M_2}{GI_2} \tag{9.145}$$

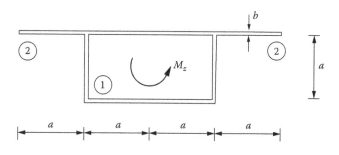

Figure 9.24

where the symbols have their obvious meanings. On the other hand, from equivalence, we have

$$M_1 + M_2 = M_z \tag{9.146}$$

Equations 9.145 and 9.146 constitute a system of linear equations in the two unknowns M_1, M_2. Once solved, these give the distribution that was sought:

$$M_1 = M_z \frac{I_1}{I_1 + I_2} \tag{9.147a}$$

$$M_2 = M_z \frac{I_2}{I_1 + I_2} \tag{9.147b}$$

Using relations (9.141) and (9.129), we then have

$$I_1 = \frac{8}{3} a^3 b \tag{9.148a}$$

$$I_2 = \frac{2}{3} a b^3 \tag{9.148b}$$

whence the ratio between rigidity factors becomes

$$\frac{I_1}{I_2} = 4 \left(\frac{a}{b} \right)^2 \tag{9.149}$$

If, for example, $a/b \simeq 10$, we obtain $I_1/I_2 \simeq 400$, and thus $M_1 \simeq M_z$ and $M_2 \simeq 0$.

Using Equations 9.130 and 9.136, we obtain the shearing stresses

$$\tau_1 = M_z \frac{I_1}{I_1 + I_2} \frac{1}{4a^2 b} \tag{9.150a}$$

$$\tau_2 = M_z \frac{I_2}{I_1 + I_2} \frac{b}{\frac{2}{3} a b^3} \tag{9.150b}$$

whence the ratio between the shearing stresses is

$$\frac{\tau_1}{\tau_2} = 4 \left(\frac{a}{b} \right)^2 \times \frac{1}{6} \left(\frac{b}{a} \right) = \frac{2}{3} \left(\frac{a}{b} \right) \tag{9.151}$$

Note that, while the ratio (9.149) between the stiffnesses is proportional to the square of the ratio (a/b), the ratio (9.151) between the stresses is proportional to the first power of (a/b). The flanges are in any case under a slight stress.

As regards multiple connected thin-walled sections, the reader is referred to Appendix G.

9.10 COMBINED SHEARING AND TORSIONAL LOADING

We intend to show how shear is energetically orthogonal to twisting moment, only if applied to the centre of torsion. To do so, let us consider a beam built-in at one of its ends and loaded on the other, in one case by a force T passing through the centre of torsion C (Figure 9.25a) and in the other by a twisting moment M_z (Figure 9.25b). If we denote the displacement in the XY plane of the point C as $\{\eta_C(M_z)\}$, caused by the moment M_z and the rotation of the cross section in the XY plane as $\varphi_z(T)$, caused by the shear T, the application of Betti's reciprocal theorem expresses the equality of the work performed by either force system acting through the displacement resulting from the other:

$$\{T\}^{\mathrm{T}}\{\eta_C(M_z)\} = M_z\varphi_z(T) \tag{9.152}$$

On the other hand, the displacement $\{\eta_C(M_z)\}$ of the centre of torsion is zero by definition, so that the work done by either system is likewise zero and in particular the rotation $\varphi_z(T)$ caused by the shear acting in point C is zero. For this reason, the **centre of torsion** is also called the **centre of shear**.

When the shearing force, then, is applied to the centre of torsion, it causes only translations and not rotations of the cross section in the XY plane. As there is no torsional deformation, i.e. $\Theta = 0$, from Equations 9.97, it may be deduced that the torsional stresses must also be zero.

Following the same line of reasoning, it is evident that the combined loading of **shear–torsion** (Figure 9.25a and b) is equivalent to a single force, parallel to the assigned shear, that presents a moment M_z with respect to the centre of torsion (Figure 9.25c). In other words, the global twisting moment M_z is evaluated as the moment of the force tangent to the cross section with respect to the centre of torsion, and not with respect to the centroid, as one might have been erroneously led to think previously.

9.11 SHEARING FORCE

Consider a Saint Venant solid loaded on the end planes by two equal and opposite shearing forces T_y, whose lines of action pass through the corresponding centres of torsion and are parallel to the central direction of inertia Y, and by a bending moment M_x which counterbalances

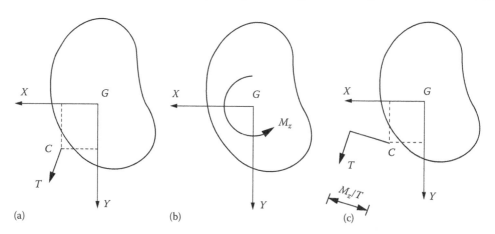

Figure 9.25

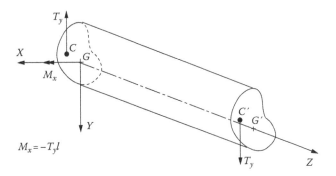

Figure 9.26

the couple $T_y l$, where l is the length of the cylindrical solid (Figure 9.26). Whereas the shear is constant, the bending moment varies linearly, from zero to the value $-T_y l$, along the axis of the beam. A typical case is that of a cantilever beam loaded by a force T_y at the end. As has already been mentioned, it is not possible to isolate the shearing characteristic, as this is the derivative of the bending moment function.

Let us now set ourselves the problem of determining the mean shearing stress acting orthogonally to a generic chord BB' (Figure 9.27a). For this purpose, let us consider a portion of the solid, bounded by two cross sections a distance dz apart and by the plane parallel to the axis of the beam, the projection of which on the cross section is represented by the chord BB' (Figure 9.27b). If we indicate the direction perpendicular to the chord BB' on the XY plane by s, the distribution of the shearing stresses τ_{sz} will be given by overturning the reciprocal shearing stresses τ_{zs} about the edge BB' (Figure 9.27b). From the equilibrium with regard to axial translation of the element of solid considered, we can then put

$$\int_{B}^{B'} \tau_{zs} \, dz \, d\xi = \int_{A'} \frac{\partial \sigma_z}{\partial z} \, dz \, dA \tag{9.153}$$

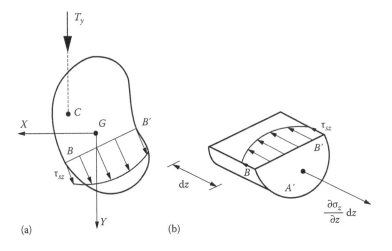

(a)   (b)

Figure 9.27

where A' is the lower portion of the cross section, and the integrand on the right-hand side of the equation represents the increment of normal stress σ_z produced by the bending moment $M_x(z)$. From Equation 9.23, we obtain

$$\frac{\partial \sigma_z}{\partial z} = \frac{\partial M_x}{\partial z} \frac{y}{I_x} \tag{9.154}$$

where M_x is the only function of z. Recalling then that the derivative of the moment is equal to the corresponding shear, we have

$$\frac{\partial \sigma_z}{\partial z} = T_y \frac{y}{I_x} \tag{9.155}$$

The relation (9.155), introduced into Equation 9.153, gives the mean shearing stresses acting orthogonally to the chord BB':

$$\bar{\tau}_{zs} = \frac{T_y S_x^{A'}}{I_x b} \tag{9.156}$$

where
  $S_x^{A'}$ is the static moment of the area A' with respect to the X axis
  b is the length of the chord

The relation (9.156) is known as the **Jourawski formula.** The components of $\{\tau_z\}$ parallel to the chord BB' are negligible in the cases in which the chord BB' is parallel to the central axis X. The mean values of these stresses, on a chord parallel to the central axis Y, are in fact proportional to the static moment $S_x^{A''}$, where A'' is one of the two portions of the section separated by the vertical chord (Figure 9.28).

From the application point of view, the Jourawski formula proves to be particularly useful, because, in addition to the fact that the aforementioned components are negligible, the local values τ_{zs} are not sensitive to variation from the mean value given by the relation (9.156). A typical example of a cross section for which the Jourawski formula provides a highly reliable estimation of the maximum shearing stress is the rectangular one (Figure 9.29a).

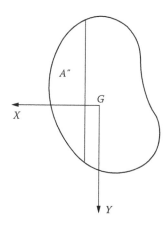

Figure 9.28

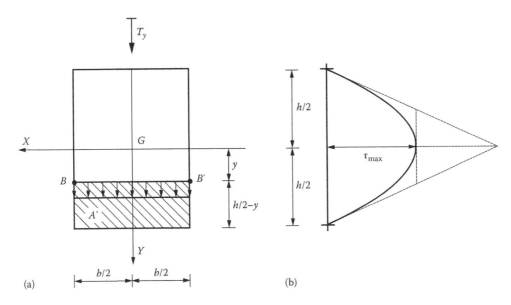

Figure 9.29

Consider a generic horizontal chord BB' for which the static moment of the area A' with respect to the X axis is

$$S_x^{A'} = b\left(\frac{h}{2} - y\right)\left[y + \frac{1}{2}\left(\frac{h}{2} - y\right)\right] = \frac{b}{2}\left(\frac{h^2}{4} - y^2\right) \tag{9.157}$$

Recalling that the moment of inertia of a rectangular area is $I_x = bh^3/12$ and applying Equation 9.156, we obtain

$$\overline{\tau}_{zy} = \frac{6T_y}{bh^3}\left(\frac{h^2}{4} - y^2\right) \tag{9.158}$$

The stress τ_{zy}, or rather its mean value on the horizontal chords, varies parabolically, vanishing on the horizontal sides and presenting one maximum on the centroidal chord (Figure 9.29b):

$$\tau_{max} = \frac{3}{2}\frac{T_y}{bh} \tag{9.159}$$

More generally, the following relation will hold good:

$$\tau_{max} = s_y \frac{T_y}{A} \tag{9.160}$$

where A stands for the area of the cross section, while s_y is a factor always greater than unity, which depends on the geometrical shape of the cross section. For the circular cross section, for instance, we have $s_y = 4/3$.

In slender beams, the shearing stresses are usually far smaller than the axial stresses due to bending moment. In the case, for instance, of a cantilever beam having rectangular cross section and loaded by a force T_y at the free end, at the built-in constraint, we have

$$\sigma_{max} = \frac{T_y l}{bh^3/12}\frac{h}{2} = \frac{6T_y l}{bh^2} \tag{9.161}$$

so that the ratio

$$\frac{\tau_{max}}{\sigma_{max}} = \frac{1}{4}\left(\frac{h}{l}\right) \tag{9.162}$$

tends to zero for $l \to \infty$.

In the case of stubby beams, the shearing stresses can be considerable, even if compared with axial stresses due to bending moment. It should be noted, however, that, below certain ratios of slenderness ($l/h \lesssim 3$), the Saint Venant theory fails to apply, since the end regions of the solid must then be neglected (the length of these regions is approximately equal to the beam depth h). Nor must it be forgotten that by the law of reciprocity of shearing stresses, longitudinal shearing stresses τ_{yz} are also present, as well as the stresses τ_{zy} acting upon the cross section (Figure 9.30a). The existence of the former stresses τ_{yz} may be inferred if we consider two beams resting one on top of the other without friction (Figure 9.30b). If the lower beam is supported at its ends and the upper one is loaded with a force, then each beam will bend, presenting fibres in tension below and fibres in compression above. At the interface between the two beams, we shall then have dilations in the upper beam and contractions in the lower one. If we now imagine that there is some form of friction acting between the two free surfaces in contact, the two beams will develop interactions that will tend to contract the upper one and dilate the lower one. These interactions, in the case where relative sliding is prevented by the presence of a bonding agent, are represented by the shearing stresses τ_{yz} (Figure 9.30c and d). In laminate composite materials, the mechanism of **delamination** is caused by the stresses τ_{yz} exceeding the limits of bonding resistance.

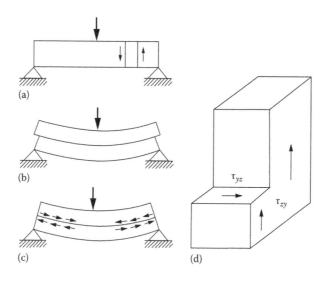

(a)

(b)

(c)   (d)

Figure 9.30

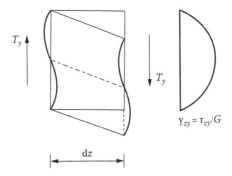

Figure 9.31

As regards deformation, the only strain component substantially different from zero is γ_{zy}. In the case of a rectangular cross section, γ_{zy} will show a parabolic variation equal, but for the factor $1/G$, to that of τ_{zy} (Figure 9.31):

$$\gamma_{zy} = \frac{\tau_{zy}}{G} = \frac{6T_y}{Gbh^3}\left(\frac{h^2}{4} - y^2\right) \tag{9.163}$$

Shearing strain is thus maximum on the centroidal plane $y = 0$ and zero on the outermost planes $y = \pm h/2$ (Figure 9.31). The result then is an inflection or warping of the cross sections out of their original planes, so as to maintain orthogonality only between the deformed section and the outermost planes.

The relative sliding between the end sections of the element of length dz (Figure 9.32) may be obtained *via* the application of Clapeyron's theorem

$$dL = \frac{1}{2}T_y\,dv = \frac{1}{2}T_y\gamma_y\,dz \tag{9.164}$$

where γ_y is called **mean shearing strain** and represents the characteristic of deformation corresponding to the shearing force T_y. The elementary work dL is expressible *via* the complementary elastic potential,

$$dL = \int_V \Psi\,dV = \frac{dz}{2G}\int_A \tau_{zy}^2\,dA \tag{9.165}$$

where the component τ_{zx} has been neglected, while the component τ_{zy} has been assumed to be given locally by its mean on the corresponding horizontal chord.

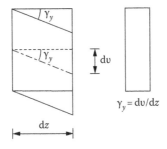

Figure 9.32

Substituting the expression (9.158) in Equation 9.165, we obtain

$$dL = \frac{dz}{2G} \int_{-b/2}^{b/2} dx \int_{-h/2}^{h/2} \frac{36T_y^2}{b^2h^6}\left(\frac{h^2}{4} - y^2\right)^2 dy \qquad (9.166)$$

from which

$$dL = \frac{dz}{2G} \frac{6}{5} \frac{T_y^2}{bh} \qquad (9.167)$$

A comparison between the foregoing expression and Equation 9.164 yields

$$\gamma_y = \frac{6}{5} \frac{T_y}{Gbh} \qquad (9.168)$$

More generally, the following relation will hold:

$$\gamma_y = t_y \frac{T_y}{GA} \qquad (9.169)$$

where t_y, called the **shear factor**, is always greater than unity and depends on the geometrical shape of the cross section. For the circular cross section, for instance, we have $t_y = 32/27$.

From Equation 9.164, we thus obtain

$$\frac{dL}{dz} = \frac{1}{2} t_y \frac{T_y^2}{GA} \qquad (9.170)$$

which shows how the work per unit length of the beam is a quadratic function also of the shearing force.

9.12 BIAXIAL SHEARING FORCE

In Section 9.10, it has been shown how the shear passing through the centre of torsion causes only translation and not rotation of the cross section in its plane. We did not, however, then specify the relation that links this translation to the shearing force components T_x, T_y. On the other hand, in the last section, we implicitly assumed that, where the axis Y is central and one of symmetry, the shearing force T_y causes exclusively a shearing strain γ_y and hence a translation of the cross section in its own Y direction. However, it is not possible to rule out *a priori* that, where the Y axis is central but not one of symmetry, the shearing force T_y may cause also a shearing strain γ_x and hence also a translation of the cross section in a direction different from its own.

To analyse the problem in a rigorous way, it is necessary to consider (open or closed) thin-walled cross sections, for which the shearing stress τ_{zs} can be considered constant on each chord orthogonal to the midline $\mathscr{C}$, while the shearing stress component orthogonal to

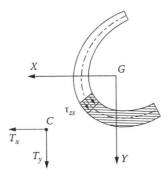

Figure 9.33

τ_{zs} can be neglected (Figure 9.33). In the case of **biaxial shear**, i.e. where there is the simultaneous presence of the two shears T_x, T_y, the stress τ_{zs} is equal to the algebraic sum of the corresponding contributions:

$$\tau_{zs}(s) = \frac{T_y}{I_x}\frac{S_x(s)}{b(s)} + \frac{T_x}{I_y}\frac{S_y(s)}{b(s)} \tag{9.171}$$

The elementary work of an element of beam is always equal to the integral of the complementary elastic energy:

$$dL = \frac{dz}{2G}\int_{\mathscr{C}}\tau_{zs}^2 b\,ds \tag{9.172}$$

From Equation 9.171, we obtain the relation

$$dL = \frac{dz}{2G}\left[\frac{T_y^2}{I_x^2}\int_{\mathscr{C}}\frac{S_x^2}{b}\,ds + \frac{T_x^2}{I_y^2}\int_{\mathscr{C}}\frac{S_y^2}{b}\,ds + 2\frac{T_xT_y}{I_xI_y}\int_{\mathscr{C}}\frac{S_xS_y}{b}\,ds\right] \tag{9.173}$$

which may be cast in the form

$$dL = \frac{dz}{2GA}\left(t_xT_x^2 + t_yT_y^2 + 2t_{xy}T_xT_y\right) \tag{9.174}$$

where

$$t_x = \frac{A}{I_y^2}\int_{\mathscr{C}}\frac{S_y^2}{b}\,ds \tag{9.175a}$$

$$t_y = \frac{A}{I_x^2}\int_{\mathscr{C}}\frac{S_x^2}{b}\,ds \tag{9.175b}$$

are the shear factors, while

$$t_{xy} = \frac{A}{I_x I_y} \int_{\mathscr{C}} \frac{S_x S_y}{b} \, ds \tag{9.175c}$$

is the **mutual shear factor**.

When in general $t_{xy} \neq 0$, the **mutual work**

$$dL(\text{mutual}) = t_{xy} \frac{T_x T_y}{GA} \, dz \tag{9.176}$$

is different from zero, and thus the shears T_x and T_y are **not energetically orthogonal**. On the other hand, t_{xy} is null when there is even only one axis of symmetry, as may be deduced from the integral expression (9.175c). Hence, only for cross sections with one or more axes of symmetry are the shears T_x and T_y energetically orthogonal.

Applying Clapeyron's theorem, we also have

$$dL = \frac{1}{2} T_x \, du + \frac{1}{2} T_y \, dv \tag{9.177}$$

And, therefore, introducing the mean shearing strains,

$$dL = \frac{1}{2} (T_x \gamma_x + T_y \gamma_y) \, dz \tag{9.178}$$

On the other hand, from the linearity of the elastic problem, the mean shearing strains are homogeneous linear functions of the shears

$$\gamma_x = a_{xx} T_x + a_{xy} T_y \tag{9.179a}$$

$$\gamma_y = a_{yx} T_x + a_{yy} T_y \tag{9.179b}$$

Substituting Equations 9.179 into Equation 9.178, and taking into account that we must have $a_{xy} = a_{yx}$ according to Betti's reciprocal theorem, we have

$$dL = \frac{dz}{2} \left(a_{xx} T_x^2 + a_{yy} T_y^2 + 2 a_{xy} T_x T_y \right) \tag{9.180}$$

A comparison between the preceding expression and Equation 9.174 furnishes, *via* the law of identity of polynomials, the linear relation that links the shearing force vector to the shearing strain (or translation) vector of the cross section:

$$\gamma_x = \frac{1}{GA} (t_x T_x + t_{xy} T_y) \tag{9.181a}$$

$$\gamma_y = \frac{1}{GA} (t_{xy} T_x + t_y T_y) \tag{9.181b}$$

The relations (9.181) hold good also for compact sections, in which case, however, the factors t_x, t_y, t_{xy} are difficult to determine.

9.13 THIN-WALLED CROSS SECTIONS SUBJECTED TO SHEAR

As has been mentioned in the foregoing section, the Jourawski formula furnishes the exact shearing stresses only in the case of thin-walled sections, with thickness b tending to zero. In this case, in fact, it is legitimate to equate the local stress with the mean stress on the chord. On the other hand, only the shearing stress component τ_{zs} (parallel to the midline) is present, by virtue of the condition of equivalence on the lateral surface of the beam.

In the sequel, we shall examine a number of typical cases, namely, the I-section, the box section and the C-section, both from the standpoint of stress and from that of deformation.

Consider the I-section of Figure 9.34a. This type of cross section is frequently used in metal constructions, because it allows the maximum moment of inertia to be achieved with the minimum use of material. A high moment of inertia implies, in fact, low axial stresses σ_z due to bending.

The static moment $S_x^{A'}$ may be represented graphically on the cross section, if we take the midline as our reference line (Figure 9.34b). The static moment varies linearly on the flanges and parabolically on the web. In the points where the flanges and the web converge, it undergoes a discontinuity, in such a way that the flux of the vector $\{\tau_{zs}\}$ is preserved. If we denote the points which immediately precede and follow the node B by B_1 and B_2, in absolute value, we have (Figure 9.34b)

$$S_x^{A'}(B_1) = b\left(\frac{h}{2}\right)\left(\frac{h}{2}\right) = \frac{1}{4}bh^2 \tag{9.182a}$$

$$S_x^{A'}(B_2) = 2S_x^{A'}(B_1) = \frac{1}{2}bh^2 \tag{9.182b}$$

The static moment in the centroid is then

$$S_x^{A'}(G) = \frac{1}{2}bh^2 + b\left(\frac{h}{2}\right)\left(\frac{h}{4}\right) = \frac{5}{8}bh^2 \tag{9.182c}$$

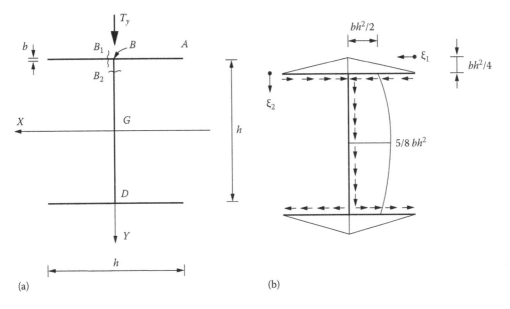

(a)   (b)

Figure 9.34

The maximum stress is in the centroid. Applying the Jourawski formula and using the moment of inertia

$$I_x = \frac{1}{12} b h^3 + 2(bh)\left(\frac{h}{2}\right)^2 = \frac{7}{12} b h^3 \tag{9.183a}$$

we find that the maximum stress is

$$\tau_{max} = \frac{T_y \left(\frac{5}{8} b h^2\right)}{\left(\frac{7}{12} b h^3\right) b} = \frac{15}{14} \frac{T_y}{bh} \tag{9.183b}$$

More particularly, considering the two coordinates ξ_1 and ξ_2 along the midline, we have (Figure 9.34b)

$$S_x^{A'}(\xi_1) = \frac{1}{2} b h \xi_1 \tag{9.184a}$$

$$S_x^{A'}(\xi_2) = \frac{1}{2} b h^2 + b\xi_2 \left(\frac{h}{2} - \frac{\xi_2}{2}\right) \tag{9.184b}$$

It is thus possible to verify that it is the web BD alone that withstands the shearing force. We have in fact

$$\int_B^D \tau_{zs} b \, ds = 2 \int_0^{h/2} \frac{T_y S_x^{A'}(\xi_2)}{I_x b} b \, d\xi_2 \tag{9.185}$$

Substituting expressions (9.183a) and (9.184b) into Equation 9.185, we obtain

$$\int_B^D \tau_{zs} b \, ds = \frac{2T_y}{\frac{7}{12} b h^3} \int_0^{h/2} \left(\frac{1}{2} b h^2 + \frac{1}{2} b h \xi_2 - \frac{1}{2} b \xi_2^2\right) d\xi_2 \tag{9.186}$$

and thus we verify that

$$\int_B^D \tau_{zs} b \, ds = T_y \tag{9.187}$$

As regards the shear deformation, the expression (9.175b) in the case of the cross section under examination becomes

$$t_y = \frac{3h}{\left(\frac{7}{12} b h^3\right)^2} \left\{ 4 \int_0^{h/2} \left[S_x^{A'}(\xi_1)\right]^2 d\xi_1 + 2 \int_0^{h/2} \left[S_x^{A'}(\xi_2)\right]^2 d\xi_2 \right\} \tag{9.188}$$

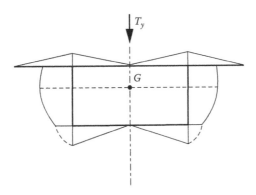

Figure 9.35

whence we obtain

$$t_y \simeq 3.38$$

Figure 9.35 gives qualitatively the diagram of shearing stresses for the box section already considered.

Finally, let us consider the C-section of Figure 9.36a. It is symmetrical only with respect to the central axis X, so that the centre of torsion will be found on that axis in a position that is unknown beforehand. In the case where there is a shear T_y passing through the centre of torsion (Figure 9.36a), the shearing stresses produced must constitute a system of forces equivalent to the only characteristic present, T_y (in addition to the bending moment M_x). Resolving then a simple problem of static equivalence, it is possible to identify the position of the centre of torsion in the case of thin-walled cross sections.

With the two reference systems ξ_1 and ξ_2 fixed on the midline, we have (Figure 9.36a)

$$S_x^{A'}(\xi_1) = bh\xi_1 \tag{9.189a}$$

$$S_x^{A'}(\xi_2) = 2bh^2 + b\xi_2\left(h - \frac{\xi_2}{2}\right) \tag{9.189b}$$

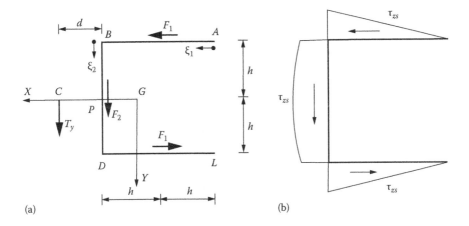

(a)   (b)

Figure 9.36

for which the application of the Jourawski formula gives the following shearing stresses:

$$\tau_{zs}(\xi_1) = \tau_{zx} = \frac{T_y}{I_x b} bh\xi_1 \tag{9.190a}$$

$$\tau_{zs}(\xi_2) = \tau_{zy} = \frac{T_y}{I_x b}\left(2bh^2 + bh\xi_2 - \frac{1}{2}b\xi_2^2\right) \tag{9.190b}$$

where $I_x = (14/3)\ bh^3$. The variation of the stress τ_{zs} is thus linear on the two horizontal plates and parabolic on the vertical plate (Figure 9.36b). The stresses τ_{zs} create a flux which runs from the upper end A to the lower one L. The magnitude of the resultant force of each horizontal distribution τ_{zs} is

$$F_1 = \int_0^{2h} \tau_{zs}(\xi_1)b\,\mathrm{d}\xi_1 = \frac{3}{14}\frac{T_y}{h^2}\left[\frac{\xi_1^2}{2}\right]_0^{2h} = \frac{3}{7}T_y \tag{9.191a}$$

while the magnitude of the vertical resultant is equal to the shear T_y:

$$F_2 = \int_0^{2h} \tau_{zs}(\xi_2)b\,\mathrm{d}\xi_2 = \frac{3}{14}\frac{T_y}{h^3}\left[2h^2\xi_2 + h\frac{\xi_2^2}{2} - \frac{1}{2}\frac{\xi_2^3}{3}\right]_0^{2h} = T_y \tag{9.191b}$$

The two horizontal resultants therefore create a counterclockwise couple $2hF_1$, and hence the shear T_y is equivalent to the system formed by this couple and by the force F_2, when it acts to the left of the vertical plate, at a distance d, such that

$$2hF_1 = dF_2 \tag{9.192}$$

from which we obtain the position of the centre of torsion C:

$$d = 2h\frac{F_1}{F_2} = \frac{6}{7}h \tag{9.193}$$

From the foregoing considerations, it may be deduced that, for thin-walled sections formed by thin rectangular elements converging to a single common point (Figure 9.37), the centre of torsion coincides with that point. In fact, the problem to be solved is a static one of

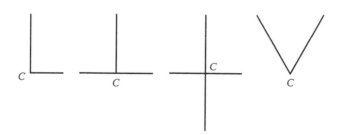

Figure 9.37

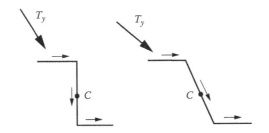

Figure 9.38

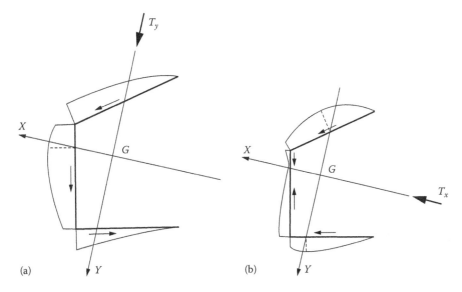

Figure 9.39

equivalence for two or more forces all passing through a single pole. Evidently, the resultant will also pass through the pole.

Also in the case of oblique polar symmetry (Figure 9.38), considerations similar to the preceding ones apply, which make it possible to identify the position of the centre of torsion in the centre of symmetry.

However, in the case of non-symmetrical sections (Figure 9.39), the task of locating the centre of torsion is far more complex and calls for the resolution of two static problems of equivalence, with respect to the shears T_x and T_y. The intersection of the lines of action of the two resultants furnishes the centre of torsion.

9.14 BEAM STRENGTH ANALYSIS

Combining all the elementary loadings acting on the Saint Venant solid, we obtain two orthogonal forces that are generally skew forces (Figure 9.40a): the axial force N, eccentric with respect to the centroid G, and the shearing force T, eccentric with respect to the centre of torsion C. The first combined load is equal to the sum of a centred axial force N and the two bending moments M_x, M_y, and generates only the axial component of stress σ_z. The second combined load is equal to the sum of a twisting moment M_z and the two shears T_x, T_y and generates only the shearing stress component τ_z. Therefore, in each point of any given

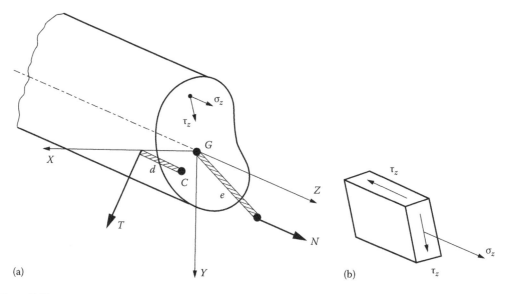

Figure 9.40

cross section, there will in general be present both the components σ_z and τ_z (Figure 9.40a), thus in every case giving rise to a plane state of stress (Figure 9.40b). It is evident that, in general, the plane of stresses varies from point to point in the cross section.

The stress tensor for the Saint Venant solid thus takes the following form:

$$[\sigma] = \begin{bmatrix} 0 & 0 & \tau_{xz} \\ 0 & 0 & \tau_{yz} \\ \tau_{zx} & \tau_{zy} & \sigma_z \end{bmatrix} \tag{9.194}$$

three of the six significant components always being zero. It should be noted that the case of the tensor (9.194) is complementary to that of the cylindrical solid of thickness t tending to zero, loaded exclusively on the lateral surface by forces contained in the middle plane. This solid, called the **Clebsch solid**, idealizes a plane plate of small thickness, loaded by forces contained in its own middle plane and thus not subject to bending.

The representation of the state of stress in a point is obtained graphically on Mohr's plane (Figure 9.41). It is sufficient to identify the two notable points $P(\sigma_z, \tau_z)$ and $P'(0, -\tau_z)$, where the stress $\tau_z > 0$ if, when applied to the plane normal to the axis Z, it tends to cause the element to rotate in a clockwise direction. The intersection of the diameter PP' with the axis σ_n gives the centre of Mohr's circle (Figure 9.41). The intersection of the horizontal through P with the vertical through P' defines the pole P^*, while the lines joining the pole P^* with the points of intersection of the circumference with the axis σ_n define the principal directions of stress. From the graphical construction, it may be noted how one of the two principal stresses is always negative and thus compressive, while the other is always positive and thus tensile.

The application of Tresca's criterion (criterion of maximum shearing stress) gives the following condition (Figure 9.41):

$$\tau_{max} = \overline{PC} = \sqrt{\left(\frac{\sigma_z}{2}\right)^2 + \tau_z^2} < \frac{1}{2}\sigma_P \tag{9.195}$$

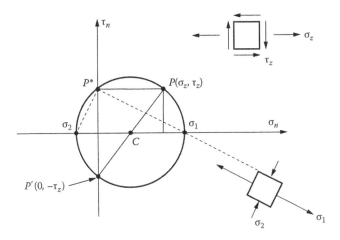

Figure 9.41

where σ_P is the uniaxial yielding stress. From Equation 9.195, we obtain then the following strength condition for the beams:

$$\sqrt{\sigma_z^2 + 4\tau_z^2} < \sigma_P \tag{9.196}$$

On the other hand, Von Mises' criterion (criterion of the maximum energy of distortion), *via* the general equation (8.114), gives

$$\sqrt{\sigma_z^2 + 3\tau_z^2} < \sigma_P \tag{9.197}$$

From a comparison between the inequalities (9.196) and (9.197), we verify again how Tresca's criterion is more conservative than that of Von Mises in that it presents the factor 4 as against the factor 3.

There follow a number of strength analyses for cross sections already introduced in Chapter 2.

Example 1

Let us reconsider the cross section of Figure 2.18, subjected to an eccentric axial force of compression (combined compression and bending). Since the eccentricities in the central reference system are (Figure 9.42)

$$e_x = 43.54 \text{ mm}, \quad e_y = 29.30 \text{ mm}$$

and the corresponding central radii of gyration

$$\rho_x^2 = \frac{I_x}{A} = 423.16 \text{ mm}^2$$

$$\rho_y^2 = \frac{I_y}{A} = 96.16 \text{ mm}^2$$

Equation 9.49 for the neutral axis is as follows:

$$1 + \frac{29.30}{423.16} y + \frac{43.54}{96.16} x = 0$$

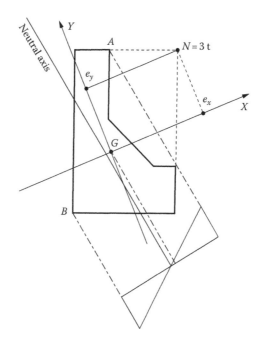

Figure 9.42

or, in segmentary form,

$$\frac{y}{14.44} + \frac{x}{2.21} = -1$$

Thus on the basis of Equation 9.48, point A, having coordinates $x_A = 15.84$ mm and $y_A = 41.08$ mm, is subjected to the compressive stress $\sigma_A = -17.88$ kg/mm², whilst point B, having coordinates $x_B = -25.00$ mm and $y_B = -17.32$ mm, is subjected to the tensile stress $\sigma_B = 18.67$ kg/mm². Both stresses are lower in absolute value than the yield stress of steel, $\sigma_P = 24$ kg/mm².

Example 2

Let us reconsider the cross section of Figure 2.19, subjected to a combination of compression and bending. Since

$$e_x = 0, \quad e_y = 35.76 \text{ mm}$$

$$\rho_x^2 = \frac{I_x}{A} = 267 \text{ mm}^2$$

Equation 9.50a of the neutral axis is

$$y = -7.5 \text{ mm}$$

The maximum stress in absolute value is thus

$$\sigma_{max} = \frac{5,000 \text{ kg}}{1,996 \text{ mm}^2} \times \frac{35.76 + 7.5}{7.5} = 14.42 \text{ kg/mm}^2$$

and proves to be less than $\sigma_P = 24$ kg/mm².

Example 3

Let us reconsider the cross section of Figure 2.20, subjected to axial force and twisting moment. The axial force produces a uniform normal stress

$$\sigma_z = \frac{N}{A} = \frac{1,500 \times 10^3 \, \text{kg}}{1,260 \, \text{cm}^2} = 12 \, \text{kg/mm}^2$$

while the twisting moment, if we neglect the contribution of the flanges, produces a uniform shearing stress

$$\tau_z = \frac{M_t}{2\Omega\delta} = \frac{40 \times 10^3 \, \text{kg} \times 10^2 \, \text{cm}}{12 \, \text{cm} \times 1082 \, \text{cm}^2} = 3 \, \text{kg/mm}^2$$

furnished by the Bredt formula (9.136).
   Applying Tresca's criterion (9.196), we obtain

$$\sqrt{\sigma_z^2 + 4\tau_z^2} = \sqrt{12^2 + 4 \times 3^2} = 13.4 \, \text{kg/mm}^2 < 24 \, \text{kg/mm}^2$$

The yield strength assessment thus gives a positive response.

Example 4

Let us consider again the cross section of Figure 2.21, subjected to a combined loading of shear–torsion. The twisting moment is $M_z = 400 \, \text{t} \times 0.4 \, \text{m} = 160 \, \text{tm}$ and produces a counterclockwise uniform shearing stress equal to

$$\tau_z(M_z) = \frac{160 \times 10^3 \, \text{kg} \times 10^2 \, \text{cm}}{10 \, \text{cm} \times (2,513 + 1,600) \, \text{cm}^2} = 3.89 \, \text{kg/mm}^2$$

On the other hand, for the calculation of stresses due to shear, it is necessary to evaluate the static moment of the part of cross section which remains above the central axis ξ.
   In Chapter 2, the static moment of the circular segment (4) with respect to the X axis has already been evaluated:

$$S_x^{(4)} = 22,600 \, \text{cm}^3$$

so that

$$S_\xi^{(4)} = S_x^{(4)} - A^{(4)} y_G = 22,600 - 444 \times 24.61 = 11,673 \, \text{cm}^3$$

The two rectilinear segments above the axis ξ present the following static moment:

$$S_\xi^{(2')} = S_x^{(3')} = \frac{1}{2}(40 - 24.61)^2 \times 5 = 592 \, \text{cm}^3.$$

Altogether, we have

$$S_\xi^{\text{max}} = (11,673 + 2 \times 592) \, \text{cm}^3 = 12,857 \, \text{cm}^3.$$

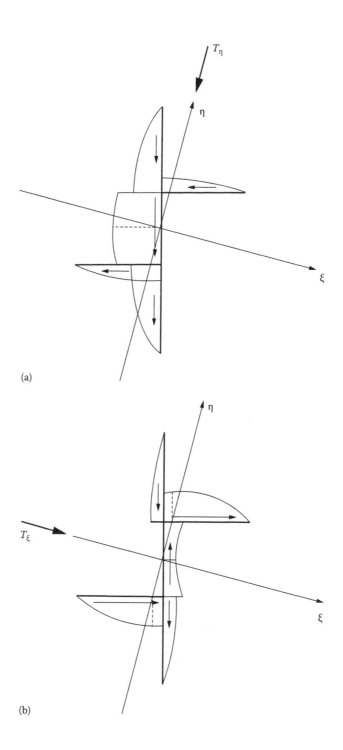

(a)

(b)

Figure 9.43

The maximum shearing stress due to shear is therefore

$$\tau_z(T) = \frac{TS_\xi^{max}}{2I_\xi \delta} = \frac{400 \times 10^3 \, kg \times 12{,}857 \, cm^3}{624{,}000 \, cm^4 \times 10 \, cm} = 8.24 \, kg/mm^2$$

The maximum shearing stress is found on the left-hand vertical segment, in correspondence with the global centroid

$$\tau_{max} = \tau_z(M_z) + \tau_z(T) = 12.13 \, kg/mm^2$$

Applying Tresca's criterion (9.196), we find

$$\sqrt{4\tau_{max}^2} = \sqrt{4 \times 12.13^2} = 24.26 \, kg/mm^2 > \sigma_P$$

whereas Von Mises' criterion (9.197) yields a positive result:

$$\sqrt{3\tau_{max}^2} = \sqrt{3 \times 12.13^2} = 21 \, kg/mm^2 < \sigma_P$$

Example 5

Let us consider again the cross section of Figure 2.22, subjected to biaxial shear. The corresponding components are equal to

$$T_\eta \simeq 193 \, t, \quad T_\xi \simeq 52.4 \, t$$

The maximum static moment with respect to the ξ axis is (Figure 9.43a)

$$S_\xi^{max} = (160 \times 20 \times 0.96 + 120 \times 13.58) \, cm^3 = 4701 \, cm^3$$

so that the maximum shearing stress due to T_η develops in the centre and is equal to

$$\tau_z(T_\eta) = \frac{T_\eta S_\xi^{max}}{I_\xi \delta} = \frac{193 \times 10^3 \, kg \times 4{,}701 \, cm^3}{204{,}000 \, cm^4 \times 4 \, cm} = 11.12 \, kg/mm^2$$

The shearing stress in the central point due to T_ξ, on the other hand, is in the opposite sense (Figure 9.43b) and tends to mitigate the previous contribution. The assessment is thus positive.

Chapter 10

Beams and plates in flexure

10.1 INTRODUCTION

In this chapter, deflected beams and plates are studied. It is shown that, for both elements, it is possible to define the deformation characteristics by derivation of the components of the generalized displacement vector, which, in addition to displacements in the strict sense, also presents rotations. Analogous with the 3D solid encountered in Chapter 8, it may be noted that, also for the 1D or 2D solid, the static operator is, but for the algebraic sign, the transpose, or rather the adjoint, of the kinematic operator. This property of **duality** proves of great utility in the case of discretization into finite elements.

The cases of curved plane beams and shells having double curvature present notable analogies with the cases of rectilinear plane beams and plates, respectively, once the rotation matrix, which converts the external reference system to the local reference system, is included in the analysis. A further noteworthy analogy is found between the differential equation of the elastic line for rectilinear beams and the equation of the elastic plane for plates. Both equations, in fact, neglect the shearing deformability and turn out to be of the fourth order in the single unknown function consisting of deflection or transverse displacement.

Two particular cases of application of the equation of the elastic line are represented by the beam on an elastic foundation and by the beam subjected to steady free oscillations.

10.2 TECHNICAL THEORY OF BEAMS

The results obtained for the Saint Venant solid on the basis of restrictive hypotheses, as regards both the geometry of the solid (rectilinear axis, constant cross section) and external loads (lateral surface not loaded, zero body forces), are usually extended in technical applications to cases in which these hypotheses are not satisfied. This means that the stresses and strains in the cross sections of beams are usually calculated using formulas obtained in the foregoing chapter, by introducing, instead of the loads acting on the end planes of the Saint Venant solid, the internal beam reactions acting in the cross sections considered. This extrapolation is also made in the case of the following:

1. Beams having non-rectilinear axes
2. Beams of variable cross section
3. Beams loaded on the lateral surface

It is required, in any case, that the radius of curvature of the geometrical axis of the beam should be much greater than the characteristic dimensions of the cross section and that the cross section should, at most, be only slightly variable.

10.3 BEAMS WITH RECTILINEAR AXES

Let us consider an elementary portion of a beam with rectilinear axis and a cross section that is symmetrical with respect to the Y axis. Let this portion be subjected to bending moment M_x and to shear T_y. Deformations due to these two characteristics will produce relative displacements between the centroids of the two extreme cross sections of the beam portion, exclusively in the direction of the Y axis. In the case of the shear T_y, we have (Figure 10.1a)

$$dv^T = \gamma_y \, dz \tag{10.1a}$$

where
  dv^T is the relative displacement in the Y direction due to the shear
  γ_y is the shearing strain dual of the shearing force
  dz is the length of the infinitesimal element of beam

In the case of bending moment, and considering the rigid rotation φ_x of the element, we have (Figure 10.1b)

$$dv^M = -\varphi_x \, dz \tag{10.1b}$$

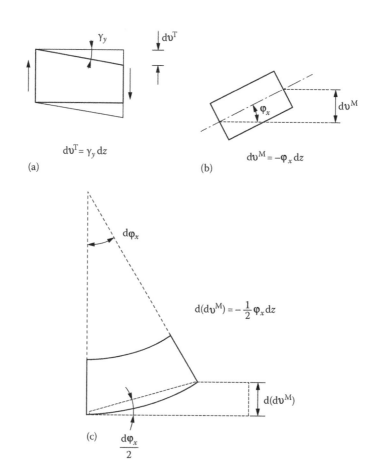

$dv^T = \gamma_y \, dz$

(a)

$dv^M = -\varphi_x \, dz$

(b)

$$d(dv^M) = -\frac{1}{2}\varphi_x \, dz$$

(c) $\dfrac{d\varphi_x}{2}$

Figure 10.1

having neglected the infinitesimals of a higher order due to the curvature, i.e. to the slope variation $d\varphi_x$ (Figure 10.1c),

$$d(dv^M) = -\frac{1}{2} d\varphi_x \, dz \qquad (10.1c)$$

Summing up the two significant contributions of the shear, Equation 10.1a, and of the bending moment, Equation 10.1b, we obtain

$$dv = dv^T + dv^M = \gamma_y \, dz - \varphi_x \, dz \qquad (10.2)$$

from which

$$\frac{dv}{dz} = \gamma_y - \varphi_x \qquad (10.3)$$

At this point, we are able to formulate the fundamental equations of the elastic problem for 1D solids with a rectilinear axis. The kinematic equations constitute, as in the case of the 3D solid, the definition of the characteristics of deformation as functions of the generalized displacements

$$\begin{bmatrix} \gamma_x \\ \gamma_y \\ \varepsilon_z \\ \chi_x \\ \chi_y \\ \Theta \end{bmatrix} = \begin{bmatrix} \dfrac{d}{dz} & 0 & 0 & 0 & -1 & 0 \\ 0 & \dfrac{d}{dz} & 0 & +1 & 0 & 0 \\ 0 & 0 & \dfrac{d}{dz} & 0 & 0 & 0 \\ 0 & 0 & 0 & \dfrac{d}{dz} & 0 & 0 \\ 0 & 0 & 0 & 0 & \dfrac{d}{dz} & 0 \\ 0 & 0 & 0 & 0 & 0 & \dfrac{d}{dz} \end{bmatrix} \begin{bmatrix} u \\ v \\ w \\ \varphi_x \\ \varphi_y \\ \varphi_z \end{bmatrix} \qquad (10.4)$$

where among the components of the deformation vector appear the shearing strains, γ_x, γ_y, the axial dilation ε_z, the curvatures, χ_x, χ_y, and the unit angle of torsion, Θ, whilst among the components of the displacement vector appear, in addition to the ordinary components, u, v, w, also the generalized components, φ_x, φ_y, φ_z, i.e. the angles of rotation about the reference axes. The transformation matrix is differential and shows on the diagonal the total derivative d/dz, while the off-diagonal terms are all zero except for two, which have absolute values of unity and derive from relation (10.3) and from its analogue

$$\frac{du}{dz} = \gamma_x + \varphi_y \qquad (10.5)$$

Also relation (10.4), like relation (8.8), may be written in compact form,

$$\{q\} = [\partial]\{\eta\} \tag{10.6}$$

where
    $\{q\}$ indicates the vector of the deformation characteristics
    $\{\eta\}$ the vector of generalized displacements
    $[\partial]$ the matrix operator of relation (10.4)

On the other hand, also the indefinite equations of equilibrium, if relations (5.12) are taken into account, can be presented in matrix form,

$$
\begin{bmatrix}
\dfrac{d}{dz} & 0 & 0 & 0 & 0 & 0 \\[2mm]
0 & \dfrac{d}{dz} & 0 & 0 & 0 & 0 \\[2mm]
0 & 0 & \dfrac{d}{dz} & 0 & 0 & 0 \\[2mm]
0 & -1 & 0 & \dfrac{d}{dz} & 0 & 0 \\[2mm]
+1 & 0 & 0 & 0 & \dfrac{d}{dz} & 0 \\[2mm]
0 & 0 & 0 & 0 & 0 & \dfrac{d}{dz}
\end{bmatrix}
\begin{bmatrix} T_x \\ T_y \\ N \\ M_x \\ M_y \\ M_z \end{bmatrix}
+
\begin{bmatrix} q_x \\ q_y \\ p \\ m_x \\ m_y \\ m_z \end{bmatrix}
=
\begin{bmatrix} 0 \\ 0 \\ 0 \\ 0 \\ 0 \\ 0 \end{bmatrix}
\tag{10.7}
$$

where among the components of the vector of static characteristics appear the shearing forces, T_x, T_y, the axial force N, the bending moments, M_x, M_y, and the twisting moment, M_z, whilst among the components of the vector of external generalized forces appear, in addition to the transverse distributed loads, q_x, q_y, and the axial distributed load, p, also the bending distributed moments, m_x, m_y and the twisting distributed moment, m_z. The matrix operator presents the total derivative d/dz in all diagonal positions, while the off-diagonal terms are all zero, except for two which are equal to unity and express the equality of the shear with the derivative of the corresponding bending moment (neglecting the distributed moments, m_x, m_y). The matrix operator (10.7) is equal to the transpose of the operator (10.4) but for the finite terms which change algebraic sign. This is said to be the **adjoint operator** of the previous one, and *vice versa*. In compact form, we can write

$$[\partial]^*\{Q\} + \{\mathscr{F}\} = \{0\} \tag{10.8}$$

where
    $\{Q\}$ is the vector of static characteristics
    $\{\mathscr{F}\}$ is the vector of external forces

It should be noted that, unlike in the case of the 3D solid, the matrices $[\partial]$ and $[\partial]^*$ are square (6×6). This means, as we already know, that it is possible to determine the internal reactions of a statically determinate beam by using only the relations of equilibrium. This, unfortunately, does not occur in the case of a 3D body constrained isostatically, the stress field of which may be determined only by applying, in addition to the static equations, also the constitutive and kinematic equations.

On the other hand, also in the case of beams constrained in a redundant manner (i.e. statically indeterminate beams), it is necessary to have recourse to the constitutive equations to define their static characteristics, in addition, of course, to the deformations and the displacements. The relations which link the static characteristics with the dual characteristics of deformation may be presented in matrix form:

$$
\begin{bmatrix} \gamma_x \\ \gamma_y \\ \varepsilon_z \\ \chi_x \\ \chi_y \\ \Theta \end{bmatrix} =
\begin{bmatrix}
\dfrac{t_x}{GA} & \dfrac{t_{xy}}{GA} & 0 & 0 & 0 & 0 \\[2mm]
\dfrac{t_{xy}}{GA} & \dfrac{t_y}{GA} & 0 & 0 & 0 & 0 \\[2mm]
0 & 0 & \dfrac{1}{EA} & 0 & 0 & 0 \\[2mm]
0 & 0 & 0 & \dfrac{1}{EI_x} & 0 & 0 \\[2mm]
0 & 0 & 0 & 0 & \dfrac{1}{EI_y} & 0 \\[2mm]
0 & 0 & 0 & 0 & 0 & \dfrac{1}{GI_t}
\end{bmatrix}
\begin{bmatrix} T_x \\ T_y \\ N \\ M_x \\ M_y \\ M_z \end{bmatrix}
\tag{10.9}
$$

Expressed in compact form, they are

$$
\{q\} = [H]^{-1}\{Q\}
\tag{10.10a}
$$

where $[H]^{-1}$ represents the inverse of the Hessian matrix of the elastic potential of the beam. On the other hand, the inverse relation of Equation 10.10a also holds:

$$
\{Q\} = [H]\{q\}
\tag{10.10b}
$$

Applying Clapeyron's theorem, we obtain the work of deformation per unit length of the beam, i.e. the elastic potential of the beam,

$$
\frac{dL}{dz} = \frac{1}{2}\{Q\}^T\{q\}
\tag{10.11}
$$

which, on the basis of Equation 10.10a, becomes

$$
\frac{dL}{dz} = \frac{1}{2}\{Q\}^T[H]^{-1}\{Q\}
\tag{10.12a}
$$

or, on the basis of Equation 10.10b

$$
\frac{dL}{dz} = \frac{1}{2}\{q\}^T[H]\{q\}
\tag{10.12b}
$$

where we have used the relation of symmetry $[H]^T = [H]$. Rendering relation (10.12a) explicit, we obtain

$$\frac{dL}{dz} = \frac{1}{2}\left(t_x \frac{T_x^2}{GA} + t_y \frac{T_y^2}{GA} + 2t_{xy} \frac{T_x T_y}{GA} + \frac{N^2}{EA} + \frac{M_x^2}{EI_x} + \frac{M_y^2}{EI_y} + \frac{M_z^2}{GI_t} \right) \tag{10.13}$$

which is a quadratic form of the static characteristics. In the case where the cross section of the beam presents at least one axis of symmetry, the mutual factor of shear, t_{xy}, vanishes and the total work is equal to the sum of the contributions of the single characteristics, the principle of superposition being applicable in this case.

Having now at our disposal the equations of kinematics and statics and the constitutive equations for the beam, we can obtain **Lamé's equation in operator form**

$$([\partial]^*[H][\partial])\{\eta\} = -\{\mathscr{F}\} \tag{10.14}$$

The matrix and differential operator of the second order in round brackets can be called the **Lamé operator:**

$$\underset{(6\times 6)}{[\mathscr{L}]} = \underset{(6\times 6)}{[\partial]^*} \underset{(6\times 6)}{[H]} \underset{(6\times 6)}{[\partial]} \tag{10.15}$$

It turns out to be a (6×6) matrix and, in non-homogeneous problems, in which the matrix $[H]$ is a function of the axial coordinate z, is also a function of z.

Finally, the boundary conditions may be conditions of equivalence at the ends,

$$[\mathscr{N}]^T\{Q\} = \{Q_0\} \tag{10.16}$$

where the matrix $[\mathscr{N}]^T$ coincides with the identity matrix $[1]$ and causes a value of unity to correspond to each differential term of the matrix $[\partial]^*$. The boundary conditions can, on the other hand, also represent displacements imposed at the ends:

$$\{\eta\} = \{\eta_0\} \tag{10.17}$$

In conclusion, the **elastic problem of the rectilinear beam** can be summarized as follows:

$$[\mathscr{L}]\{\eta\} = -\{\mathscr{F}\}, \qquad \text{for } 0 < z < l \tag{10.18a}$$

$$([H][\partial])\{\eta\} = \{Q_0\}, \quad \text{for } z = 0, l \tag{10.18b}$$

$$\{\eta\} = \{\eta_0\}, \qquad \text{for } z = 0, l \tag{10.18c}$$

where the static boundary condition (10.18b) holds good in the end point or points that are subjected to loading, as does the kinematic boundary condition (10.18c) in the constrained end point or points. In the case where there are no conditions on the displacements, the loads $\{Q_0\}$ must constitute a self-balancing system.

The previously developed formulation is notably simplified in the case where the beam is **loaded in the plane.** If we assume a cross section symmetrical with respect to the Y axis, the characteristics of deformation reduce to the shearing strain γ_y, the axial dilation ε_z and the curvature χ_x, so that of the relations (10.4), (10.7) and (10.9), only those corresponding to the second, third and fourth rows and columns remain significant. In particular, the **kinematic equations** simplify as follows:

$$
\begin{bmatrix} \gamma_y \\ \varepsilon_z \\ \chi_x \end{bmatrix} = \begin{bmatrix} \dfrac{d}{dz} & 0 & +1 \\[2mm] 0 & \dfrac{d}{dz} & 0 \\[2mm] 0 & 0 & \dfrac{d}{dz} \end{bmatrix} \begin{bmatrix} \upsilon \\ w \\ \varphi_x \end{bmatrix}
\tag{10.19}
$$

The **static equations** likewise simplify thus:

$$
\begin{bmatrix} \dfrac{d}{dz} & 0 & 0 \\[2mm] 0 & \dfrac{d}{dz} & 0 \\[2mm] -1 & 0 & \dfrac{d}{dz} \end{bmatrix} \begin{bmatrix} T_y \\ N \\ M_x \end{bmatrix} + \begin{bmatrix} q \\ p \\ m \end{bmatrix} = \begin{bmatrix} 0 \\ 0 \\ 0 \end{bmatrix}
\tag{10.20}
$$

The **constitutive equations** reduce then to a diagonal relation

$$
\begin{bmatrix} \gamma_y \\ \varepsilon_z \\ \chi_x \end{bmatrix} = \begin{bmatrix} \dfrac{t_y}{GA} & 0 & 0 \\[2mm] 0 & \dfrac{1}{EA} & 0 \\[2mm] 0 & 0 & \dfrac{1}{EI_x} \end{bmatrix} \begin{bmatrix} T_y \\ N \\ M_x \end{bmatrix}
\tag{10.21}
$$

10.4 PLANE BEAMS WITH CURVILINEAR AXES

Let us consider once more the element of beam with curvilinear axis of Figure 5.3. The curvilinear coordinate s is considered as increasing as we proceed from left to right along the beam, while the angle $d\vartheta$ is considered to be positive if it is counterclockwise. In accordance with the aforementioned conventions, also the radius of curvature r acquires an algebraic sign on the basis of the relation

$$
ds = r d\vartheta
\tag{10.22}
$$

As regards the generalized displacements of the generic cross section, the radial displacement υ is positive if it is in the positive direction of the Y^* axis (where Y^*Z^* is a system of right-handed axes travelling along the axis of the beam), the axial displacement w is positive if it is in the positive direction of the curvilinear coordinate s and finally the **variation of the angle** φ is positive if it is counterclockwise (Figure 10.2a).

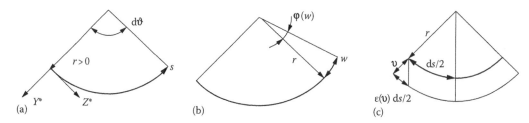

Figure 10.2

We shall now show how the kinematic equation (10.19) must be modified to take into account the **intrinsic curvature** of the beam. The axial displacement w produces, in fact, a slope variation $\varphi(w)$, which, in accordance with the scheme of Figure 10.2b and neglecting infinitesimals of a higher order, equals

$$\varphi(w) = \frac{w}{r} \tag{10.23}$$

On the other hand, the radial displacement υ produces an axial dilation $\varepsilon(\upsilon)$ which, in accordance with the scheme of Figure 10.2c and neglecting infinitesimals of a higher order, is given by

$$\varepsilon(\upsilon) = \frac{\upsilon}{r} \tag{10.24}$$

As a consequence of an infinitesimal relative rotation $d\varphi$ of the extreme cross sections of the beam element of Figure 5.3, the angle between the sections can be obtained as the sum $(d\vartheta + d\varphi)$ of the initial and intrinsic relative rotation and the elastic and flexural relative rotation. The new curvature is then

$$\chi_{\text{total}} = \frac{(d\vartheta + d\varphi)}{ds} \tag{10.25}$$

so that the **variation of curvature** is

$$\chi = \chi_{\text{total}} - \frac{1}{r} = \frac{d\varphi}{ds} \tag{10.26}$$

Substituting the relations (10.23), (10.24) and (10.26), which furnish respectively the slope variation to be deducted, the additional axial dilation and the variation of curvature, the **kinematic equations** (10.19) transform as follows:

$$\begin{bmatrix} \gamma \\ \varepsilon \\ \chi \end{bmatrix} = \begin{bmatrix} \dfrac{d}{ds} & -\dfrac{1}{r} & +1 \\ \dfrac{1}{r} & \dfrac{d}{ds} & 0 \\ 0 & 0 & \dfrac{d}{ds} \end{bmatrix} \begin{bmatrix} \upsilon \\ w \\ \varphi \end{bmatrix} \tag{10.27}$$

The **indefinite equations of equilibrium**, or **static equations**, have previously been derived in Section 5.2. Equations 5.4 may be reproposed in matrix form:

$$\begin{bmatrix} \dfrac{d}{ds} & -\dfrac{1}{r} & 0 \\[2mm] \dfrac{1}{r} & \dfrac{d}{ds} & 0 \\[2mm] -1 & 0 & \dfrac{d}{ds} \end{bmatrix} \begin{bmatrix} T \\ N \\ M \end{bmatrix} + \begin{bmatrix} q \\ p \\ m \end{bmatrix} = \begin{bmatrix} 0 \\ 0 \\ 0 \end{bmatrix} \tag{10.28}$$

It should be noted that, but for the algebraic signs of the non-differential terms, the static matrix is the transpose of the kinematic one.

Finally, as regards the **constitutive equations**, if the radius of curvature is much greater than the characteristic dimensions of the cross section, Equation 10.21 can be used to very good approximation.

The rotation matrix which transforms the global reference system YZ into the local reference system Y^*Z^* is the following:

$$[N] = \begin{bmatrix} \cos\vartheta & \sin\vartheta & 0 \\ -\sin\vartheta & \cos\vartheta & 0 \\ 0 & 0 & 1 \end{bmatrix} \tag{10.29}$$

so that the vectors of the external forces and of the generalized displacements in the local reference system may be expressed by premultiplying the respective vectors evaluated in the global reference system by the matrix $[N]$:

$$\{\mathscr{F}^*\} = [N]\{\mathscr{F}\} \tag{10.30a}$$

$$\{\eta^*\} = [N]\{\eta\} \tag{10.30b}$$

The static and kinematic equations can thus be expressed as follows:

$$[\partial]^*\{Q\} + \{\mathscr{F}^*\} = \{0\} \tag{10.31a}$$

$$\{q\} = [\partial]\{\eta^*\} \tag{10.31b}$$

which, on the basis of Equations 10.30, become

$$[\partial]^*\{Q\} + [N]\{\mathscr{F}\} = \{0\} \tag{10.32a}$$

$$\{q\} = [\partial][N]\{\eta\} \tag{10.32b}$$

Substituting Equations 10.10b and 10.32b in Equation 10.32a, we have

$$[\partial]^*[H][\partial][N]\{\eta\} + [N]\{\mathscr{F}\} = \{0\} \tag{10.33}$$

Premultiplying both sides of Equation 10.33 by $[N]^T$, we obtain finally

$$([N]^T[\partial]^*[H][\partial][N])\{\eta\} = -\{\mathscr{F}\} \tag{10.34}$$

which is Lamé's equation for curved beams and arches.

The **elastic problem for curved beams and arches** can then be summarized as follows:

$$[\mathscr{L}]\{\eta\} = -\{\mathscr{F}\}, \qquad \text{for } 0 < s < l \tag{10.35a}$$

$$([N]^T[H][\partial][N])\{\eta\} = \{Q_0\}, \quad \text{for } s = 0, l \tag{10.35b}$$

$$\{\eta\} = \{\eta_0\}, \qquad \text{for } s = 0, l \tag{10.35c}$$

10.5 DIFFERENTIAL EQUATION OF THE ELASTIC LINE

As regards beams with rectilinear axes and cross sections that are symmetrical with respect to the Y axis, loaded in the plane of symmetry YZ, we shall arrive at a differential equation in the unknown function $\upsilon(z)$, called **deflection** or **transverse displacement**, by neglecting the contributions of deformation due to shear, which, for sufficiently slender beams, are much less than the contributions of deformation due to bending moment.

Let us consider a cantilever beam of length l, loaded at the free end by a concentrated force F (Figure 10.3a). The increment of vertical displacement due to shear is (Figure 10.3b)

$$d\upsilon^T = \gamma_y\, dz \tag{10.36}$$

so that the vertical displacement at the free end due to shear is

$$\upsilon^T(l) = \gamma_y l \tag{10.37}$$

On the other hand, the increment of vertical displacement due to bending moment is (Figure 10.3c)

$$d\upsilon^M = -\varphi_x\, dz \tag{10.38}$$

Differentiating both sides of Equation 10.38 with respect to z, we obtain

$$\frac{d^2\upsilon^M}{dz^2} = -\frac{d\varphi_x}{dz} \tag{10.39}$$

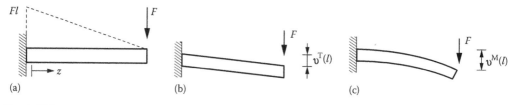

Figure 10.3

Recalling the expressions of elastic curvature (9.40) and (9.41), we obtain a second-order differential equation:

$$\frac{d^2 v^M}{dz^2} = -\frac{M_x}{EI_x} \qquad (10.40)$$

The bending moment M_x, as well as the elastic characteristic E and the geometrical characteristic I_x, can be functions of the z coordinate. Usually, however, cases are considered where the material and the cross section do not vary along the axis of the beam. The bending moment, in the case of the cantilever beam of Figure 10.3a, instead varies linearly,

$$M_x = F(z - l) \qquad (10.41)$$

so that the differential equation (10.40) becomes

$$\frac{d^2 v^M}{dz^2} = \frac{F}{EI_x}(l - z) \qquad (10.42)$$

and produces the complete integral

$$v^M(z) = -\frac{F}{6EI_x}z^3 + \frac{Fl}{2EI_x}z^2 + C_1 z + C_2 \qquad (10.43)$$

The two constants C_1 and C_2 are obtained by applying the boundary conditions. The built-in constraint does not allow any translations or rotations of the fixed-end section of the beam, so that

$$v^M(0) = C_2 = 0 \qquad (10.44a)$$

$$\frac{dv^M}{dz}(0) = C_1 = 0 \qquad (10.44b)$$

and the vertical displacement at the free end is therefore (Figure 10.3c)

$$v^M(l) = \frac{Fl^3}{3EI_x} \qquad (10.45)$$

From relations (10.37) and (10.45), it is already evident that the vertical displacement due to shear is a linear function of the length l of the beam, while the vertical displacement due to bending moment is proportional to the third power of l. More particularly, it is possible to sum up the two contributions and write

$$v(l) = \frac{Fl^3}{3EI_x}\left[1 + 6(1 + v)t_y\left(\frac{\rho_x}{l}\right)^2\right] \qquad (10.46)$$

where ρ_x indicates the radius of gyration of the cross section with respect to the X axis. It is hence possible to understand how, since we normally have $\rho_x \ll l$, the contribution of shear is negligible.

The **differential equation of the elastic line** thus coincides substantially with Equation 10.40:

$$\frac{d^2 v}{dz^2} = -\frac{M_x}{EI_x} \tag{10.47}$$

If the bending moment vanishes in a cross section within the beam, also the curvature vanishes and, in accordance with Equation 10.47, the second derivative of the vertical displacement evanishes. In correspondence with that cross section, the elastic deformed configuration (or elastic line) of the axis of the beam will thus show a point of inflection.

Differentiating both sides of Equation 10.47 with respect to z, we obtain

$$\frac{d^3 v}{dz^3} = -\frac{T_y}{EI_x} \tag{10.48}$$

which is a third-order differential equation that has as its known term the shear function. Differentiating again, we finally obtain

$$\frac{d^4 v}{dz^4} = \frac{q}{EI_x} \tag{10.49}$$

which is the alternative version of the equation of the elastic line, of the fourth order, with the known term proportional to the transverse distributed load $q(z)$. This version requires a more laborious integration, with the identification of four arbitrary constants, on the basis of the static and kinematic boundary conditions. To compensate for this, on the other hand, it is not necessary to determine beforehand the bending moment function $M_x(z)$.

From Equation 10.49, it may be deduced how a discontinuity in the transverse distributed load function $q(z)$ causes a discontinuity on the fourth derivative of the transverse

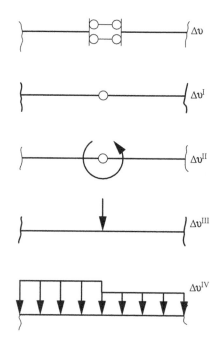

Figure 10.4

displacement $\upsilon(z)$, and thus a relatively minor discontinuity. In like manner and considering Equation 10.48, we may deduce how a discontinuity in the shear function $T_y(z)$, and thus a concentrated transverse load, causes a discontinuity on the third derivative of the transverse displacement $\upsilon(z)$. From Equation 10.47, we can deduce how a discontinuity in the bending moment function $M_x(z)$, and thus a concentrated moment, causes a discontinuity on the second derivative of the transverse displacement $\upsilon(z)$. To have, instead, the discontinuities more pronounced in the transverse displacement $\upsilon(z)$, i.e. the discontinuity on the first derivative, or relative rotation, and the discontinuity on the function $\upsilon(z)$ itself, or relative sliding, the appropriate disconnections are necessary, represented in one case by the hinge and in the other by the double rod parallel to the axis of the beam. Figure 10.4 provides a synthesis of the various cases presented previously.

10.6 NOTABLE DISPLACEMENTS AND ROTATIONS IN ELEMENTARY SCHEMES

Here we shall discuss displacements and rotations of notable cross sections in elementary structural schemes. To this end, we shall proceed to integrate the differential equation of the elastic line introduced previously, considering the specific boundary conditions of constraint at the same time.

As a first case, let us examine the cantilever beam AB, built-in at A and subjected to a concentrated moment at point B (Figure 10.5a). Since the bending moment is constant along the axis of the cantilever and is also negative, Equation 10.47 takes the following form:

$$\frac{d^2\upsilon}{dz^2} = \frac{m}{EI} \tag{10.50}$$

the flexural rigidity EI also being considered constant. The complete integral of Equation 10.50 is thus

$$\upsilon(z) = \frac{m}{EI}\frac{z^2}{2} + C_1 z + C_2 \tag{10.51}$$

the two constants C_1 and C_2 being obtainable by means of the two boundary conditions

$$\upsilon(0) = \upsilon'(0) = 0 \tag{10.52}$$

which express the vanishing of the vertical displacement and of the rotation at the built-in constraint, respectively. We obtain

$$C_1 = C_2 = 0 \tag{10.53}$$

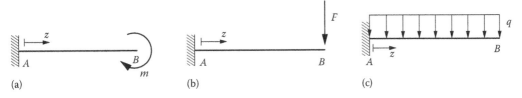

(a)   (b)   (c)

Figure 10.5

and thus the vertical displacement at the end B is

$$v_B = v(l) = \frac{ml^2}{2EI} \tag{10.54a}$$

while the rotation at the same end is

$$\varphi_B = -v'(l) = -\frac{ml}{EI} \tag{10.54b}$$

A slightly more complex case is that of the same cantilever beam, loaded by a force perpendicular to the axis at the end B (Figure 10.5b). In this case, the moment is a function of the axial coordinate z, and therefore the differential equation of the elastic line (10.47) is particularized as follows:

$$\frac{d^2v}{dz^2} = \frac{F(l-z)}{EI} \tag{10.55}$$

A first integration gives

$$v'(z) = -\frac{Fz^2}{2EI} + \frac{Fl}{EI}z + C_1 \tag{10.56}$$

and a second one

$$v(z) = -\frac{Fz^3}{6EI} + \frac{Fl}{2EI}z^2 + C_1z + C_2 \tag{10.57}$$

As the boundary conditions of Equation 10.52 still hold, we obtain finally

$$v_B = \frac{Fl^3}{3EI} \tag{10.58a}$$

$$\varphi_B = -\frac{Fl^2}{2EI} \tag{10.58b}$$

As our last elementary case for the cantilever beam, let us take a uniform distributed load q (Figure 10.5c). In this case, the differential equation is

$$\frac{d^2v}{dz^2} = \frac{q(l-z)^2}{2EI} \tag{10.59}$$

which, once integrated, yields

$$v'(z) = \frac{q}{2EI}\left(\frac{z^3}{3} - lz^2 + l^2z + C_1 \right) \tag{10.60}$$

$$v(z) = \frac{q}{2EI}\left(\frac{z^4}{12} - l\frac{z^3}{3} + l^2\frac{z^2}{2} + C_1z + C_2 \right) \tag{10.61}$$

The application of the conditions of Equations 10.52 confirms Equations 10.53, and thus

$$v_B = \frac{ql^4}{8EI} \tag{10.62a}$$

$$\varphi_B = -\frac{ql^3}{6EI} \tag{10.62b}$$

Let us now look at the case of a beam supported at both ends subjected to a concentrated moment at one of the two ends (Figure 10.6a). The equation of the elastic line is

$$\frac{d^2 v}{dz^2} = -\frac{mz}{EIl} \tag{10.63}$$

with the boundary conditions

$$v(0) = v(l) = 0 \tag{10.64}$$

Integrating Equation 10.63, we obtain

$$v'(z) = \frac{1}{EI}\left(-\frac{m}{2l}z^2 + C_1\right) \tag{10.65a}$$

$$v(z) = \frac{1}{EI}\left(-\frac{m}{6l}z^3 + C_1 z + C_2\right) \tag{10.65b}$$

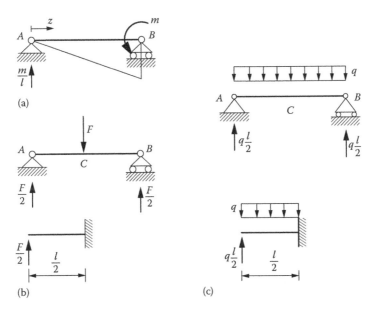

Figure 10.6

from which, applying Equations 10.64, there follows

$$C_1 = \frac{ml}{6} \tag{10.66a}$$

$$C_2 = 0 \tag{10.66b}$$

The rotation function (10.65a) thus takes on the following final aspect:

$$\upsilon'(z) = \frac{1}{EI}\left(-\frac{m}{2l}z^2 + \frac{ml}{6}\right) \tag{10.67}$$

whence it is possible to determine the rotations of the end sections:

$$\varphi_A = -\upsilon'(0) = -\frac{ml}{6EI} \tag{10.68a}$$

$$\varphi_B = -\upsilon'(l) = \frac{ml}{3EI} \tag{10.68b}$$

When the supported beam is loaded symmetrically, it is possible to consider only one half of it, with a built-in constraint which represents kinematically the condition of symmetry and the reactive force in place of the support. This is the case, for instance, of the vertical force in the centre (Figure 10.6b). In this way, we find ourselves once again faced with the case of the cantilever beam of Figure 10.5b, for which Equations 10.58 provide, respectively, the vertical displacement in the centre and the rotations at the ends:

$$\upsilon_C = \frac{F/2(l/2)^3}{3EI} = \frac{Fl^3}{48EI} \tag{10.69a}$$

$$\varphi_B = -\varphi_A = \frac{F/2(l/2)^2}{2EI} = \frac{Fl^2}{16EI} \tag{10.69b}$$

In like manner, in the case of uniform distributed load (Figure 10.6c), using Equations 10.58 and 10.62, we obtain

$$\upsilon_C = \frac{q(l/2)(l/2)^3}{3EI} - \frac{q(l/2)^4}{8EI} = \frac{5}{384}\frac{ql^4}{EI} \tag{10.70a}$$

$$\varphi_B = -\varphi_A = \frac{q(l/2)(l/2)^2}{2EI} - \frac{q(l/2)^3}{6EI} = \frac{ql^3}{24EI} \tag{10.70b}$$

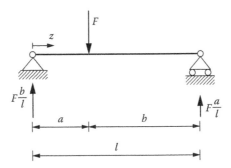

Figure 10.7

If the vertical force F is not applied in the centre (Figure 10.7), it is not possible to exploit the properties of symmetry. The bending moment function is piecewise linear:

$$M(z) = \frac{Fb}{l}z, \qquad \text{for } 0 \leq z \leq a \qquad (10.71a)$$

$$M(z) = \frac{Fb}{l}z - F(z - a), \quad \text{for } a \leq z \leq l \qquad (10.71b)$$

Two distinct differential equations must therefore be considered:

$$\frac{d^2 v_1}{dz^2} = -\frac{Fb}{EIl}z, \qquad \text{for } 0 \leq z \leq a \qquad (10.72a)$$

$$\frac{d^2 v_2}{dz^2} = -\frac{Fb}{EIl}z + \frac{F}{EI}(z - a), \quad \text{for } a \leq z \leq l \qquad (10.72b)$$

Integrating Equation 10.72a, we obtain

$$\frac{dv_1}{dz} = -\frac{Fb}{2EIl}z^2 + C_1 \qquad (10.73a)$$

$$v_1 = -\frac{Fb}{6EIl}z^3 + C_1 z + C_2 \qquad (10.73b)$$

whilst, integrating Equation 10.72b, we get

$$\frac{dv_2}{dz} = -\frac{Fb}{2EIl}z^2 + \frac{F}{2EI}(z - a)^2 + C_3 \qquad (10.74a)$$

$$v_2 = -\frac{Fb}{6EIl}z^3 + \frac{F}{6EI}(z - a)^3 + C_3 z + C_4 \qquad (10.74b)$$

The two conditions of continuity for the rotation and for the displacement in $z = a$

$$\left(\frac{dv_1}{dz}\right)_{z=a} = \left(\frac{dv_2}{dz}\right)_{z=a} \tag{10.75a}$$

$$v_1(z = a) = v_2(z = a) \tag{10.75b}$$

give

$$C_1 = C_3 \tag{10.76a}$$

$$C_2 = C_4 \tag{10.76b}$$

while the conditions at the ends

$$v_1(0) = 0 \tag{10.77a}$$

$$v_2(l) = 0 \tag{10.77b}$$

determine the values

$$C_1 = \frac{Fb}{6EIl}(l^2 - b^2) \tag{10.78a}$$

$$C_2 = 0 \tag{10.78b}$$

In general, it is necessary to resort to the conditions of continuity expressed by Equations 10.75 whenever there is a discontinuity or non-regularity in the moment function (Figure 10.8a and b) or a discontinuity in the flexural rigidity EI of the beam (Figure 10.8c).

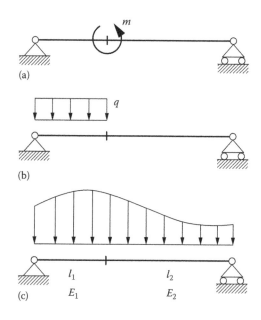

(a)

(b)

(c)

Figure 10.8

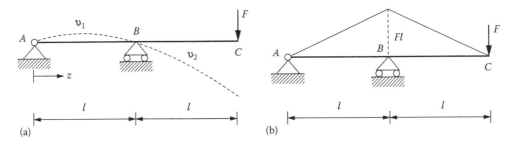

Figure 10.9

The latter eventually occurs in the case of an abrupt variation of the cross section or in the case of a sharp variation in the material of which the beam is made. When these variations occur with continuity, it is instead necessary to integrate the differential equation of the elastic line, considering the continuous functions $E(z)$, $I(z)$.

Let us consider finally the beam with overhanging end of Figure 10.9a. Also in this case, the bending moment is piecewise linear, and thus it is expedient to consider two unknown functions v_1 (portion AB) and v_2 (portion BC). The differential equations of the elastic line, corresponding to the two portions, are (Figure 10.9b)

$$\frac{d^2 v_1}{dz^2} = \frac{Fz}{EI}, \qquad \text{for } 0 \le z \le l \tag{10.79a}$$

$$\frac{d^2 v_2}{dz^2} = \frac{Fz}{EI} - \frac{2F(z-l)}{EI}, \qquad \text{for } l \le z \le 2l \tag{10.79b}$$

with the boundary conditions

$$v_1(0) = 0 \tag{10.80a}$$

$$v_1(l) = 0 \tag{10.80b}$$

$$v_2(l) = 0 \tag{10.80c}$$

$$v_1'(l) = v_2'(l) \tag{10.80d}$$

Integrating Equation 10.79a, we obtain

$$\frac{dv_1}{dz} = \frac{F}{2EI} z^2 + C_1 \tag{10.81a}$$

$$v_1 = \frac{F}{6EI} z^3 + C_1 z + C_2 \tag{10.81b}$$

while integrating Equation 10.79b, we find

$$\frac{d\upsilon_2}{dz} = \frac{F}{2EI}z^2 - \frac{F}{EI}(z-l)^2 + C_3 \tag{10.82a}$$

$$\upsilon_2 = \frac{F}{6EI}z^3 - \frac{F}{3EI}(z-l)^3 + C_3 z + C_4 \tag{10.82b}$$

By applying Equations 10.80, we obtain four equations in the four arbitrary constants C_i, $i = 1, 2, 3, 4$:

$$C_2 = 0 \tag{10.83a}$$

$$\frac{F}{6EI}l^3 + C_1 l + C_2 = 0 \tag{10.83b}$$

$$\frac{F}{6EI}l^3 + C_3 l + C_4 = 0 \tag{10.83c}$$

$$\frac{F}{2EI}l^2 + C_1 = \frac{F}{2EI}l^2 + C_3 \tag{10.83d}$$

The solution is readily obtained:

$$C_1 = C_3 = -\frac{Fl^2}{6EI} \tag{10.84a}$$

$$C_2 = C_4 = 0 \tag{10.84b}$$

The rotation and the vertical displacement at the end of the overhang can be found on the basis of Equations 10.82 and 10.84:

$$\varphi_C = -\upsilon_2'(2l) = -\frac{5}{6}\frac{Fl^2}{EI} \tag{10.85a}$$

$$\upsilon_C = \upsilon_2(2l) = \frac{2}{3}\frac{Fl^3}{EI} \tag{10.85b}$$

10.7 COMPOSITION OF ROTATIONS AND DISPLACEMENTS

As can be seen in the last section, the resolution of the equation of the elastic line is not always immediate and indeed often involves very laborious calculations. In some cases, as in that of Figure 10.9, it is certainly more convenient to apply the principle of superposition, considering separately the effects of the force on the cantilever beam BC (Figure 10.10a) and the effects of the internal reactions on the supported beam AB (Figure 10.10b). As regards the cantilever beam BC, we have

$$\varphi_A^{(1)} = 0 \tag{10.86a}$$

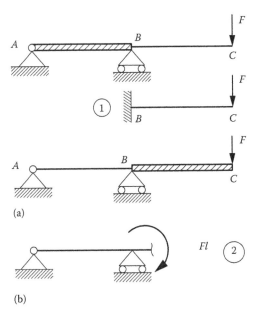

Figure 10.10

$$\varphi_B^{(1)} = 0 \tag{10.86b}$$

$$\varphi_C^{(1)} = -\frac{Fl^2}{2EI} \tag{10.86c}$$

$$\upsilon_C^{(1)} = \frac{Fl^3}{3EI} \tag{10.86d}$$

while for the supported beam, subjected to the moment Fl at the end B, we have

$$\varphi_A^{(2)} = \frac{(Fl)l}{6EI} \tag{10.87a}$$

$$\varphi_B^{(2)} = -\frac{(Fl)l}{3EI} \tag{10.87b}$$

$$\varphi_C^{(2)} = \varphi_B^{(2)} \tag{10.87c}$$

$$\upsilon_C^{(2)} = \left|\varphi_B^{(2)}\right| l \tag{10.87d}$$

whereby, finally, summing up the contributions of Equations 10.86 and 10.87, we obtain

$$\varphi_A = \frac{Fl^2}{6EI} \tag{10.88a}$$

$$\varphi_B = -\frac{Fl^2}{3EI} \tag{10.88b}$$

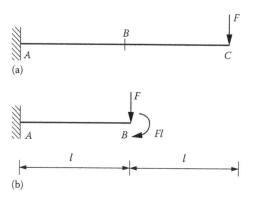

Figure 10.11

$$\varphi_C = -\frac{5}{6}\frac{Fl^2}{EI} \tag{10.88c}$$

$$\upsilon_C = \frac{2}{3}\frac{Fl^3}{EI} \tag{10.88d}$$

Note how Equations 10.88c and 10.88d coincide with Equations 10.85a and 10.85b.

If we intend to determine the rotation and displacement of the midpoint of a cantilever beam of length $2l$, loaded by a force at the end (Figure 10.11a), it is sufficient to consider a cantilever beam halved and loaded by the internal characteristics of shear and bending moment (Figure 10.11b). Summation of the contributions yields

$$\varphi_B = -\frac{Fl^2}{2EI} - \frac{(Fl)l}{EI} = -\frac{3}{2}\frac{Fl^2}{EI} \tag{10.89a}$$

$$\upsilon_B = \frac{Fl^3}{3EI} + \frac{(Fl)l^2}{2EI} = \frac{5}{6}\frac{Fl^3}{EI} \tag{10.89b}$$

If, instead, the same cantilever beam is loaded in its midpoint (Figure 10.12), and we wish to determine the rotation and displacement of the end cross section, taking into account that the portion BC is not subjected to bending but only to rigid rotation, we have

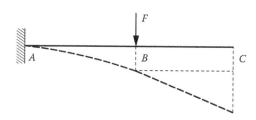

Figure 10.12

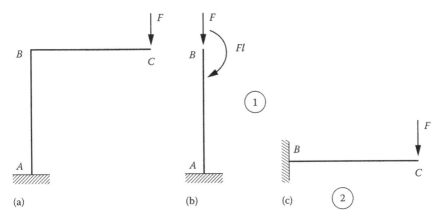

Figure 10.13

$$\varphi_C = \varphi_B = -\frac{Fl^2}{2EI} \tag{10.90a}$$

$$\upsilon_C = \upsilon_B + |\varphi_B|l = \frac{Fl^3}{3EI} + \frac{Fl^2}{2EI}l = \frac{5}{6}\frac{Fl^3}{EI} \tag{10.90b}$$

The displacements (10.89b) and (10.90b) are identical on the basis of Betti's reciprocal theorem.

Let us consider the L-shaped cantilever beam of Figure 10.13a, subjected to a force acting at the end C, and let us seek to determine the translation and rotation of this cross section. We shall therefore make the summation of the contributions that result from the scheme of Figure 10.13b, in which the characteristics of axial force and bending moment act on the vertical cantilever AB, with the contributions that emerge from the scheme of Figure 10.13c, consisting of the overhang BC, built-in at B and loaded with the external force F at C:

$$u_B^{(1)} = \frac{(Fl)l^2}{2EI} \tag{10.91a}$$

$$\varphi_B^{(1)} = -\frac{(Fl)l}{EI} \tag{10.91b}$$

$$\varphi_C^{(1)} = \varphi_B^{(1)} \tag{10.91c}$$

$$\upsilon_C^{(1)} = |\varphi_B^{(1)}|l \tag{10.91d}$$

$$u_C^{(1)} = u_B^{(1)} \tag{10.91e}$$

$$\upsilon_C^{(2)} = \frac{Fl^3}{3EI} \tag{10.91f}$$

$$\varphi_C^{(2)} = -\frac{Fl^2}{2EI} \tag{10.91g}$$

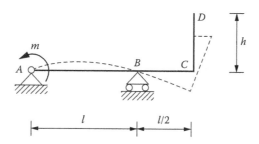

Figure 10.14

Summing up the corresponding terms, we obtain

$$\varphi_C = \varphi_C^{(1)} + \varphi_C^{(2)} = -\frac{3}{2}\frac{Fl^2}{EI} \tag{10.92a}$$

$$\upsilon_C = \upsilon_C^{(1)} + \upsilon_C^{(2)} = \frac{4}{3}\frac{Fl^3}{EI} \tag{10.92b}$$

$$u_C = u_C^{(1)} = \frac{Fl^3}{2EI} \tag{10.92c}$$

As our final example, let us now examine the beam of Figure 10.14 subjected to a concentrated moment at A. The overhang is devoid of loads and hence is not deflected, but only rotates rigidly, under the action of the remainder of the structure. On the hypothesis of small displacements, and applying the rules of kinematics of rigid bodies, we have

$$\varphi_D = \varphi_B = -\frac{ml}{6EI} \tag{10.93a}$$

$$\upsilon_D = \upsilon_C = |\varphi_B|\frac{l}{2} = \frac{ml^2}{12EI} \tag{10.93b}$$

$$u_D = |\varphi_B|h = \frac{mlh}{6EI} \tag{10.93c}$$

10.8 BEAM ON ELASTIC FOUNDATION

We shall now examine the case of a beam supported on a foundation which reacts elastically and bilaterally. Referring to the fact that the reaction of the foundation is assumed to be proportional to the vertical displacement υ, we represent this foundation usually as a bed of springs (Figure 10.15). This model provides a significant representation of the case of a beam set in the ground or of a rail fastened to sleepers.

If we denote the elastic rigidity of the foundation by K, the differential equation (10.49), which holds in the absence of the foundation, is modified as follows:

$$\frac{d^4\upsilon}{dz^4} = \frac{q}{EI} - \frac{K}{EI}\upsilon \tag{10.94}$$

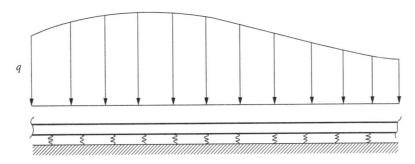

Figure 10.15

where the index x of the moment of inertia has been omitted. In the case where $q = 0$, we have

$$\frac{d^4 v}{dz^4} + 4\beta^4 v = 0 \qquad (10.95)$$

with

$$\beta = \sqrt[4]{\frac{K}{4EI}} \qquad (10.96)$$

The complete integral of Equation 10.95 is

$$v(z) = e^{\beta z}(C_1 \cos \beta z + C_2 \sin \beta z) + e^{-\beta z}(C_3 \cos \beta z + C_4 \sin \beta z) \qquad (10.97)$$

where the constants C_i with $i = 1, 2, 3, 4$, are to be identified *via* the boundary conditions.

When instead, q is constant and different from zero, the complete integral (10.97) must be supplemented by the particular solution $v = q/K$.

Once the analytical expression of the displacement v is known, it is possible to obtain the rotation φ, the bending moment M and the shear T by derivation.

In the case of an infinitely long beam resting on an elastic foundation, loaded by a concentrated force F (Figure 10.16), the terms of Equation 10.97 that contain the factor $e^{\beta z}$ must vanish, because at infinity the displacement vanishes. We shall therefore have $C_1 = C_2 = 0$, so that

$$v(z) = e^{-\beta z}(C_3 \cos \beta z + C_4 \sin \beta z) \qquad (10.98)$$

At the point of application of the force, from symmetry, we have then

$$\left(\frac{dv}{dz}\right)_{z=0} = -\beta(C_3 - C_4) = 0 \qquad (10.99)$$

and hence the displacement will be determined but for a single factor

$$v(z) = Ce^{-\beta z}(\cos \beta z + \sin \beta z) \qquad (10.100)$$

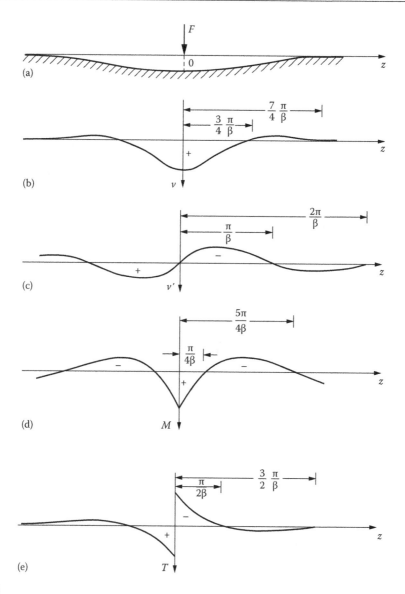

Figure 10.16

This factor C can finally be determined on the basis of the condition of equilibrium to the vertical translation of the entire beam:

$$F = 2 \int_0^\infty K v \, dz \qquad (10.101)$$

Substituting Equation 10.100 in Equation 10.101, we obtain

$$F = \frac{2KC}{\beta} \qquad (10.102)$$

and hence the elastic line is represented by the function

$$\upsilon(z) = \frac{F\beta}{2K} e^{-\beta z}(\cos\beta z + \sin\beta z) = \frac{F\beta}{2K} A_{\beta z} \tag{10.103a}$$

By derivation, we obtain

$$\varphi = -\frac{d\upsilon}{dz} = \frac{F\beta^2}{K} e^{-\beta z}\sin\beta z = \frac{F\beta^2}{K} B_{\beta z} \tag{10.103b}$$

$$M = -EI\frac{d^2\upsilon}{dz^2} = \frac{F}{4\beta} e^{-\beta z}(\cos\beta z - \sin\beta z) = \frac{F}{4\beta} C_{\beta z} \tag{10.103c}$$

$$T = -EI\frac{d^3\upsilon}{dz^3} = -\frac{F}{2} e^{-\beta z}\cos\beta z = -\frac{F}{2} D_{\beta z} \tag{10.103d}$$

Table 10.1 gives the values of the functions $A_{\beta z}$, $B_{\beta z}$, $C_{\beta z}$, $D_{\beta z}$, as the argument βz varies. The maximum values attained by the transverse displacement, the bending moment and the shear are found for $z = 0$ (Figure 10.16). The functions (10.103) represent all exponentially smoothing sinusoids. The wavelength λ is defined by the relation

$$\beta\lambda = 2\pi \tag{10.104}$$

from which we deduce

$$\lambda = \frac{2\pi}{\beta} = 2\pi\sqrt[4]{\frac{4EI}{K}} \tag{10.105}$$

In the case of multiple concentrated loads, the principle of superposition can furnish the displacement and the internal characteristics at each point. For example, when two equal forces are applied, at a distance apart of 1500 mm, on a beam with $\beta = 10^{-3}$ mm^{-1}, we have (Table 10.1)

$$A_{\beta z}(1.5) \simeq 0.23$$

$$C_{\beta z}(1.5) \simeq -0.20$$

whereby the total displacement under each force is increased by approximately 23%,

$$\upsilon = \frac{F\beta}{2K}(1 + 0.23)$$

while the total bending moment under each force is reduced by approximately 20%,

$$M = \frac{F}{4\beta}(1 - 0.20)$$

Table 10.1

βz	$A_{\beta z}$	$B_{\beta z}$	$C_{\beta z}$	$D_{\beta z}$	βz	$A_{\beta z}$	$B_{\beta z}$	$C_{\beta z}$	$D_{\beta z}$
0	1.0000	0	1.0000	1.0000	3.6	-0.0366	-0.0121	-0.0124	-0.0245
0.1	0.9907	0.0903	0.8100	0.9003	3.7	-0.0341	-0.0131	-0.0079	-0.0210
0.2	0.9651	0.1627	0.6398	0.8024	3.8	-0.0314	-0.0137	-0.0040	-0.0177
0.3	0.9267	0.2189	0.4888	0.7077	3.9	-0.0286	-0.0140	-0.0008	-0.0147
0.4	0.8784	0.2610	0.3564	0.6174	$5\pi/4$	-0.0278	-0.0140	0	-0.0139
0.5	0.8231	0.2908	0.2415	0.5323	4.0	-0.0258	-0.0139	0.0019	-0.0120
0.6	0.7628	0.3099	0.1431	0.4530	4.1	-0.0231	-0.0136	0.0040	-0.0095
0.7	0.6997	0.3199	0.0599	0.3798	4.2	-0.0204	-0.0131	0.0057	-0.0074
$\pi/4$	0.6448	0.3224	0	0.3224	4.3	-0.0179	-0.0125	0.0070	-0.0054
0.8	0.6354	0.3223	-0.0093	0.3131	4.4	-0.0155	-0.0117	0.0079	-0.0038
0.9	0.5712	0.3185	-0.0657	0.2527	4.5	-0.0132	-0.0108	0.0085	-0.0023
1.0	0.5083	0.3096	-0.1108	0.1988	4.6	-0.0111	-0.0100	0.0089	-0.0011
1.1	0.4476	0.2967	-0.1457	0.1510	4.7	-0.0092	-0.0091	0.0090	-0.0001
1.2	0.3899	0.2807	-0.1716	0.1091	$3\pi/2$	-0.0090	-0.0090	0.0090	0
1.3	0.3355	0.2626	-0.1897	0.0729	4.8	-0.0075	-0.0082	0.0089	0.0007
1.4	0.2849	0.2430	-0.2011	0.0419	4.9	-0.0059	-0.0073	0.0087	0.0014
1.5	0.2384	0.2226	-0.2068	0.0158	5.0	-0.0046	-0.0065	0.0084	0.0019
$\pi/2$	0.2079	0.2079	-0.2079	0	5.1	-0.0033	-0.0057	0.0080	0.0023
1.6	0.1959	0.2018	-0.2077	-0.0059	5.2	-0.0023	-0.0049	0.0075	0.0026
1.7	0.1576	0.1812	-0.2047	-0.0235	5.3	-0.0014	-0.0042	0.0069	0.0028
1.8	0.1234	0.1610	-0.1985	-0.0376	5.4	-0.0006	-0.0035	0.0064	0.0029
1.9	0.0932	0.1415	-0.1899	-0.0484	$7\pi/4$	0	-0.0029	0.0058	0.0029
2.0	0.0667	0.1230	-0.1794	-0.0563	5.5	0.0000	-0.0029	0.0058	0.0029
2.1	0.0439	0.1057	-0.1675	-0.0618	5.6	0.0005	-0.0023	0.0052	0.0029
2.2	0.0244	0.0895	-0.1548	-0.0652	5.7	0.0010	-0.0018	0.0046	0.0028
2.3	0.0080	0.0748	-0.1416	-0.0668	5.8	0.0013	-0.0014	0.0041	0.0027
$3\pi/4$	0	0.0671	-0.1342	-0.0671	5.9	0.0015	-0.0010	0.0036	0.0026
2.4	-0.0056	0.0613	-0.1282	-0.0669	6.0	0.0017	-0.0007	0.0031	0.0024
2.5	-0.0166	0.0492	-0.1149	-0.0658	6.1	0.0018	-0.0004	0.0026	0.0022
2.6	-0.0254	0.0383	-0.1019	-0.0636	6.2	0.0019	-0.0002	0.0022	0.0020
2.7	-0.0320	0.0287	-0.0895	-0.0608	2π	0.0019	0	0.0019	0.0019
2.8	-0.0369	0.0204	-0.0777	-0.0573	6.3	0.0019	+0.0001	0.0018	0.0018
2.9	-0.0403	0.0132	-0.0666	-0.0534	6.4	0.0018	0.0003	0.0015	0.0017
3.0	-0.0423	0.0070	-0.0563	-0.0493	6.5	0.0018	0.0004	0.0012	0.0015
3.1	-0.0431	0.0019	-0.0469	-0.0450	6.6	0.0017	0.0005	0.0009	0.0013
π	-0.0432	0	-0.0432	-0.0432	6.7	0.0016	0.0006	0.0006	0.0011
3.2	-0.0431	-0.0024	-0.0383	-0.0407	6.8	0.0015	0.0006	0.0004	0.0010
3.3	-0.0422	-0.0058	-0.0306	-0.0364	6.9	0.0014	0.0006	0.0002	0.0008
3.4	-0.0408	-0.0085	-0.0237	-0.0323	7.0	0.0013	0.0006	0.0001	0.0007
3.5	-0.0389	-0.0106	-0.0177	-0.0283	$9\pi/4$	0.0012	0.0006	0	0.0006

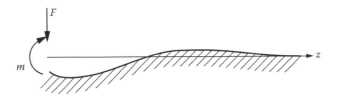

Figure 10.17

Let us consider finally a semi-infinite beam, loaded at its end by a force F and a moment m (Figure 10.17). Since also in this case the displacement at infinity must be zero, the relation (10.98) will hold with the boundary conditions

$$\left(\frac{\mathrm{d}^2 v}{\mathrm{d}z^2}\right)_{z=0} = -\frac{m}{EI} \tag{10.106a}$$

$$\left(\frac{\mathrm{d}^3 v}{\mathrm{d}z^3}\right)_{z=0} = \frac{F}{EI} \tag{10.106b}$$

From Equations 10.98 and 10.106, we obtain

$$C_3 = \frac{F - \beta m}{2\beta^3 EI} \tag{10.107a}$$

$$C_4 = \frac{m}{2\beta^2 EI} \tag{10.107b}$$

Equation 10.98 therefore transforms as follows:

$$v(z) = \frac{e^{-\beta z}}{2\beta^3 EI}[F\cos\beta z - \beta m(\cos\beta z - \sin\beta z)]$$

$$= \frac{2F\beta}{K} D_{\beta z} - \frac{2m\beta^2}{K} C_{\beta z} \tag{10.108a}$$

By derivation, we then obtain

$$\varphi(z) = \frac{2F\beta^2}{K} A_{\beta z} - \frac{4m\beta^3}{K} D_{\beta z} \tag{10.108b}$$

The vertical displacement and the rotation at the end of the beam are therefore

$$v(0) = \frac{2\beta}{K}(F - m\beta) \tag{10.109a}$$

$$\varphi(0) = \frac{2\beta^2}{K}(F - 2m\beta) \tag{10.109b}$$

10.9 DYNAMICS OF DEFLECTED BEAMS

With the purpose of analysing the free flexural oscillations of beams, let us consider the differential equation of the elastic line (10.49), replacing the distributed load $q(z)$ with the force of inertia,

$$\frac{\partial^4 v}{\partial z^4} = -\frac{\mu}{EI}\frac{\partial^2 v}{\partial t^2} \tag{10.110}$$

where μ denotes the linear density of the beam (mass per unit length).

Equation 10.110 is an equation with separable variables, the solution being representable as the product of two different functions, each one having a single variable:

$$v(z,t) = \eta(z)f(t) \tag{10.111}$$

Substituting Equation 10.111 in Equation 10.110, we obtain

$$\frac{d^4\eta}{dz^4}f + \frac{\mu}{EI}\eta\frac{d^2f}{dt^2} = 0 \tag{10.112}$$

Dividing Equation 10.112 by the product ηf, we have

$$-\frac{\left(\dfrac{d^2f}{dt^2}\right)}{f} = \frac{EI}{\mu}\frac{\left(\dfrac{d^4\eta}{dz^4}\right)}{\eta} = \omega^2 \tag{10.113}$$

where ω^2 represents a positive constant, the two first terms of Equation 10.113 being at the most functions of the time t and the coordinate z, respectively.

From Equation 10.113, there follow two ordinary differential equations:

$$\frac{d^2f}{dt^2} + \omega^2 f = 0 \tag{10.114a}$$

$$\frac{d^4\eta}{dz^4} - \alpha^4\eta = 0 \tag{10.114b}$$

with

$$\alpha = \sqrt[4]{\frac{\mu\omega^2}{EI}} \tag{10.115}$$

Whereas Equation 10.114a is the equation of the harmonic oscillator, with the well-known complete integral

$$f(t) = A\cos\omega t + B\sin\omega t \tag{10.116a}$$

Equation 10.114b has the complete integral

$$\eta(z) = C\cos\alpha z + D\sin\alpha z + E\cosh\alpha z + F\sinh\alpha z \tag{10.116b}$$

As we shall see later on, the constants A, B may be determined on the basis of the initial conditions, while the constants C, D, E, F may be determined on the basis of the boundary conditions. However, the parameter ω remains for the moment undetermined, and so also the parameter α according to Equation 10.115. This represents the eigenvalue of the problem, from the mathematical standpoint, or the angular frequency of the system, from the mechanical point of view. In the sequel, we shall see how the angular frequency ω may also be obtained on the basis of the boundary conditions. We shall obtain in fact an infinite number of eigenvalues ω_i and thus α_i, just as also an infinite number of eigenfunctions, f_i and thus η_i. The complete integral of the differential equation (10.110) may therefore be given the following form, on the basis of the principle of superposition:

$$v(z,t) = \sum_{i=1}^{\infty} \eta_i(z) f_i(t) \tag{10.117}$$

with

$$f_i(t) = A_i \cos \omega_i t + B_i \sin \omega_i t \tag{10.118a}$$

$$\eta_i(z) = C_i \cos \alpha_i z + D_i \sin \alpha_i z + E_i \cosh \alpha_i z + F_i \sinh \alpha_i z \tag{10.118b}$$

The eigenfunctions η_i are **orthonormal functions.** We may in fact write Equation 10.114b for two different eigensolutions:

$$\eta_j^{IV} = \alpha_j^4 \eta_j \tag{10.119a}$$

$$\eta_k^{IV} = \alpha_k^4 \eta_k \tag{10.119b}$$

Multiplying the first of Equations 10.119 by η_k and the second by η_j, and integrating over the length of the beam, we obtain

$$\int_0^l \eta_k \eta_j^{IV}\, dz = \alpha_j^4 \int_0^l \eta_k \eta_j\, dz \tag{10.120a}$$

$$\int_0^l \eta_j \eta_k^{IV}\, dz = \alpha_k^4 \int_0^l \eta_j \eta_k\, dz \tag{10.120b}$$

Integrating by parts the left-hand sides, the foregoing equations transform as follows:

$$\left[\eta_k \eta_j'''\right]_0^l - \left[\eta_k' \eta_j''\right]_0^l + \int_0^l \eta_k'' \eta_j''\, dz = \alpha_j^4 \int_0^l \eta_k \eta_j\, dz \tag{10.121a}$$

$$\left[\eta_j \eta_k'''\right]_0^l - \left[\eta_j' \eta_k''\right]_0^l + \int_0^l \eta_j'' \eta_k''\, dz = \alpha_k^4 \int_0^l \eta_j \eta_k\, dz \tag{10.121b}$$

When each of the two ends of the beam is constrained by a built-in support ($\eta = \eta' = 0$), or by a hinge ($\eta = \eta'' = 0$), or by a double rod ($\eta''' = \eta' = 0$), or yet again is unconstrained ($\eta''' = \eta'' = 0$), in this last case the remaining end of the beam being built-in, the quantities in square brackets vanish. Subtracting member by member, we thus have

$$\left(\alpha_j^4 - \alpha_k^4\right) \int_0^l \eta_j \eta_k \, dz = 0 \tag{10.122}$$

from which there follows the condition of orthonormality,

$$\int_0^l \eta_j \eta_k dz = \delta_{jk} \tag{10.123}$$

where δ_{jk} is the **Kronecker delta**. Thus, when the eigenvalues are distinct, the integral of the product of the corresponding eigenfunctions vanishes. When, instead, the indices j and k coincide, the condition of normality reminds us that the eigenfunctions are defined but for a factor of proportionality, as follows from the homogeneity of Equation 10.114b.

As we have already had occasion to mention, the constants A_i, B_i of Equation 10.116a are determined *via* the initial conditions

$$v(z,0) = v_0(z) \tag{10.124a}$$

$$\frac{\partial v}{\partial t}(z,0) = \dot{v}_0(z) \tag{10.124b}$$

which, on the basis of Equations 10.117 and 10.118a, become

$$\sum_{i=1}^{\infty} A_i \eta_i(z) = v_0(z) \tag{10.125a}$$

$$\sum_{i=1}^{\infty} \omega_i B_i \eta_i(z) = \dot{v}_0(z) \tag{10.125b}$$

Multiplying by any desired eigenfunction η_j, and integrating over the length of the beam, we obtain

$$\sum_{i=1}^{\infty} A_i \int_0^l \eta_i \eta_j \, dz = \int_0^l \eta_j v_0 dz \tag{10.126a}$$

$$\sum_{i=1}^{\infty} \omega_i B_i \int_0^l \eta_i \eta_j dz = \int_0^l \eta_j \dot{v}_0 dz \tag{10.126b}$$

Taking into account the condition of orthonormality, Equation 10.123, finally we have

$$A_j = \int_0^l \eta_j \upsilon_0 dz \tag{10.127a}$$

$$B_j = \frac{1}{\omega_j} \int_0^l \eta_j \dot{\upsilon}_0 \, dz \tag{10.127b}$$

When the system is initially perturbed, by assigning to the beam a deformed configuration proportional to one of the eigenfunctions, with an initial zero velocity, the beam, once left free to oscillate, continues to do so in proportion to the initial deformed configuration. In this case, we have

$$\upsilon_0(z) = a\eta_i(z) \tag{10.128a}$$

$$\dot{\upsilon}_0(z) = 0 \tag{10.128b}$$

where a is an arbitrary constant of proportionality. Equations 10.127 then furnish

$$A_j = a\delta_{ij} \tag{10.129a}$$

$$B_j = 0 \tag{10.129b}$$

and hence the complete integral (10.117) takes the following form:

$$\upsilon(z,t) = a\eta_i(z)\cos\omega_i t = \upsilon_0(z)\cos\omega_i t \tag{10.130}$$

The beam therefore oscillates in proportion to the initial deformation and with an angular frequency that corresponds to the same eigenfunction. These oscillations are called the **natural modes of vibration of the system.**

More particularly, as regards a beam supported at both ends, of length l (Figure 10.18a), the boundary conditions imposed on the expression (10.118b) in correspondence to the end A yield

$$\eta(0) = C + E = 0 \tag{10.131a}$$

$$\eta''(0) = -\alpha^2(C - E) = 0 \tag{10.131b}$$

whence we obtain

$$C = E = 0 \tag{10.132}$$

The boundary conditions corresponding to the end B yield, on the other hand

$$\eta(l) = D\sin\alpha l + F\sinh\alpha l = 0 \tag{10.133a}$$

$$\eta''(l) = -\alpha^2(D\sin\alpha l - F\sinh\alpha l) = 0 \tag{10.133b}$$

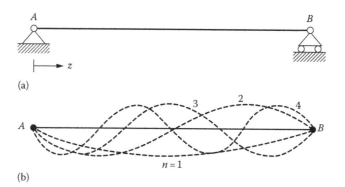

Figure 10.18

from which it follows that

$$D \sin \alpha l = 0 \tag{10.134a}$$

$$F \sinh \alpha l = 0 \tag{10.134b}$$

From Equation 10.134b, we obtain

$$F = 0 \tag{10.135}$$

whilst from Equation 10.134a, once the trivial solution $D = 0$ has been ruled out, it follows that

$$\alpha = n \frac{\pi}{l} \tag{10.136}$$

where n is a natural number.
From relation (10.115), we deduce

$$\alpha_n^4 = n^4 \frac{\pi^4}{l^4} = \frac{\mu \omega_n^2}{EI} \tag{10.137}$$

whereby we obtain the **natural angular frequencies** of the system

$$\omega_n = n^2 \frac{\pi^2}{l^2} \sqrt{\frac{EI}{\mu}} \tag{10.138}$$

and hence the proper periods thereof:

$$T_n = \frac{2\pi}{\omega_n} = \frac{2l^2}{\pi n^2} \sqrt{\frac{\mu}{EI}} \tag{10.139}$$

The normalized eigenfunctions are thus represented by the following succession (Figure 10.18b):

$$\eta_n(z) = \sqrt{\frac{2}{l}} \sin n\pi \frac{z}{l} \tag{10.140}$$

(a)   (b)   (c)

Figure 10.19

In the case of the cantilever beam of Figure 10.19, the boundary conditions imposed on the expression (10.118b) are

$$\eta(0) = C + E = 0 \tag{10.141a}$$

$$\eta'(0) = \alpha(D + F) = 0 \tag{10.141b}$$

$$\eta''(l) = -\alpha^2(C \cos \alpha l + D \sin \alpha l - E \cosh \alpha l - F \sinh \alpha l) = 0 \tag{10.141c}$$

$$\eta'''(l) = \alpha^3(C \sin \alpha l - D \cos \alpha l + E \sinh \alpha l + F \cosh \alpha l) = 0 \tag{10.141d}$$

Whilst from the first two equations we obtain

$$E = -C, \quad F = -D \tag{10.142}$$

from the last two there follows

$$C(\cos \alpha l + \cosh \alpha l) + D(\sin \alpha l + \sinh \alpha l) = 0 \tag{10.143a}$$

$$C(\sin \alpha l - \sinh \alpha l) - D(\cos \alpha l + \cosh \alpha l) = 0 \tag{10.143b}$$

The system of algebraic equations (10.143) gives, on the other hand, a solution different from the trivial one, if and only if the determinant of the coefficients is zero:

$$(\cos \alpha l + \cosh \alpha l)^2 + (\sin^2 \alpha l - \sinh^2 \alpha l) = 0 \tag{10.144}$$

Computing expression (10.144), we obtain the trigonometric equation which provides the set or **spectrum** of eigenvalues:

$$\cos \alpha_n l \cosh \alpha_n l = -1 \tag{10.145}$$

The first three roots of Equation 10.145 are

$$\alpha_1 l = 1.875, \quad \alpha_2 l = 4.694, \quad \alpha_3 l = 7.885$$

The angular frequencies and the proper periods of the cantilever beam are given by

$$\omega_n = \alpha_n^2 \sqrt{\frac{EI}{\mu}} \tag{10.146a}$$

$$T_n = \frac{2\pi}{\alpha_n^2} \sqrt{\frac{\mu}{EI}} \tag{10.146b}$$

The fundamental period is therefore

$$T_1 = 1.79l^2 \sqrt{\frac{\mu}{EI}} \qquad (10.147)$$

and is approximately three times as long as that of the supported beam given by Equation 10.139:

$$T_1 = 0.64l^2 \sqrt{\frac{\mu}{EI}} \qquad (10.148)$$

The first three eigenfunctions have the aspect shown in Figure 10.19.

In the case of a rope in tension (Figure 10.18), the flexural rigidity EI is vanishingly small, so that the bending moment, in the case of large displacements, is given by the product of the axial force and the transverse displacement:

$$M = -N\upsilon \qquad (10.149)$$

Applying the relation (4.25), it is possible to obtain the equivalent transverse load, so that the equation of equilibrium to vertical translation is the following:

$$N\frac{\partial^2 \upsilon}{\partial z^2} = \mu \frac{\partial^2 \upsilon}{\partial t^2} \qquad (10.150)$$

Equation 10.150 transforms into the wave equation

$$\frac{\partial^2 \upsilon}{\partial t^2} = c^2 \frac{\partial^2 \upsilon}{\partial z^2} \qquad (10.151)$$

where

$$c^2 = \frac{N}{\mu} \qquad (10.152)$$

is the square of the velocity of the transverse wave in the rope in tension.

Equation 10.151 is formally identical to the equation of longitudinal waves in elastic bars. If in fact we replace the distributed longitudinal force $\mathcal{F}_x(x)$ in the static equation (8.53a) with the force of inertia $-\mu(\partial^2 u/\partial t^2)$, we obtain

$$EA\frac{\partial^2 u}{\partial x^2} = \mu \frac{\partial^2 u}{\partial t^2} \qquad (10.153)$$

and thus

$$\frac{\partial^2 u}{\partial t^2} = c^2 \frac{\partial^2 u}{\partial x^2} \qquad (10.154)$$

where in this case

$$c^2 = \frac{EA}{\mu} \qquad\qquad (10.155)$$

is the square of the velocity of the longitudinal wave in the elastic bar.

10.10 FLEXURAL BUCKLING OF SLENDER BEAMS LOADED IN COMPRESSION

The solution to an elastic problem may result in a state of stable, neutral or unstable equilibrium, depending on the magnitude of the load applied. The state of a system is said to be stable when the system, having been displaced slightly by a perturbation, upon being released returns of its own in accord with its original configuration. A similar situation is that of a ball in a condition of (stable) equilibrium at the bottom of a hole (Figure 10.20a).

If the ball is removed from its equilibrium point and then released again to gravity, after a few oscillations, it will return to its initial position. Conversely, the condition is unstable if an elastic system, after a perturbation, tends to move farther and farther away from its original configuration. A similar situation is that of a ball in (unstable) equilibrium at the top of a protuberance (Figure 10.20b). Finally, the condition is neutral when the application of a perturbation results in the elastic system remaining in a condition of equilibrium in the new configuration, without tending either to return to its original position or to move farther away from it (Figure 10.20c).

Loss of stability of an elastic system, commonly referred to as **buckling**, generally occurs in slender structural members loaded in compression, e.g. the columns of a building, truss struts, machine shafts, bridge arches and piers, dome and roof vaults, submarine shells.

Let us consider a beam of length l, hinged at one end and simply supported at the other, consisting of two rigid parts connected by an elastic hinge having stiffness k (Figure 10.21). If the beam is subjected to an axial force N, applied onto the roller support, equilibrium is restored by an equal force applied in the opposite direction by the restraint at the opposite end. We shall see later how this state of equilibrium can be stable, unstable or neutral as a function of the magnitude of the load applied, N. Let us assume for instance that the elastic system is perturbed by lowering the elastic hinge by an infinitesimal quantity, υ. Under the assumption of linearization of the constraints, the two rigid parts will rotate by the same infinitesimal angle $\varphi = 2\upsilon/l$, but in opposite directions. In the

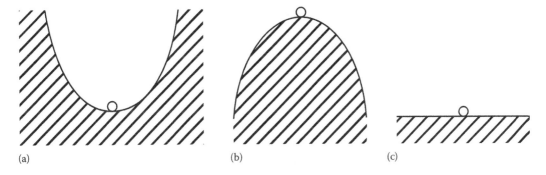

(a)   (b)   (c)

Figure 10.20

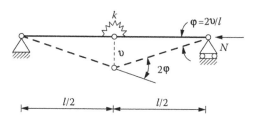

Figure 10.21

new deflected configuration, the elastic hinge will be subject to two opposite moments: a stabilizing moment, proportional, according to stiffness k, to the angles 2φ formed by the two parts (Figure 10.21):

$$M_s = k\frac{4v}{l} \tag{10.156}$$

and an unstabilizing moment, proportional to the arm v of the applied force with respect to the elastic hinge:

$$M_i = Nv \tag{10.157}$$

Obviously, when N is small enough, we get $M_i < M_s$ and the equilibrium condition is stable. Conversely, when N is sufficiently large, we get $M_i > M_s$ and the system becomes unstable. Notice, in fact, that both M_s and M_i turn out to be proportional to v and the hence this kinematic parameter does not affect the stability condition:

$$N < k\frac{4}{l} \Leftrightarrow \text{stable equilibrium} \tag{10.158}$$

$$N > k\frac{4}{l} \Leftrightarrow \text{unstable equilibrium} \tag{10.159}$$

$$N = k\frac{4}{l} \Leftrightarrow \text{neutral equilibrium} \tag{10.160}$$

The force which separates stable from unstable conditions, $N_c = 4k/l$, is called **critical load**. It should be noted that this load depends solely on the geometric and mechanical characteristics of the system and is totally independent of the amount v of the perturbation envisaged. In other words, if $N > N_c$, even the smallest perturbation will suffice to disrupt system stability, whereas, if $N < N_c$, even larger perturbations will not engender a crisis.

So far, we have considered a case of what is commonly referred to as concentrated elasticity. Let us now consider a more general case of a distributed elasticity beam loaded in compression (Figure 10.22). In this case, the perturbation will be a function $v(z)$ of imposed displacements. Hence, each beam cross section will be subjected both to a stabilizing moment, given by equation:

$$M_s = -EI\frac{d^2v}{dz^2} \tag{10.161}$$

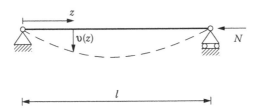

Figure 10.22

and hence proportional, according to stiffness EI, to the second derivative of $v(z)$, and to an unstabilizing moment, given by Equation 10.157, where v is now a function of the axial coordinate z.

By imposing that, in each section of the beam, the stabilizing moment equates the unstabilizing moment, we get the following differential equation:

$$\frac{\mathrm{d}^2 v}{\mathrm{d}z^2} + \alpha^2 v = 0 \tag{10.162}$$

where

$$\alpha^2 = \frac{N}{EI} \tag{10.163}$$

The perturbation, $v(z)$, is called **eigenfunction**, while parameter α is the **eigenvalue** of the problem. The general integral of Equation 10.162 is

$$v(z) = C_1 \sin \alpha z + C_2 \cos \alpha z \tag{10.164}$$

and through the boundary conditions

$$v(0) = v(l) = 0 \tag{10.165}$$

we get

$$C_2 = 0 \tag{10.166a}$$

$$C_1 \sin \alpha l = 0 \tag{10.166b}$$

From Equation 10.166b, we get

$$\alpha l = n\pi \tag{10.167}$$

with n is natural number and, therefore, by applying Equation 10.163,

$$N_{cn} = n^2 \pi^2 \frac{EI}{l^2} \tag{10.168}$$

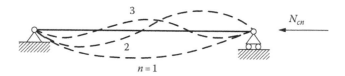

Figure 10.23

For each eigenvalue N_{cn}, there corresponds an eigenfunction

$$v_n(z) = C_1 \sin \alpha_n z \tag{10.169}$$

which represents the critical deformation mode of the beam for that load. This deformation configuration consists of a number n of sinusoidal half-waves (Figure 10.23). Of course, if there are no further constraints on the beam apart from the two external supports, the critical load occurs for $n = 1$:

$$N_{c1} = \pi^2 \frac{EI}{l^2} \tag{10.170}$$

This force, called **Euler's critical load**, is the load which determines the buckling of the beam. For $N < N_{c1}$, the equilibrium is stable; for $N = N_{c1}$, the equilibrium is neutral; and for $N > N_{c1}$, the equilibrium is unstable. It should be noted that Euler's critical load increases in proportion to the rigidity EI of the beam and decreases in inverse proportion to the square of the beam length.

However, Euler's formula shows its limitations in dealing with not sufficiently slender beams, where the inelastic behaviour of the material can interact with the buckling mechanism.

Let us denote with

$$\sigma_c = \frac{N_{c1}}{A} \tag{10.171}$$

Euler's critical pressure, which, according to (10.170), may be expressed as

$$\sigma_c = \pi^2 \frac{EI}{l^2 A} = \pi^2 E \frac{\rho^2}{l^2} \tag{10.172}$$

where ρ stands for the radius of inertia of the section in the direction of the bending axis. If we denote the slenderness, l/ρ, with λ, Equation 10.172 can be expressed as

$$\sigma_c = \frac{\pi^2 E}{\lambda^2} \tag{10.173}$$

By plotting Equation 10.173 on the σ–λ^2 plane, we get the so-called **Euler's hyperbola** (Figure 10.24). This hyperbola envisages critical pressures approaching zero for slenderness ratios approaching infinity, and, conversely, critical pressures approaching infinity for slenderness ratios approaching zero. The latter tendency is unlikely since for stubby beams the failure due to yielding

$$\sigma_c = \sigma_P \tag{10.174}$$

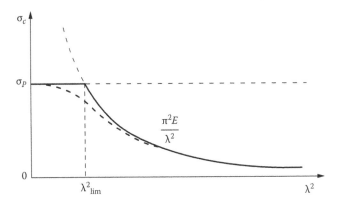

Figure 10.24

can precede, even markedly, that due to buckling (see Equation 10.173). If there were no interaction between the two critical phenomena, the transition between the two would occur, in a discontinuous manner, for a **limit slenderness**:

$$\lambda_{\lim} = \pi\sqrt{\frac{E}{\sigma_P}} \tag{10.175}$$

i.e. a function of the elastic modulus, E, and the material's yielding stress σ_P. For steel, it is $E/\sigma_P \sim 10^3$ and hence $\lambda_{\lim} \sim 10^2$.

In actual fact, the two critical phenomena interact, and hence there is a gradual transition from one to the other as the slenderness of the beam varies. The critical pressure is therefore supplied by the dashed curve shown in Figure 10.24, which links the two critical curves corresponding to Equations 10.173 and 10.174, smoothing the cuspidal point formed at their intersection.

This connecting curve is normally given in tabulated form, by setting

$$\sigma < \frac{\sigma_P}{\omega} \tag{10.176}$$

where ω is a safety factor higher than one, which is a function of the material and the slenderness ratio of the beam.

So far, we have examined the case of a beam restrained by a hinge and a roller support. Equation 10.162 shows, on the other hand, the equilibrium equation of a beam, whatever the means of constraint. The boundary conditions instead vary according to the constraints at the ends. Table 10.2 shows the different possible cases: a beam supported at either end, a cantilever beam, a beam built-in at one end and supported at the other, a beam with one end built-in and the other constrained with a transverse double rod, a beam with one end build-in and the other constrained by an axial double rod, and a beam supported at one end and constrained at the other by an axial double rod. For each case, Table 10.2 gives the critical load, which can always be expressed in the following form:

$$N_{c1} = \pi^2 \frac{EI}{l_0^2} \tag{10.177}$$

Table 10.2

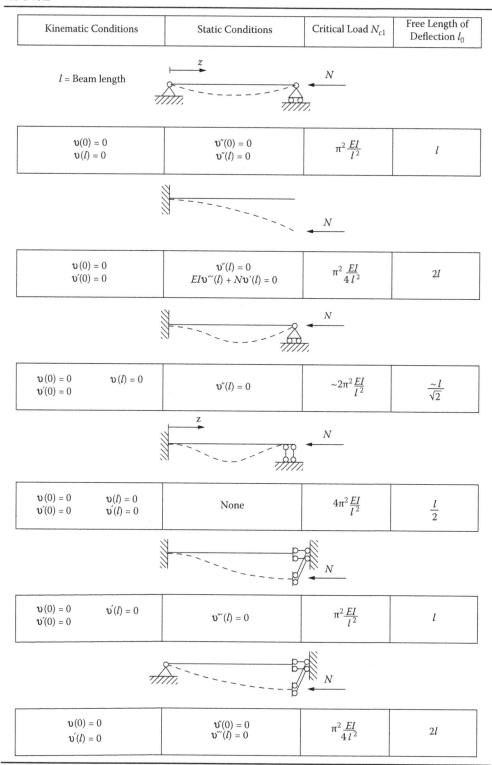

Kinematic Conditions	Static Conditions	Critical Load N_{c1}	Free Length of Deflection l_0
$\upsilon(0) = 0$ $\upsilon(l) = 0$	$\upsilon''(0) = 0$ $\upsilon''(l) = 0$	$\pi^2 \dfrac{EI}{l^2}$	l
$\upsilon(0) = 0$ $\upsilon'(0) = 0$	$\upsilon''(l) = 0$ $EI\upsilon'''(l) + N\upsilon'(l) = 0$	$\pi^2 \dfrac{EI}{4\,l^2}$	$2l$
$\upsilon(0) = 0 \qquad \upsilon(l) = 0$ $\upsilon'(0) = 0$	$\upsilon''(l) = 0$	$\sim 2\pi^2 \dfrac{EI}{l^2}$	$\dfrac{\sim l}{\sqrt{2}}$
$\upsilon(0) = 0 \qquad \upsilon(l) = 0$ $\upsilon'(0) = 0 \qquad \upsilon'(l) = 0$	None	$4\pi^2 \dfrac{EI}{l^2}$	$\dfrac{l}{2}$
$\upsilon(0) = 0 \qquad \upsilon'(l) = 0$ $\upsilon'(0) = 0$	$\upsilon'''(l) = 0$	$\pi^2 \dfrac{EI}{l^2}$	l
$\upsilon(0) = 0$ $\upsilon'(l) = 0$	$\upsilon''(0) = 0$ $\upsilon'''(l) = 0$	$\pi^2 \dfrac{EI}{4\,l^2}$	$2l$

The dimension l_0 is the so-called **free length of deflection**, which is the distance between the two successive points of inflection in the critical deformed configuration.

In some cases, on account of their simplicity, beam systems may correspond to the elementary schemes shown in Table 10.2. In particular, portal frames with rigid cross members may be referred directly to the last four cases, according to whether or not the windbracing is present, and whether the feet of the columns are hinged or built-in (Figure 10.25).

In other cases, the axial redundant reactions, obtainable with the usual equations of congruence, can cause instability of equilibrium. A classical case is that of bars hinged or

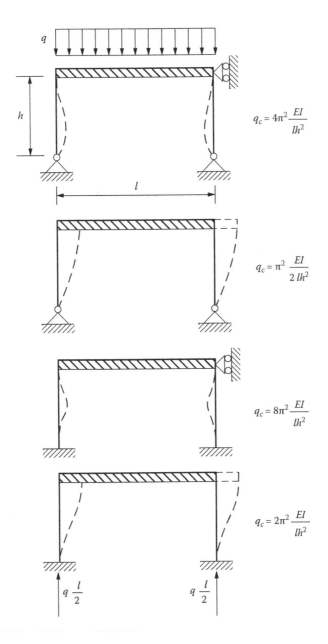

$$q_c = 4\pi^2 \frac{EI}{lh^2}$$

$$q_c = \pi^2 \frac{EI}{2\,lh^2}$$

$$q_c = 8\pi^2 \frac{EI}{lh^2}$$

$$q_c = 2\pi^2 \frac{EI}{lh^2}$$

Figure 10.25

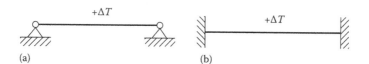

Figure 10.26

built-in at the ends (Figure 10.26a and b), subjected to an increase in temperature and hence to prevented dilation. If the bar is only hinged at the ends, the critical temperature increase is (Figure 10.26a)

$$\Delta T_c = \frac{\pi^2}{\alpha\lambda^2} \tag{10.178}$$

while it is quadrupled if the bar is built-in (Figure 10.26b).

10.11 PLATES IN FLEXURE

Plates are structural elements where one dimension is negligible in comparison with the other two. This dimension is termed **thickness. Plane plates**, in particular, are cylindrical solids whose generators are at least one order of magnitude smaller than the dimensions of the faces (the reverse of the situation we have in the case of the Saint Venant solid).

Let us consider a plate of thickness h, loaded by distributed forces orthogonal to the faces and constrained at the edge (Figure 10.27). Let XY be the middle plane of the plate and Z the orthogonal axis. The so-called **Kirchhoff kinematic hypothesis** assumes that the segments orthogonal to the middle plane, after deformation has occurred, remain orthogonal to the deformed middle plane (Figure 10.28). Denoting then the angle of rotation about the Y axis as φ_x and the angle of rotation about the X axis as φ_y, the displacement of a generic point P of coordinates x, y, z will present the following three components:

$$u = \varphi_x z = -\frac{\partial w}{\partial x} z \tag{10.179a}$$

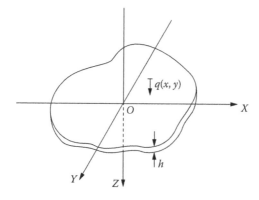

Figure 10.27

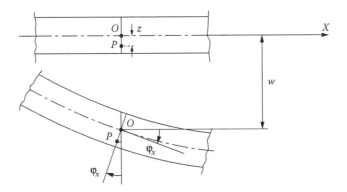

Figure 10.28

$$v = \varphi_y z = -\frac{\partial w}{\partial y} z \tag{10.179b}$$

$$w = w(x, y) \tag{10.179c}$$

The relation (10.179c) indicates that all the points belonging to one and the same segment orthogonal to the middle plane are displaced in that direction by the same quantity.

From the kinematic hypothesis (10.179), it follows, by simple derivation, that the strain field is

$$\varepsilon_x = \frac{\partial u}{\partial x} = \frac{\partial \varphi_x}{\partial x} z = -\frac{\partial^2 w}{\partial x^2} z \tag{10.180a}$$

$$\varepsilon_y = \frac{\partial v}{\partial y} = \frac{\partial \varphi_y}{\partial y} z = -\frac{\partial^2 w}{\partial y^2} z \tag{10.180b}$$

$$\varepsilon_z = \frac{\partial w}{\partial z} = 0 \tag{10.180c}$$

$$\gamma_{xy} = \frac{\partial u}{\partial y} + \frac{\partial v}{\partial x} = \left(\frac{\partial \varphi_x}{\partial y} + \frac{\partial \varphi_y}{\partial x} \right) z = -2 \frac{\partial^2 w}{\partial x \partial y} z \tag{10.180d}$$

$$\gamma_{xz} = \frac{\partial u}{\partial z} + \frac{\partial w}{\partial x} = 0 \tag{10.180e}$$

$$\gamma_{yz} = \frac{\partial v}{\partial z} + \frac{\partial w}{\partial y} = 0 \tag{10.180f}$$

Kirchhoff's kinematic hypothesis generates therefore a condition of plane strain. The three significant components of strain may be expressed as follows:

$$\varepsilon_x = \chi_x z \tag{10.181a}$$

$$\varepsilon_y = \chi_y z \tag{10.181b}$$

$$\gamma_{xy} = \chi_{xy} z \tag{10.181c}$$

where

χ_x and χ_y are the flexural curvatures of the middle plane in the respective directions
χ_{xy} is twice the unit angle of torsion of the middle plane in the X and Y directions

For the condition of plane stress, the constitutive relations (8.73) become

$$\varepsilon_x = \frac{1}{E}(\sigma_x - \nu\sigma_y) \tag{10.182a}$$

$$\varepsilon_y = \frac{1}{E}(\sigma_y - \nu\sigma_x) \tag{10.182b}$$

$$\gamma_{xy} = \frac{1}{G}\tau_{xy} \tag{10.182c}$$

It is important, however, to note that a condition cannot, at the same time, be both one of plane strain and one of plane stress. The thickness h is, on the other hand, assumed to be so small as to enable very low, and consequently negligible, stresses σ_z to develop. From Equations 10.182a and b, we find

$$\sigma_x - \nu\sigma_y = E\varepsilon_x \tag{10.183a}$$

$$\nu\sigma_y - \nu^2\sigma_x = E\nu\varepsilon_y \tag{10.183b}$$

whence, by simple addition, we obtain the expressions

$$\sigma_x = \frac{E}{1-\nu^2}(\varepsilon_x + \nu\varepsilon_y) \tag{10.184a}$$

$$\sigma_y = \frac{E}{1-\nu^2}(\varepsilon_y + \nu\varepsilon_x) \tag{10.184b}$$

$$\tau_{xy} = \frac{E}{2(1+\nu)}\gamma_{xy} \tag{10.184c}$$

From Equations 10.181, there thus follows the stress field of the plate

$$\sigma_x = \frac{E}{1-\nu^2}(\chi_x + \nu\chi_y)z \tag{10.185a}$$

$$\sigma_y = \frac{E}{1-\nu^2}(\chi_y + \nu\chi_x)z \tag{10.185b}$$

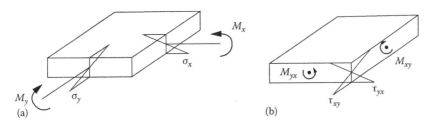

Figure 10.29

$$\tau_{xy} = \frac{E}{2(1+\nu)}\chi_{xy}z \tag{10.185c}$$

Integrating, over the thickness, the stresses expressed by Equations 10.185, we obtain the **characteristics of the internal reaction**, which are bending and twisting moments per unit length (Figure 10.29):

$$M_x = \int_{-h/2}^{h/2} \sigma_x z\,dz \tag{10.186a}$$

$$M_y = \int_{-h/2}^{h/2} \sigma_y z\,dz \tag{10.186b}$$

$$M_{xy} = M_{yx} = \int_{-h/2}^{h/2} \tau_{xy} z\,dz \tag{10.186c}$$

Substituting Equations 10.185 in Equations 10.186, we obtain finally the **constitutive equations** of the plane plate,

$$M_x = D(\chi_x + \nu\chi_y) \tag{10.187a}$$

$$M_y = D(\chi_y + \nu\chi_x) \tag{10.187b}$$

$$M_{xy} = M_{yx} = \frac{1-\nu}{2}D\chi_{xy} \tag{10.187c}$$

where

$$D = \frac{Eh^3}{12(1-\nu^2)} \tag{10.188}$$

is the flexural rigidity of the plate.

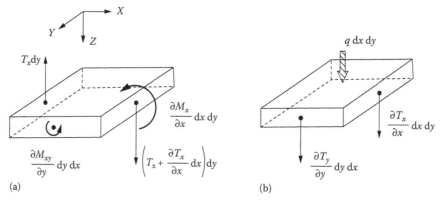

Figure 10.30

Let us then determine the **indefinite equations of equilibrium,** considering an infinitesimal element of the plate submitted to the external load and to the static characteristics. The condition of equilibrium with regard to rotation about the Y axis (Figure 10.30a) yields

$$\left(\frac{\partial M_x}{\partial x}dx\right)dy + \left(\frac{\partial M_{xy}}{\partial y}dy\right)dx - (T_x dy)dx = 0 \tag{10.189}$$

from which we deduce

$$\frac{\partial M_x}{\partial x} + \frac{\partial M_{xy}}{\partial y} - T_x = 0 \tag{10.190a}$$

Likewise, we have

$$\frac{\partial M_{xy}}{\partial x} + \frac{\partial M_y}{\partial y} - T_y = 0 \tag{10.190b}$$

The condition of equilibrium with regard to translation in the direction of the Z axis (Figure 10.30b) gives

$$\left(\frac{\partial T_x}{\partial x}dx\right)dy + \left(\frac{\partial T_y}{\partial y}dy\right)dx + q\,dx\,dy = 0 \tag{10.191}$$

from which we deduce

$$\frac{\partial T_x}{\partial x} + \frac{\partial T_y}{\partial y} + q = 0 \tag{10.192}$$

The remaining three conditions of equilibrium, that with regard to rotation about the Z axis and those with regard to translation in the X and Y directions, are identically satisfied in that the plate has been assumed to be loaded by forces not exerted on the middle plane.

If γ_x and γ_y denote the shearing strains due to the shearing forces, T_x and T_y, respectively, the kinematic equations define the characteristics of deformation as functions of the generalized displacements, in the following way:

$$
\begin{bmatrix} \gamma_x \\ \gamma_y \\ \chi_x \\ \chi_y \\ \chi_{xy} \end{bmatrix} =
\begin{bmatrix}
\dfrac{\partial}{\partial x} & +1 & 0 \\[2mm]
\dfrac{\partial}{\partial y} & 0 & +1 \\[2mm]
0 & \dfrac{\partial}{\partial x} & 0 \\[2mm]
0 & 0 & \dfrac{\partial}{\partial y} \\[2mm]
0 & \dfrac{\partial}{\partial y} & \dfrac{\partial}{\partial x}
\end{bmatrix}
\begin{bmatrix} w \\ \varphi_x \\ \varphi_y \end{bmatrix}
\tag{10.193}
$$

It is to be noted that the shearing strains γ_x and γ_y have so far been neglected, starting from Equations 10.179a and 10.179b, as also in Equations 10.180e and 10.180f.

The **static equations** (10.190) and (10.192), on the other hand, in matrix form are presented as follows:

$$
\begin{bmatrix}
\dfrac{\partial}{\partial x} & \dfrac{\partial}{\partial y} & 0 & 0 & 0 \\[2mm]
-1 & 0 & \dfrac{\partial}{\partial x} & 0 & \dfrac{\partial}{\partial y} \\[2mm]
0 & -1 & 0 & \dfrac{\partial}{\partial y} & \dfrac{\partial}{\partial x}
\end{bmatrix}
\begin{bmatrix} T_x \\ T_y \\ M_x \\ M_y \\ M_{xy} \end{bmatrix}
+ \begin{bmatrix} q \\ 0 \\ 0 \end{bmatrix}
= \begin{bmatrix} 0 \\ 0 \\ 0 \end{bmatrix}
\tag{10.194}
$$

Also in the case of plates, **static–kinematic duality** is expressed by the fact that the static matrix, neglecting the algebraic sign of the unity terms, is the transpose of the kinematic matrix, and *vice versa*.

The **constitutive equations** (10.187), finally, can also be cast in matrix form,

$$
\begin{bmatrix} T_x \\ T_y \\ M_x \\ M_y \\ M_{xy} \end{bmatrix} =
\begin{bmatrix}
\dfrac{5}{6}Gh & 0 & 0 & 0 & 0 \\[2mm]
0 & \dfrac{5}{6}Gh & 0 & 0 & 0 \\[2mm]
0 & 0 & D & \nu D & 0 \\[2mm]
0 & 0 & \nu D & D & 0 \\[2mm]
0 & 0 & 0 & 0 & \dfrac{1-\nu}{2}D
\end{bmatrix}
\begin{bmatrix} \gamma_x \\ \gamma_y \\ \chi_x \\ \chi_y \\ \chi_{xy} \end{bmatrix}
\tag{10.195}
$$

where the factor 5/6 is the inverse of the shear factor corresponding to a rectangular cross section of unit base and thickness h. Note that, while the thickness h appears in the first two rows raised to the first power, in the remaining rows, it appears raised to the third power, in agreement with Equation 10.188. The shearing stiffness appears therefore more important than the flexural stiffness, by as much as two orders of magnitude, and this explains why the shearing strains are often neglected.

The equations of kinematics (10.193) and statics (10.194) and the constitutive equations (10.195) may be cast in compact form,

$$\{q\} = [\partial]\{\eta\} \tag{10.196a}$$

$$[\partial]^*\{Q\} + \{\mathcal{F}\} = \{0\} \tag{10.196b}$$

$$\{Q\} = [H]\{q\} \tag{10.196c}$$

so that if, as we have already done in the case of the 3D solid and the beam, we denote Lamé's matrix operator by

$$\underset{(3\times3)}{[\mathcal{L}]} = \underset{(3\times5)}{[\partial]^*}\underset{(5\times5)}{[H]}\underset{(5\times3)}{[\partial]} \tag{10.197}$$

the elastic problem of the deflected plane plate is represented by the following operator equation furnished with the corresponding boundary conditions:

$$[\mathcal{L}]\{\eta\} = -\{\mathcal{F}\}, \quad \forall P \in S \tag{10.198a}$$

$$[\mathcal{N}]^\mathrm{T}\{Q\} = \{p\}, \quad \forall P \in \mathscr{C}_p \tag{10.198b}$$

$$\{\eta\} = \{\eta_0\}, \quad \forall P \in \mathscr{C}_\eta \tag{10.198c}$$

In the preceding, $\mathscr{C}_p$ denotes the portion of the edge $\mathscr{C}$ on which the static conditions are assigned, while $\mathscr{C}_\eta$ denotes the complementary portion on which the kinematic (or constraint) conditions are assigned.

Designating the unit vector normal to the boundary portion $\mathscr{C}_p$ as $\{n\}$ (Figure 10.31), it is simple to render the boundary condition of equivalence explicitly (10.198b):

$$
\begin{bmatrix}
n_x & n_y & 0 & 0 & 0 \\
0 & 0 & n_x & 0 & n_y \\
0 & 0 & 0 & n_y & n_x
\end{bmatrix}
\begin{bmatrix}
T_x \\ T_y \\ M_x \\ M_y \\ M_{xy}
\end{bmatrix}
=
\begin{bmatrix}
T_n \\ M_{nx} \\ M_{ny}
\end{bmatrix}
\tag{10.199}
$$

M_{nx} and M_{ny} being the components of the moment vector acting on the section of normal n. As in the case of the 3D solid, also in the case of the plate, the matrix $[\mathcal{N}]^\mathrm{T}$ is linked to the

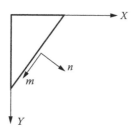

Figure 10.31

matrix operator $[\partial]^*$ of the static equation (10.194), matching each partial derivative with the corresponding direction cosine of the normal to the boundary.

Finally, the **elastic problem of the deflected plane plate** can be summarized as follows:

$$[\mathcal{L}]\{\eta\} = -\{\mathcal{F}\}, \quad \forall P \in S \tag{10.200a}$$

$$([\mathcal{N}]^{\mathrm{T}}[H]\{\partial\})\{\eta\} = \{p\}, \quad \forall P \in \mathscr{C}_p \tag{10.200b}$$

$$\{\eta\} = \{\eta_0\}, \quad \forall P \in \mathscr{C}_\eta \tag{10.200c}$$

Equations 10.200 are formally identical to Equations 8.52, once the 3D domain V has been replaced by the 2D one S of the plate and the external surface S by the boundary $\mathscr{C}$ of the plate.

10.12 SOPHIE GERMAIN EQUATION

Neglecting the shearing deformability of the plate, it is possible to arrive at a differential equation in the single kinematic unknown w. Deriving Equations 10.190, we find

$$\frac{\partial T_x}{\partial x} = \frac{\partial^2 M_x}{\partial x^2} + \frac{\partial^2 M_{xy}}{\partial x \partial y} \tag{10.201a}$$

$$\frac{\partial T_y}{\partial y} = \frac{\partial^2 M_{xy}}{\partial x \partial y} + \frac{\partial^2 M_y}{\partial y^2} \tag{10.201b}$$

and substituting the foregoing Equations 10.201 in Equation 10.192, we obtain

$$\frac{\partial^2 M_x}{\partial x^2} + 2\frac{\partial^2 M_{xy}}{\partial x \partial y} + \frac{\partial^2 M_y}{\partial y^2} + q = 0 \tag{10.202}$$

The constitutive equations (10.187), if we neglect the shearing strains, become

$$M_x = -D\left(\frac{\partial^2 w}{\partial x^2} + \nu\frac{\partial^2 w}{\partial y^2}\right) \tag{10.203a}$$

$$M_y = -D\left(\frac{\partial^2 w}{\partial y^2} + \nu\frac{\partial^2 w}{\partial x^2}\right) \tag{10.203b}$$

$$M_{xy} = -D(1-\nu)\frac{\partial^2 w}{\partial x \partial y} \tag{10.203c}$$

Substituting Equations 10.203 in Equation 10.202, we deduce finally the **Sophie Germain equation,**

$$\frac{\partial^4 w}{\partial x^4} + 2\frac{\partial^4 w}{\partial x^2 \partial y^2} + \frac{\partial^4 w}{\partial y^4} = \frac{q}{D} \tag{10.204}$$

which is the fourth-order differential equation corresponding to the **elastic plane.** If we indicate the Laplacian by

$$\nabla^2 = \frac{\partial^2}{\partial x^2} + \frac{\partial^2}{\partial y^2} \tag{10.205}$$

Equation 10.181 can also be written as

$$\nabla^2(\nabla^2 w) = \frac{q}{D} \tag{10.206}$$

or, even more synthetically,

$$\nabla^4 w = \frac{q}{D} \tag{10.207}$$

Note the formal analogy between Equation 10.207 and the equation of the elastic line (10.49).

The term **principal directions of moment** relative to a point of the deflected plate is given to the two orthogonal directions along which the twisting moment M_{xy} vanishes and consequently the shearing stresses τ_{xy} likewise vanish. These directions thus coincide with the principal ones of stress. The term **principal directions of curvature** is applied to the two orthogonal directions along which the unit angle of torsion $\chi_{xy}/2$ vanishes. In the case where the material is assumed to be isotropic, the constitutive equation (10.187c) shows how the principal directions of moment and the principal directions of curvature must coincide.

The **Finite Difference Method** for the approximate numerical solution of the Sophie Germain equation is proposed in Appendix F, while the multilayer plates of **composite materials** are discussed in Appendix E.

Chapter 11

Statically indeterminate beam systems

Method of forces

11.1 INTRODUCTION

Redundant beam systems, i.e. ones that contain a surplus number of constraints, are statically indeterminate. As we have seen in Chapter 3, this means that they can be balanced by ∞^{v-g} different sets of reactive forces, $v - g$ being the degree of redundancy. We therefore need to identify the particular single set of reactive forces which, in addition to equilibrium, also implies **congruence**, or rather the respect of the internal and external constraints, notwithstanding the deformations induced in the structural elements.

From the operative viewpoint, the **method of forces** consists of eliminating $v - g$ degrees of constraint, so as to reduce the given structure to a statically determinate beam system, and applying to this system, in addition to the external forces, the unknown constraint reactions exerted by the constraints that have been removed. The $v - g$ equations of congruence will then impose abeyance of the kinematic conditions corresponding to the suppressed constraints and, once resolved, will yield the $v - g$ elementary reactions exerted by these constraints, which are called **hyperstatic unknowns**.

When resolving a statically indeterminate structure, one is therefore confronted with the problem of finding a suitable way of disconnecting it so as to obtain the statically determinate scheme on which to impose the conditions of congruence. In principle, the disconnection can be performed in an infinite number of different ways, as it is possible to reduce the degrees of constraint both externally and internally, and, among the internal constraints, it is possible to reduce the infinite internal fixed-joint constraints which guarantee the continuity of the beam. Normally, however, it is convenient to reduce or suppress the external constraints, or to interrupt the continuity of the structure, by inserting hinges in points of concurrence of two or more beams (fixed-joint nodes). In the first case, the equations of congruence will impose the annihilation of the displacements of the points that are sites of statically indeterminate reactions, whereas in the second case, the so-called **angular congruence** will be imposed, i.e. an equal elastic rotation at all the beam ends which converge at the same fixed-joint node.

11.2 AXIAL INDETERMINACY

Consider a rectilinear beam of length l, hinged at the ends A and B and subjected to an axial force F, acting at a distance a from the end A and b from the end B (Figure 11.1a). Thus loaded, the beam is statically indeterminate, since the pairs of reactions H_A and H_B, which together with the force F make up a balanced system, are infinite. Replacing the hinge B with a roller support having a horizontal plane of movement (Figure 11.1b) and applying the hyperstatic unknown X at the same end B, we obtain the equivalent statically determinate

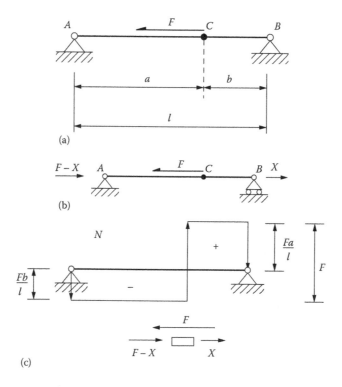

Figure 11.1

scheme. The equation of congruence must express the existence of the suppressed constraint, i.e. the displacement of the roller support is zero:

$$w_B = \frac{X}{EA}l - \frac{F}{EA}a = 0 \tag{11.1}$$

where the first term represents the contribution of the reaction X, while the second represents the contribution of the external force. Notice that we have implicitly made use of the principle of superposition. The force X, in fact, generates a characteristic of tension on the entire beam, while the force F generates a characteristic of compression only on the portion AC. The force F thus contracts the portion AC, while the portion CB is drawn along by a rigid translation.

From Equation 11.1, we derive

$$X = F\frac{a}{l} \tag{11.2a}$$

$$F - X = F\frac{b}{l} \tag{11.2b}$$

so that the force F is supported by the two end constraints in direct proportion to the reciprocal distances from the point of application. The portion CB is thus subjected to tension, while the portion AC is subjected to compression.

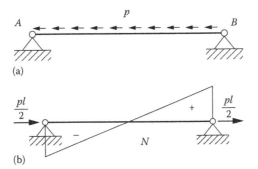

(a)

(b)

Figure 11.2

The axial force diagram (Figure 11.1c) thus shows a discontinuity at the point of application of the force. The element of beam straddling this point is in equilibrium under the action of the external force and of two internal reactions having the same sense.

If the beam considered previously were submitted to a uniform distribution of axial forces p (Figure 11.2a), the skew-symmetry of the structural scheme would make it possible to recognize two equal constraint reactions having the same sense, so that, on the basis of Equation 5.12a, the axial force diagram would be linear and skew-symmetrical, with a zero in the centre and the extreme values equal to $-\frac{1}{2}pl$ in A and $+\frac{1}{2}pl$ in B (Figure 11.2b).

Finally, let us consider a case of double axial redundancy: a beam of length $2l$ hinged at the ends and in the centre, loaded by a concentrated axial force acting in the centre of the left-hand span (Figure 11.3a). The equivalent statically determinate scheme may be obtained by transforming two of the three hinges into as many horizontally moving roller supports. Figure 11.3b depicts the scheme with the roller supports in B and C and the respective redundant reactions X_1 and X_2. The two equations of congruence express the immovability of the points B and C,

$$w_B = \frac{(X_1 + X_2)}{EA}l - \frac{F}{EA}\frac{l}{2} = 0 \tag{11.3a}$$

$$w_C = \frac{X_2}{EA}2l + \frac{X_1}{EA}l - \frac{F}{EA}\frac{l}{2} = 0 \tag{11.3b}$$

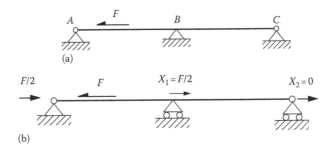

(a)

(b)

Figure 11.3

from which we obtain

$$X_1 = \frac{F}{2}, \quad X_2 = 0 \tag{11.4}$$

The solution found expresses the fact that only the two hinges between which the external force F is applied react.

11.3 ELEMENTARY STATICALLY INDETERMINATE SCHEMES

Consider the rectilinear beam of length l, built-in at the end A and supported at the end B (Figure 11.4a), subjected to the distributed load q. This structure has one degree of redundancy. The equivalent statically determinate scheme is obtained by eliminating one of the three external constraints (excluding the axial one) and imposing congruence, i.e. abeyance of the constraint that has been suppressed. Alternatively, though only in principle, it would also be possible to disconnect the beam internally, but this would not prove to be a convenient approach in actual operative terms.

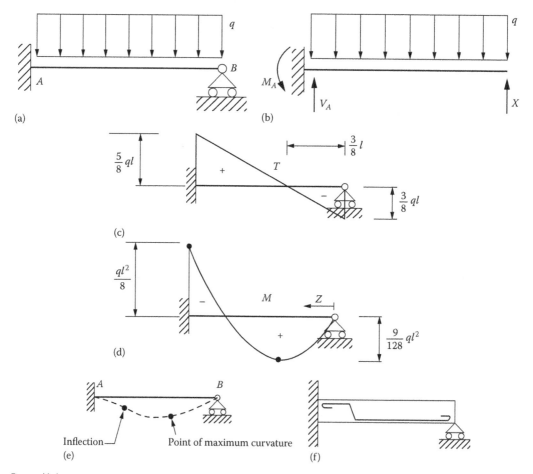

Figure 11.4

A first equivalent statically determinate scheme is obtained by eliminating the roller support in B and subjecting the cantilever beam AB, not only to the distributed load q, but also to the redundant reaction X, which is an unknown vertical force acting at the end B (Figure 11.4b). Superposing the effects, the condition of congruence becomes

$$v_B = \frac{ql^4}{8EI} - \frac{Xl^3}{3EI} = 0 \tag{11.5}$$

This equation contains the single unknown X. From Equation 11.5, we obtain

$$X = \frac{3}{8}ql \tag{11.6}$$

The reactions at the built-in end are then equal to (Figure 11.4b)

$$V_A = ql - \frac{3}{8}ql = \frac{5}{8}ql \tag{11.7a}$$

$$M_A = \frac{1}{2}ql^2 - \frac{3}{8}ql^2 = \frac{1}{8}ql^2 \tag{11.7b}$$

The shear diagram is thus linear with extreme values equal to $\frac{5}{8}ql$ in A and $-\frac{3}{8}ql$ in B (Figure 11.4c). The bending moment diagram may be plotted by points. The moment at the built-in constraint is equal in fact to $\frac{1}{8}ql^2$, while the moment is zero at the hinge B (Figure 11.4d) and at the point where the function

$$M(z) = \frac{3}{8}qlz - \frac{1}{2}qz^2 = -\frac{1}{2}qz\left(z - \frac{3}{4}l\right) \tag{11.8}$$

vanishes, i.e. at a distance $z = \frac{3}{4}l$ from the end B. Another notable value of the moment is the maximum one at the point of zero shear:

$$M_{\max} = M\left(\frac{3}{8}l\right) = \frac{9}{128}ql^2 \tag{11.9}$$

On the basis of the four notable points discussed earlier, the diagram can be immediately plotted (Figure 11.4d). The elastic deformed configuration of the beam can be plotted qualitatively, with respect to the constraints and the moment diagram (Figure 11.4e). The built-in constraint A imposes a deformed configuration with zero vertical displacement and zero rotation at A, just as the roller support B imposes a deformed configuration with zero vertical displacement at B, while it allows rotation of the end section B. The deformed configuration of course presents a point of inflection where the moment becomes zero and then undergoes a change in sign. There will be extended fibres in the upper portion between the end A and the inflection, and, *vice versa*, in the lower portion between the inflection and the end B.

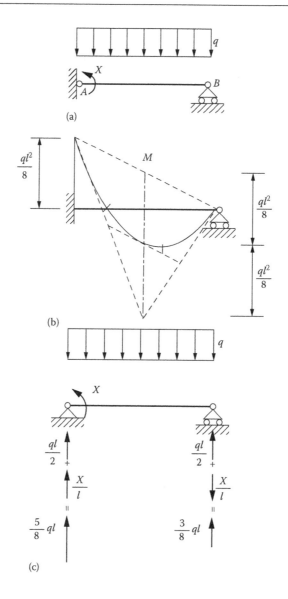

Figure 11.5

Summarizing, we can say that the deformed configuration will present the point of inflection at a distance $\frac{3}{4}l$ from the end B, and the point of maximum curvature at the distance $\frac{3}{8}l$ once again from point B (Figure 11.4e). In the case of the beam made of reinforced concrete, the reinforcement must follow the path of the stretched fibres (Figure 11.4f).

   A second equivalent statically determinate scheme is obtained by eliminating the degree of constraint with regard to rotation of the built-in end A, i.e. by replacing the built-in constraint with a hinge and by applying an unknown redundant moment X at the end A itself (Figure 11.5a). Summing up the elastic rotations of the end section A, due respectively to the external load q and to the redundant reaction X, we obtain the equation of congruence

$$\varphi_A = -\frac{ql^3}{24EI} + \frac{Xl}{3EI} = 0 \tag{11.10}$$

from which there follows

$$X = \frac{1}{8}ql^2 \tag{11.11}$$

This result coincides with the moment at the built-in constraint, deduced in the previous solution.

The moment diagram, in the framework of the present scheme of resolution, may be obtained graphically (Figure 11.5b). The partial diagram due to the redundant reaction is linear with extreme values equal to $-\frac{1}{8}ql^2$ in A and zero in B. On the other hand, the partial diagram due to the distributed load is parabolic with the maximum which again equals $\frac{1}{8}ql^2$. The graphical sum of the two partial diagrams may be obtained by following the usual procedure, outlined in Chapters 4 through 6, based on the properties of the arcs of parabola. From the mid-point of the triangular diagram, two consecutive vertical segments are drawn, each having a length of $\frac{1}{8}ql^2$, so as to obtain the third point of the parabola and the point of intersection of the end tangents, respectively. The third tangent is parallel to the line joining the end points. The total diagram thus obtained coincides with that of Figure 11.4d.

As regards the shear diagram, the equivalent statically determinate scheme must be balanced by two vertical forces at either end, directed upwards and equal to $\frac{1}{2}ql$, and by a clockwise couple of vertical forces $X/l = \frac{1}{8}ql$ (Figure 11.5c). Making the vector summation of the partial reactions, we obtain once more the extreme values of shearing force shown in Figure 11.4c.

Let us now consider the same beam, built-in at one end and supported at the other, loaded in this case by a couple m at the end B (Figure 11.6a). Even though it is possible to accommodate the case within the equivalent statically determinate scheme consisting of the cantilever beam and obtained by eliminating the roller support, in the ensuing treatment, we

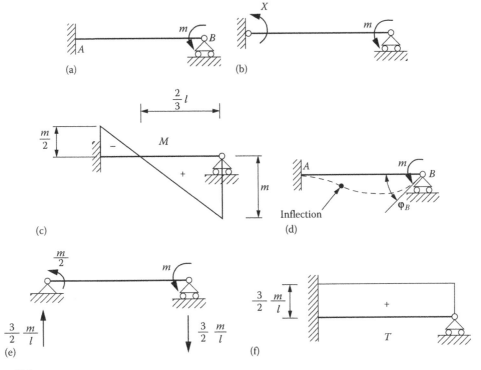

Figure 11.6

shall consider the second scheme used previously: that of the beam supported at either end (Figure 11.6b). The condition of congruence is

$$\varphi_A = \frac{Xl}{3EI} - \frac{ml}{6EI} = 0 \tag{11.12}$$

from which we obtain

$$X = \frac{m}{2} \tag{11.13}$$

The moment at the built-in constraint is then half that applied. The moment diagram is linear and hence vanishes at the distance $\frac{2}{3}l$ from the support (Figure 11.6c). At that point, the elastic deformed configuration of the beam undergoes an inflection (Figure 11.6d). It is, moreover, possible to calculate the elastic rotation of the end section B:

$$\varphi_B = \frac{ml}{3EI} - \frac{(m/2)l}{6EI} = \frac{ml}{4EI} \tag{11.14}$$

It is important to introduce the concept of rotational stiffness of the beam built-in at one end and supported at the other, for moments applied at the supported end:

$$k = \frac{m}{\varphi_B} = \frac{4EI}{l} \tag{11.15}$$

This stiffness is directly proportional to the elastic modulus of the material and to the moment of inertia of the cross section, and inversely proportional to the length of the beam.

The vertical reactions must balance the two moments having the same sense, m and $X = m/2$. They are thus equal and opposite forces of magnitude $(3/2)\,m/l$ (Figure 11.6e). The shear diagram is thus constant and positive (Figure 11.6f).

Note that, if in the cases hitherto examined we were to replace the roller support with a hinge, the solutions would not vary at all, on account of the absence of the axial force. Even if a redundant reaction were supposed, this would be zero, thus yielding the only contribution to the equation of axial congruence $w_B = 0$ (Figure 11.1b).

In the case of a rectilinear beam built in at both ends, loaded in any manner whatsoever (Figure 11.7a), the degree of static indeterminacy is three and can be eliminated by removing one of the two built-in constraints (Figure 11.7b). The three redundant unknowns, consisting of the elementary built-in constraint reactions, may be determined using the three equations of congruence with regard to horizontal translation, to vertical translation and to rotation, respectively:

$$w_B = 0, \quad v_B = 0, \quad \varphi_B = 0 \tag{11.16}$$

In the particular case of a vertical load concentrated in the centre (Figure 11.8a), the scheme has only one degree of redundancy. The other two degrees of redundancy, which are potentially present, do not appear owing to symmetry and the lack of horizontal components in

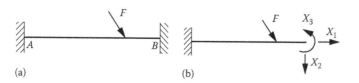

(a)   (b)

Figure 11.7

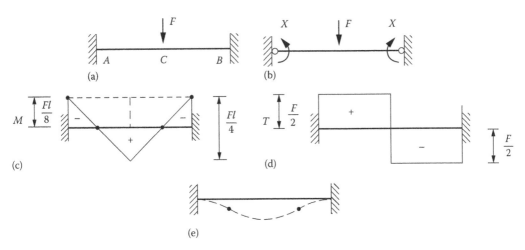

Figure 11.8

the external loading. The equivalent statically determinate scheme is then found by inserting two hinges at the ends and applying two equal and opposite redundant moments X (Figure 11.8b). Just as there is only one degree of redundancy, there is also a single equation of congruence with regard to rotation:

$$\varphi_A = -\varphi_B = \frac{Xl}{3EI} + \frac{Xl}{6EI} - \frac{Fl^2}{16EI} = 0 \tag{11.17}$$

whence we obtain

$$X = \frac{1}{8}Fl \tag{11.18}$$

The bending moment diagram can be constructed graphically by superposition (Figure 11.8c). The two redundant moments furnish a constant partial diagram which is equal in absolute value to $1/8Fl$, whilst the central force F produces a bilinear partial diagram with the maximum equal to $1/4Fl$. The total diagram thus intersects the axis of the beam at two symmetrical points at a distance $l/4$ from the built-in constraints.

The shear diagram is that produced exclusively by the force F, the two redundant moments constituting a self-balanced system (Figure 11.8d).

Since the elastic deformed configuration of the beam has to satisfy the constraint conditions as well as the deflections suggested by the moment diagram, it will appear as in Figure 11.8e, with two inflections corresponding to the points where the bending moment becomes zero.

It is interesting to see how, by using the properties of symmetry of the structure, it is possible to reduce it even to a statically determinate scheme and so resolve it with the use of equilibrium equations alone. One may consider just the half-beam on the left, once the centre C is constrained with a double connecting rod (Figure 11.9a). The vertical force $F/2$ is countered by the built-in constraint, so that it is possible to transform the scheme of Figure 11.9a into that of Figure 11.9b, where the built-in constraint has been replaced by a second double rod loaded by the vertical reaction $F/2$. This latter scheme is skew-symmetrical and can be reduced to that of Figure 11.9c, which presents a roller support in the new centre and thus

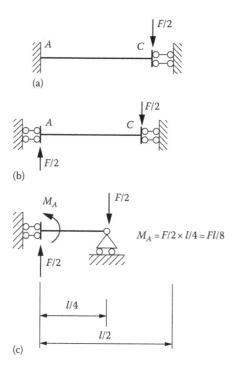

Figure 11.9

emerges as statically determinate. It is clear that the reactive moment exerted by the double rod A, in the scheme of Figure 11.9c, is

$$M_A = \frac{F}{2} \times \frac{l}{4} = \frac{1}{8} Fl$$

and represents the fixed-end moment already defined following another procedure.

Adopting the same synthetic approach seen hitherto, it is not possible, on the other hand, to resolve the case of a beam built in at both ends and subjected to a distributed load q (Figure 11.10a). The equivalent statically determinate scheme is that obtained, as we have

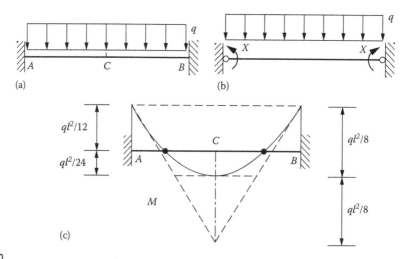

Figure 11.10

already seen, by replacing the built-in constraints with hinges and by applying two equal and opposite redundant moments (Figure 11.10b). The equation of congruence is

$$\varphi_A = -\varphi_B = \frac{Xl}{3EI} + \frac{Xl}{6EI} - \frac{ql^3}{24EI} = 0 \tag{11.19}$$

from which we obtain

$$X = \frac{ql^2}{12} \tag{11.20}$$

Whereas then the shear diagram is equal to one of the scheme of a beam supported at both ends and subjected to a distributed load q (Figure 5.20c), the moment diagram is obtained from the graphical addition of a constant diagram with a value equal to $-\frac{1}{12}ql^2$ and a parabolic diagram with a maximum of $\frac{1}{8}ql^2$ (Figure 11.10c). The moment in the centre is thus $M_C = \frac{1}{24}ql^2$.

The scheme of a beam constrained by a built-in support and a double rod, loaded by a vertical force F applied to the double rod (Figure 11.11a), from symmetry is equivalent to the beam of twice the length, i.e. $2l$, built in at both ends and subjected to twice the force, $2F$, in the centre (Figure 11.11b), or to the beam of length $l/2$, constrained by a double rod and a roller support (Figure 11.11c). In either case, the bending moment and shearing force diagrams are those represented in Figure 11.11d and e, respectively. The elastic displacement of point B can be determined in various ways. The simplest is, however,

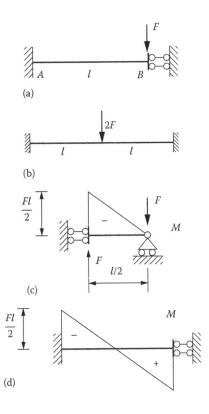

Figure 11.11

(continued)

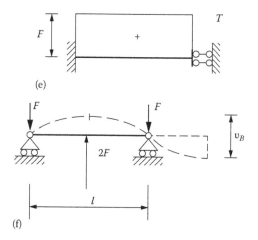

(e)

(f)

Figure 11.11 (continued)

the one based on the scheme of Figure 11.11c, or rather, by symmetry, on the scheme of Figure 11.11f:

$$v_B = 2\left(\frac{2Fl^3}{48EI}\right) = \frac{Fl^3}{12EI} \tag{11.21}$$

Likewise, the scheme of a beam constrained by a hinge and a double rod, acted upon by a vertical force F applied at the double rod (Figure 11.12a), from symmetry is equivalent to the beam of twice the length, $2l$, supported at both ends and subjected in the centre to twice the force, $2F$ (Figure 11.12b). The bending moment and shear diagrams are those represented in Figure 11.12c and d. The vertical displacement of the point B constrained by the double rod is

$$v_B = \frac{2F(2l)^3}{48EI} = \frac{Fl^3}{3EI} \tag{11.22}$$

The aforementioned structure, loaded by a moment at the hinged end (Figure 11.13a), is equivalent to the beam of twice the length, $2l$, supported at both ends and subjected to two opposite moments at the ends (Figure 11.13b). The moment is thus constant and the shear absent. The rotation at the extreme section A is

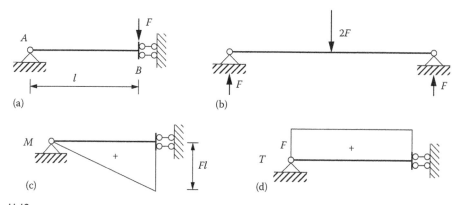

(a)

(b)

(c)

(d)

Figure 11.12

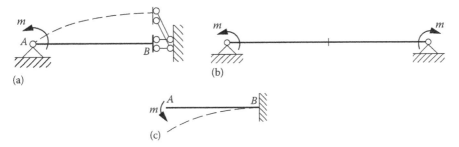

Figure 11.13

$$\varphi_A = \frac{m(2l)}{3EI} + \frac{m(2l)}{6EI} = \frac{ml}{EI} \tag{11.23}$$

Notice that the rotation φ_A is the same as that undergone by the end section of a cantilever beam, built in at B and loaded by the moment m at A (Figure 11.13c). In fact, the moment diagrams of the schemes of Figure 11.13a and c coincide, just as the same boundary condition, $v'_B = 0$, applies in both cases. The deformed configuration, on the other hand, remains the same but for one additional constant (vertical translation).

11.4 ELASTIC CONSTRAINTS

Up to now, the constraints, whether internal or external, have been considered as rigid, i.e. as conditions of congruence, where the displacements or rotations vanish. In practice, however, the constraints cannot always be treated simply as rigid. They are said to settle elastically when the reaction of the constraint is proportional to the displacement undergone by the constraint itself. In what follows, we shall compare the results for rigidly constrained redundant structures with those for the same structures constrained elastically. In statically determinate structures, on the other hand, the constraint reactions and the diagrams of characteristics do not depend on the stiffness of the constraints, since in any case the equilibrium equations are the same.

Let us consider the continuous beam on three supports shown in Figure 11.14a, subjected to the moment m acting at the end C. There exists an infinite number of triads of constraint reactions V_A, V_B, V_C, equilibrants of the moment m. The one that also ensures congruence could be obtained by eliminating any one of the three supports, by applying the corresponding unknown redundant reaction and by imposing the condition that the corresponding vertical displacement should become zero. An even faster approach is that of interrupting the continuity of the beam by inserting a hinge on the support B (Figure 11.14b) and applying the redundant reactions transmitted by the removed constraint, viz. two equal and opposite moments. The condition of congruence at this point will concern the continuity of the elastic line, which will not be able to present cusps in B. That is, the section B thought of as belonging to the beam AB must rotate by the same amount as that by which the same cross section, thought of as belonging to the beam BC, rotates:

$$\varphi_{BA} = \varphi_{BC} \tag{11.24}$$

Rendering both sides explicit, we have

$$\frac{Xl}{3EI} = -\frac{Xl}{3EI} + \frac{ml}{6EI} \tag{11.25}$$

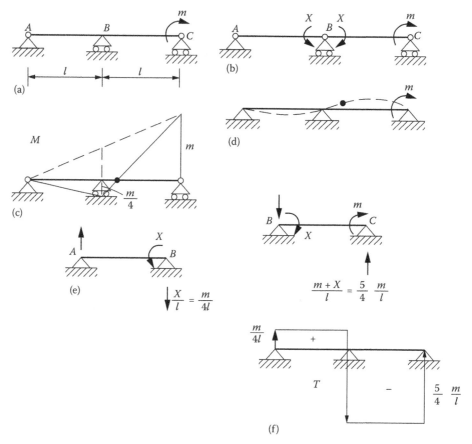

Figure 11.14

from which we obtain

$$X = \frac{m}{4} \qquad\qquad (11.26)$$

Knowing now the moments at the three supports, the moment diagram can at once be drawn (Figure 11.14c). It will suffice to lay out to scale a segment of length m above the support C, a segment of length $m/4$ beneath the support B and then join the notable points of the diagram thus defined with straight line segments. A point of annihilation of the moment in the right-hand span is then identified at the distance $l/5$ from the central support. At this point, the elastic deformed configuration possesses an inflection (Figure 11.14d).

The constraint reactions, and hence the shear diagram, may be determined by isolating the supported beams AB and BC, acted upon by the external and the redundant loads (Figure 11.14e). The two end reactions are directed upwards and are $V_A = m/4l$, $V_C = 5m/4l$, while the reaction of the intermediate support is the sum of the reactions that apply to the two schemes of Figure 11.14e, $V_B = 3m/2l$. It is a force directed downwards, which produces a discontinuity of the first kind in the shear diagram (Figure 11.14f).

Consider again the foregoing scheme, assuming, however, an intermediate elastically compliant support (Figure 11.15a). This compliance can, for instance, represent the axial compliance of a connecting rod. In this case, the stiffness of the equivalent spring is $k = EA/l$ (Figure 11.15b). When the external constraints are elastically compliant, it is

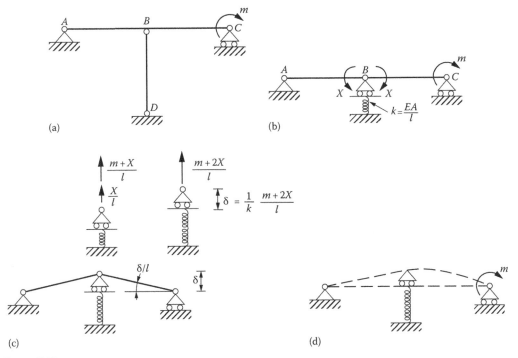

Figure 11.15

necessary to consider, not only, as in the usual case, the action of the constraint on the structure, but also the action of the structure on the compliant constraint. The scheme of Figure 11.15c shows how the spring is loaded by a force X/l transmitted by the left-hand beam and by a force $(m + X)/l$ transmitted by the right-hand beam, both directed upwards in accordance with the conventions assumed. The support will then rise by the quantity

$$\delta = \frac{m + 2X}{kl} \tag{11.27}$$

which is a function of the redundant unknown X and will induce a rigid rotation in both beams, with an absolute value equal to δ/l (Figure 11.15c).

The equation of congruence, in implicit form, is once more Equation 11.24, where, on the other hand, also the contributions of rigid rotation must appear, as well as those of elastic rotation previously considered. We shall therefore have

$$\frac{Xl}{3EI} + \frac{m + 2X}{kl^2} = -\frac{Xl}{3EI} + \frac{ml}{6EI} - \frac{m + 2X}{kl^2} \tag{11.28}$$

whence we obtain

$$X = m \frac{[l/(6EI) - 2/(kl^2)]}{[2l/(3EI) + 4/(kl^2)]} \tag{11.29}$$

When the stiffness of the spring tends to infinity, we find again the previous result (Figure 11.14c)

$$\lim_{k \to \infty} X = \frac{m}{4} \tag{11.30}$$

On the other hand, when the stiffness of the spring tends to zero, we find again the value of moment in the centre for a beam supported at both ends, having a length of $2l$ loaded by a moment at the end:

$$\lim_{k \to 0} X = -\frac{m}{2}$$ (11.31)

Note that the moment X vanishes when

$$k = \frac{12EI}{l^3}$$ (11.32)

Hence, for smaller stiffness, the inflection disappears in the elastic line, and only the upper fibres are stretched. For $k = 12EI/l^3$, the moment is zero in the left-hand span, since only the supports B and C react, and the deformed configuration of the beam is rigid between the supports A and B (Figure 11.15d), i.e. the left-hand span rotates rigidly counterclockwise, drawn along by the deflection of the right-hand span.

As a second example of elastic constraint, consider again the beam built in at one end and supported at the other, acted upon by a distributed load q (Figure 11.16a). In this case, assume that the built-in constraint A is rotationally compliant. The equivalent statically determinate scheme is the same as that of Figure 11.5a, but, in the equation of congruence, the action of the beam on the built-in constraint must be taken into account (Figure 11.16b). Whereas, that is, the built-in constraint acts on the beam with a counterclockwise moment X, the beam will act on the built-in constraint with a clockwise moment X and will cause it to rotate by an angle $-X/k$. The equation of congruence is thus modified as follows:

$$\varphi_A = \frac{Xl}{3EI} - \frac{ql^3}{24EI} = -\frac{X}{k}$$ (11.33)

from which we obtain the redundant unknown

$$X = \frac{[ql^3/(24EI)]}{[l/(3EI) + 1/k]}$$ (11.34)

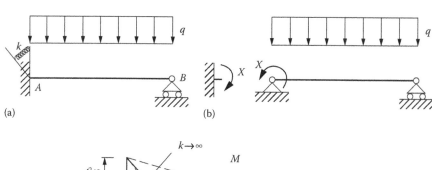

(a)   (b)

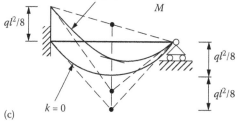

(c)

Figure 11.16

When the rotational stiffness of the built-in constraint tends to infinity, we find again the **rigid joint moment**

$$\lim_{k \to \infty} X = \frac{1}{8} q l^2 \qquad (11.35a)$$

On the other hand, when the rotational stiffness of the built-in constraint tends to zero, the moment vanishes because the constraint A turns into a hinge:

$$\lim_{k \to 0} X = 0 \qquad (11.35b)$$

The moment diagram will thus in general be contained between the two limit diagrams of Figure 11.16c.

11.5 INELASTIC CONSTRAINTS (IMPOSED DISPLACEMENTS)

The cases of inelastic constraint settlements, which will be dealt with in the ensuing discussion, can be more appropriately termed **imposed displacements**. In fact, it is not a question of modifying in some way the reactive properties of the constraint, but rather of imposing a predetermined displacement on the constraint itself. Such a displacement will therefore not be a function of the loads and the redundant unknowns, but will itself perform the function of an external load, it being a datum of the problem.

In the case where infinitesimal displacements are imposed on the constraints of a statically determinate beam system, the system adapts by undergoing only rigid rototranslations. Consequently, no external or internal reactions develop. On the other hand, with the exception of particular cases, it may be stated that the displacements imposed on a statically indeterminate beam system generate reactions and deflections in the beams. Evidently, this is due to the redundant degree of constraint, which may be said to oppose the deformation of the system. A rational explanation of this may be found in the matrix treatment of the kinematics of rigid systems, presented in Chapter 3.

The kinematic matrix of a statically determinate beam system is square, and hence the augmented matrix, containing the imposed displacements, will have the same rank, whatever the displacements imposed on the constraints. By virtue of the Rouché–Capelli theorem, it is thus possible to state that there exists a single kinematic solution, represented by the rigid deformed configuration of the system.

The kinematic matrix of a statically indeterminate beam system is rectangular and is augmented by the imposed displacements on its longer side. This allows the augmented matrix to have a rank greater than that of the kinematic matrix, unless the imposed displacements are so particular as to produce a column of known terms linearly dependent on the columns of the kinematic matrix. If we exclude the latter case, the Rouché–Capelli theorem makes it possible to assert that there does not exist a solution in the framework of the kinematics of the rigid body. The beam system will thus have to adapt by undergoing deformations (dilations and deflections). The solution is determinable only in the context of the kinematics of the deformable body.

Consider a beam built in at one end and supported at the other, and let the vertical displacement η_0 of the built-in constraint be imposed (Figure 11.17a). Let the equivalent statically determinate scheme be that of a beam supported at both ends (Figure 11.17b). This scheme rotates only rigidly, after the imposition of the displacement η_0 (Figure 11.17c). The equation of congruence will consider both the elastic rotation induced by the redundant moment X, and the rigid rotation η_0/l,

$$\varphi_A = -\frac{Xl}{3EI} + \frac{\eta_0}{l} = 0 \qquad (11.36)$$

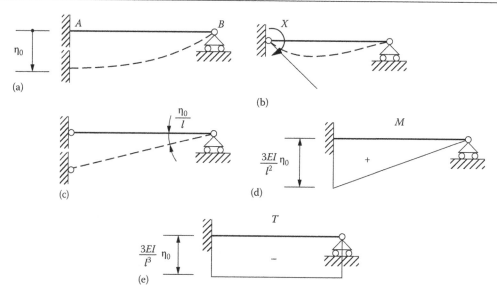

Figure 11.17

from which we obtain

$$X = \frac{3EI}{l^2} \eta_0 \tag{11.37}$$

The reactive moment of the built-in constraint is thus proportional to the magnitude η_0 of the inelastic settlement. The bending moment diagram is thus linear with a maximum at the built-in constraint (Figure 11.17d), whilst the shear is constant and equal to $-X/l$ (Figure 11.17e).

A different equivalent statically determinate scheme could be that of a cantilever beam, with the reaction V_B at the free end determinable using the equation of congruence

$$\upsilon_B = \eta_0 - \frac{V_B l^3}{3EI} = 0 \tag{11.38}$$

whence

$$V_B = \frac{3EI}{l^3} \eta_0 \tag{11.39}$$

On the other hand, the reaction V_B may be obtained also by inverting the formula (11.22) corresponding to the scheme of a beam with a hinge and a double rod (Figure 11.12a).

Let a rotation φ_0 of the built-in constraint be imposed on the scheme of the beam built in at one end and supported at the other (Figure 11.18a). Let the equivalent statically determinate scheme consist of a beam supported at either end, loaded by the unknown moment X (Figure 11.18b). The condition of rotational congruence thus reads

$$\varphi_A = \frac{Xl}{3EI} = \varphi_0 \tag{11.40}$$

from which we deduce

$$X = \frac{3EI}{l} \varphi_0 \tag{11.41}$$

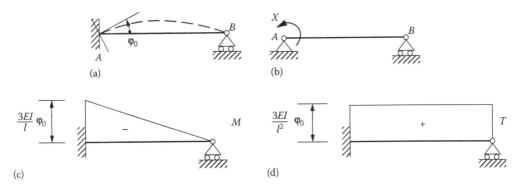

Figure 11.18

Also in this case, the moment diagram is linear (Figure 11.18c) with a maximum at the built-in constraint, while the shear is constant and equal to X/l (Figure 11.18d).

Let us now pass on to the beam built in at both ends, and let a vertical displacement and a rotation be imposed separately to one of the two built-in constraints. In the case of the imposed displacement η_0 (Figure 11.19a), the scheme can be split up into one symmetrical scheme plus one skew-symmetrical scheme (Figure 11.19b). Whereas the symmetrical scheme represents a simple rigid translation $\eta_0/2$ downwards of the entire beam, the skew-symmetrical scheme can be brought back to the one previously studied of a beam built in at one end and supported at the other (Figure 11.19c), for which the moment at the built-in constraint is

$$X = \frac{3EI}{(l/2)^2}\frac{\eta_0}{2} = \frac{6EI}{l^2}\eta_0 \tag{11.42}$$

The moment diagram is thus linear and skew-symmetrical (Figure 11.19d), while the shear diagram is constant and equal to (Figure 11.19e)

$$T = -\frac{2X}{l} = -\frac{12EI}{l^3}\eta_0 \tag{11.43}$$

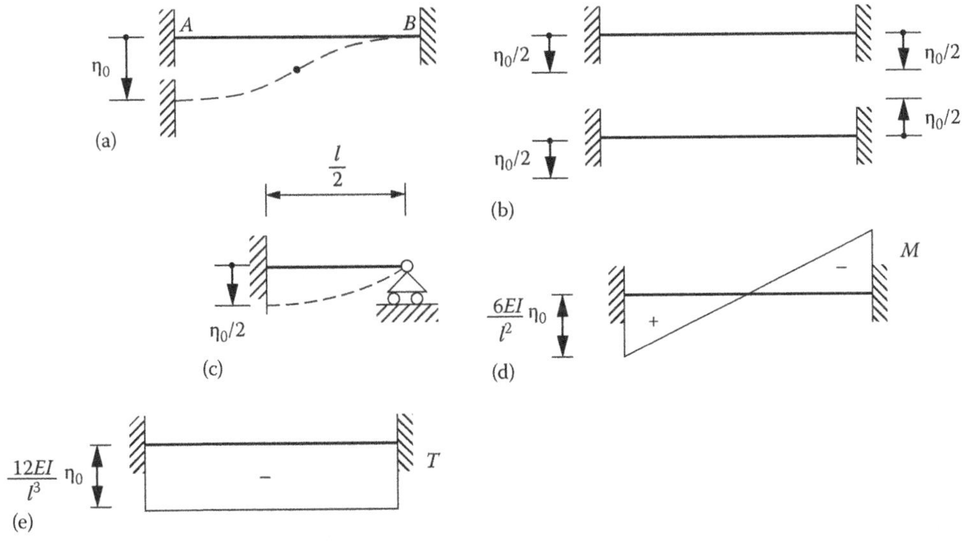

Figure 11.19

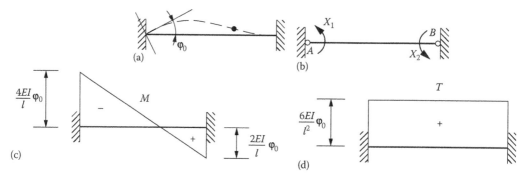

Figure 11.20

In the case of imposed rotation φ_0 (Figure 11.20a), the equivalent statically determinate scheme is that of a beam supported at either end with unknown redundant moments at the ends (Figure 11.20b). The equations of congruence will impose the rotations φ_0 in A and zero in B

$$\varphi_A = \frac{X_1 l}{3EI} - \frac{X_2 l}{6EI} = \varphi_0 \tag{11.44a}$$

$$\varphi_B = \frac{X_2 l}{3EI} - \frac{X_1 l}{6EI} = 0 \tag{11.44b}$$

From the foregoing equations, we can deduce the two redundant unknowns:

$$X_1 = \frac{4EI}{l}\varphi_0 \tag{11.45a}$$

$$X_2 = \frac{2EI}{l}\varphi_0 \tag{11.45b}$$

The moment diagram is thus linear and asymmetrical, with a value at the end at which the built-in constraint rotates, which is twice that at the other end (Figure 11.20c). The shear is constant (Figure 11.20d), having a value of

$$T = \frac{X_1 + X_2}{l} = \frac{6EI}{l^2}\varphi_0 \tag{11.46}$$

Note that it would have been possible to disconnect just the built-in constraint A, in which case the equation of congruence would have been expressible on the basis of Equation 11.14:

$$\varphi_A = \frac{X_1 l}{4EI} = \varphi_0 \tag{11.47}$$

In this way, the solution outlined previously is once more obtained.

Finally, let us consider the scheme of the beam built in at one end and constrained with a double rod at the other. Whereas the vertical displacement of the built-in constraint does not generate reactions and deflections (it is one of those cases in which the augmented matrix has the same rank as the kinematic matrix), an imposed rotation φ_0 generates the

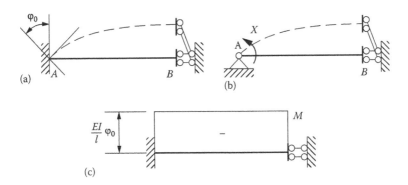

Figure 11.21

elastic deformation of Figure 11.21a. The equivalent statically determinate scheme consists of the hinged beam constrained by a double rod, loaded by the redundant moment X (Figure 11.21b). This scheme has already been studied in Section 11.3 and, on the basis of Equation 11.23, yields the following equation of congruence:

$$\varphi_A = \frac{Xl}{EI} = \varphi_0 \tag{11.48}$$

from which we deduce

$$X = \frac{EI}{l} \varphi_0 \tag{11.49}$$

The moment diagram is constant (Figure 11.21c), whereas the shear is zero.

11.6 THERMAL DISTORTIONS

Thermal distortions are the deformations induced in beams by variations in temperature within the depth h. In the present section, only linear variations of temperature within the depth h will be considered. Each **linear variation of temperature** can then be divided into a **uniform thermal variation** and a **butterfly-shaped thermal variation** (Figure 11.22). In the context of small displacements, it will be possible to apply the principle of superposition.

A uniform thermal variation ΔT, acting on an infinitesimal beam element of length $\mathrm{d}z$ (Figure 11.23), induces an elongation of the element given by

$$\mathrm{d}w = \alpha \Delta T \mathrm{d}z \tag{11.50}$$

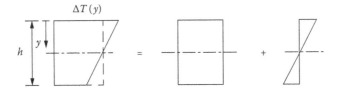

Figure 11.22

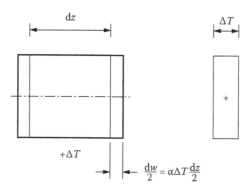

Figure 11.23

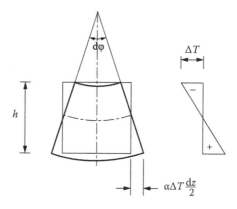

Figure 11.24

where α is the **coefficient of thermal expansion** of the material. Thus, a **thermal dilation** will be produced:

$$\varepsilon_T = \frac{\mathrm{d}w}{\mathrm{d}z} = \alpha \Delta T \tag{11.51}$$

The butterfly-shaped thermal variation of Figure 11.24 brings about lengthening of the lower fibres and shortening of the upper ones, so that, if the cross sections are assumed to be orthogonal to the axis of the beam even after thermal deformation has occurred, a relative rotation is produced between the end sections of the element,

$$\mathrm{d}\varphi = 2\frac{\alpha \Delta T(\mathrm{d}z/2)}{h/2} = 2\alpha \Delta T \frac{\mathrm{d}z}{h} \tag{11.52}$$

and hence a **thermal curvature:**

$$\chi_T = \frac{\mathrm{d}\varphi}{\mathrm{d}z} = 2\alpha \frac{\Delta T}{h} \tag{11.53}$$

Since, on the other hand, the rotation is equal, but for the algebraic sign, to the derivative of the vertical displacement υ,

$$\varphi = -\frac{d\upsilon}{dz} \tag{11.54}$$

the Equation 11.53 becomes

$$\frac{d^2\upsilon}{dz^2} = -2\alpha\frac{\Delta T}{h} \tag{11.55}$$

The differential equation (11.55), which governs the thermal deflection of the beam, is formally identical to Equation 10.47, which governs the elastic deflection and, in place of the temperature variation ΔT, presents the bending moment M. Putting together the two contributions, it is possible to write the **equation of the thermoelastic line**:

$$\frac{d^2\upsilon}{dz^2} = -\left(\frac{M}{EI} + 2\alpha\frac{\Delta T}{h}\right) \tag{11.56}$$

Let us now consider some elementary statically determinate schemes subject to thermal distortions. For the same reasons pointed out in the case of imposed displacements, the thermal distortions do not generate additional reactions and elastic deformations in statically determinate structures. The latter are thus freely deformed without forcing of any sort. The cantilever beam of Figure 11.25, subjected to a uniform thermal variation, is lengthened, for example, by the amount

$$w_B = \int_A^B \varepsilon_T\,dz = \alpha\Delta Tl \tag{11.57}$$

The same cantilever beam of Figure 11.25, subjected to a butterfly-shaped thermal variation (Figure 11.26), will undergo a rotation of thermal origin at the end section B,

$$\varphi_B = \int_A^B \chi_T\,dz = -2\alpha\frac{\Delta T}{h}l \tag{11.58a}$$

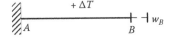

Figure 11.25

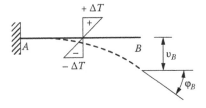

Figure 11.26

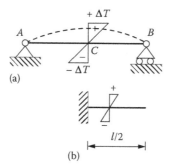

Figure 11.27

and a deflection

$$v_B = \int_A^B \chi_T z \, dz = \alpha \frac{\Delta T}{h} l^2 \tag{11.58b}$$

which can be determined also by resolving the differential equation (11.55), with the boundary conditions: $v(A) = v'(A) = 0$.

The supported beam subjected to a butterfly-shaped thermal variation (Figure 11.27a) constitutes a symmetrical structural scheme, whereby, using the previous results for the cantilever beam, we have (Figure 11.27b)

$$\varphi_A = -\varphi_B = \alpha \frac{\Delta T}{h} l \tag{11.59a}$$

$$v_C = -\alpha \frac{\Delta T}{h} (l/2)^2 = -\alpha \frac{\Delta T}{h} \frac{l^2}{4} \tag{11.59b}$$

In the case of the beam built in at both ends, subjected to a uniform thermal variation (Figure 11.28a), the equivalent statically determinate scheme is that of a cantilever beam with an unknown axial reaction X (Figure 11.28b). The equation of congruence

$$w_B = \alpha \Delta T l - \frac{X}{EA} l = 0 \tag{11.60}$$

yields the axial force

$$X = EA \, \alpha \Delta T \tag{11.61}$$

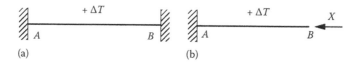

Figure 11.28

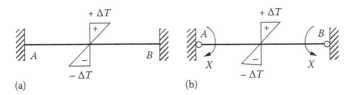

Figure 11.29

In the case of a beam built in at both ends, subjected to a butterfly-shaped thermal variation (Figure 11.29a), the equivalent statically determinate scheme is that of a beam supported at either end with redundant reactions X at the ends (Figure 11.29b). The equation of congruence

$$\varphi_A = -\varphi_B = \alpha \frac{\Delta T}{h} l - \frac{Xl}{2EI} = 0 \tag{11.62}$$

yields the built-in constraint moment, which is equal to the bending moment acting on all the sections of the beam:

$$X = 2 \frac{\alpha \Delta T EI}{h} \tag{11.63}$$

Note that since the elastic curvature is equal and opposite to the thermal curvature,

$$\chi_e = \frac{X}{EI} = 2\alpha \frac{\Delta T}{h} \tag{11.64a}$$

$$\chi_T = -2\alpha \frac{\Delta T}{h} \tag{11.64b}$$

the global thermoelastic deformed configuration is zero, and the beam remains rectilinear.

In conclusion, let us take a beam built in at one end and constrained by a vertical connecting rod at the other (Figure 11.30a). Let the beam be subjected to a butterfly-shaped thermal variation and the connecting rod to a uniform thermal variation. If we isolate the cantilever beam and the connecting rod, which exchange the redundant reaction X (Figure 11.30b), it is possible to formulate the equation of congruence, which must express the equality of the deflection of point B', belonging to the cantilever beam, and point B'', belonging to the connecting rod,

$$\upsilon_{B'} = \upsilon_{B''} \tag{11.65}$$

with

$$\upsilon_{B'} = \alpha \frac{\Delta T}{h} l^2 - \frac{Xl^3}{3EI} \tag{11.66a}$$

$$\upsilon_{B''} = -\alpha \Delta T \frac{l}{2} + \frac{X(l/2)}{EA} \tag{11.66b}$$

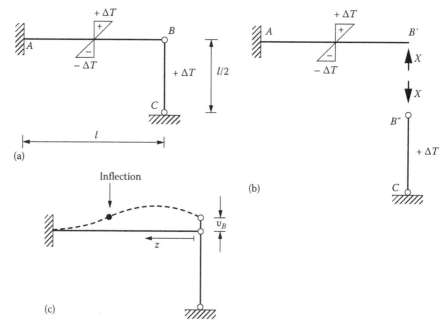

Figure 11.30

With expressions (11.66) substituted in Equation 11.65, we obtain

$$X = \frac{\alpha \Delta T E (l/h + 1/2)}{[1/(2A) + l^2/(3I)]}$$ (11.67)

The connecting rod therefore functions as a strut, as was assumed conventionally.

The beam presents a thermoelastic inflection where the elastic curvature equals, in absolute value, the thermal one,

$$\frac{Xz}{EI} = 2\alpha \frac{\Delta T}{h}$$ (11.68)

from which we obtain the coordinate of the inflection (Figure 11.30c),

$$z = \frac{2\alpha \Delta T E I}{Xh}$$ (11.69)

where X is given by Equation 11.67. Finally,

$$z = \frac{\left(I/A + (2/3)l^2 \right)}{\left(\frac{h}{2} + l \right)}$$ (11.70)

and, in the case of a rectangular section and a slender beam ($l/h \to \infty$),

$$z \simeq \frac{2}{3}l$$ (11.71)

Remaining within the restrictive hypotheses made earlier, the vertical displacement of the end B is

$$v_B = -\frac{1}{2}\alpha\Delta Tl \tag{11.72}$$

and thus, since it is negative, the point B will rise. The thermoelastic deformed configuration of the beam is depicted in Figure 11.30c.

11.7 CONTINUOUS BEAMS

The term **continuous beams** refers to rectilinear beams devoid of internal disconnections, constrained to the foundation by a series of supports and, in some cases, by built-in supports at the ends. The equivalent statically determinate structure is obtained by inserting a hinge in each support or in each external built-in support. There will thus be as many redundant unknowns and as many equations of congruence as there are disconnected nodes. The equation of rotational congruence corresponding to each node will contain the redundant moment acting on the same node, as well as the redundant moments acting on the two adjacent nodes; it is for this reason that it is called **equation of three moments**. In the case of nodes adjacent to the ends or end nodes, the number of unknowns present in the corresponding equation may prove to be less than three.

Consider the three-span continuous beam of Figure 11.31a. It is constrained by four supports (a hinge and three roller supports) and loaded by a moment m at the right-hand end D. If the two intermediate supports are disconnected with two hinges and the redundant moments X_1 and X_2 are applied (Figure 11.31b), the equations of congruence take on the following form:

$$\varphi_{BA} = \varphi_{BC} \tag{11.73a}$$

$$\varphi_{CB} = \varphi_{CD} \tag{11.73b}$$

and thus

$$-\frac{X_1l}{3EI} = \frac{X_1l}{3EI} + \frac{X_2l}{6EI} \tag{11.74a}$$

$$-\frac{X_2l}{3EI} - \frac{X_1l}{6EI} = \frac{X_2l}{3EI} + \frac{ml}{6EI} \tag{11.74b}$$

Once resolved, Equations 11.74 yield the solution $X_1 = \frac{1}{15}m$, $X_2 = -\frac{4}{15}m$, and therefore the moment diagram of Figure 11.31c. This diagram consists of a broken line which intersects the axis of the beam in two points. The points at which the bending moment becomes zero correspond to inflection points in the elastic deformed configuration (Figure 11.31d). Finally, the shear diagram is represented in Figure 11.31e. The discontinuities or jumps which the function T undergoes, proceeding from left to right, represent the vertical reactions of the supports:

$$V_A = -\frac{1}{15}\frac{m}{l}, \quad V_B = \frac{2}{5}\frac{m}{l}, \quad V_C = -\frac{8}{5}\frac{m}{l}, \quad V_D = \frac{19}{15}\frac{m}{l}$$

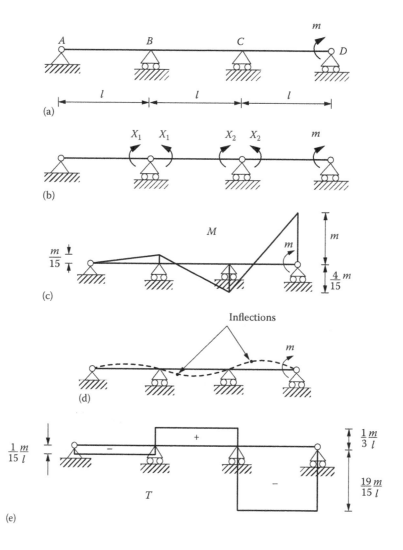

Figure 11.31

Let us now assume that the continuous beam considered previously undergoes an inelastic settlement of the support B. After disconnecting the beam as shown in Figure 11.31b, it will be necessary to consider the contributions to the rotations of the sections B and C, due both to the elastic deformability of the beams and to the rigid mechanism caused by the deflection η_0:

$$-\frac{X_1 l}{3EI} - \frac{\eta_0}{l} = \frac{X_1 l}{3EI} + \frac{X_2 l}{6EI} + \frac{\eta_0}{l} \qquad (11.75a)$$

$$-\frac{X_2 l}{3EI} - \frac{X_1 l}{6EI} + \frac{\eta_0}{l} = \frac{X_2 l}{3EI} \qquad (11.75b)$$

The solution to the system (11.75) is as follows:

$$X_1 = -\frac{18}{5}\frac{EI}{l^2}\eta_0 \qquad (11.76a)$$

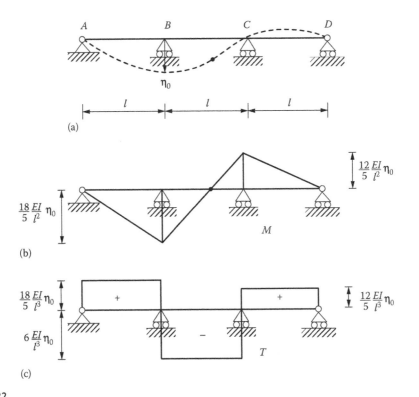

Figure 11.32

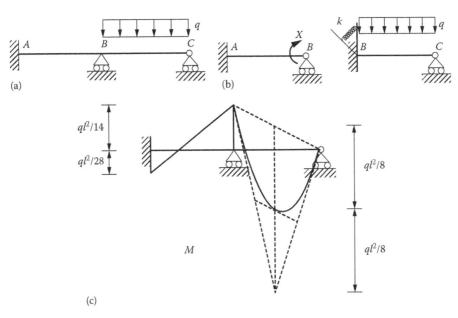

Figure 11.33

(continued)

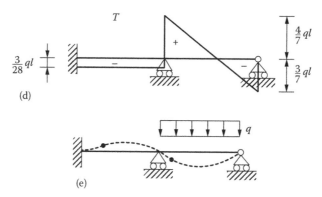

(d)

(e)

Figure 11.33 (continued)

$$X_2 = \frac{12}{5}\frac{EI}{l^2}\eta_0 \tag{11.76b}$$

The moment diagram (Figure 11.32b) intersects the axis of the beam at a single point, so that the deformed configuration presents the fibres stretched at the lower edge in the left-hand part and at the upper edge in the right-hand part (Figure 11.32a). The shear diagram is depicted in Figure 11.32c, and also in this case, the discontinuities measure the values of the constraint reactions.

As our last example of a continuous beam, we shall examine the one depicted in Figure 11.33a, which consists of two spans with a built-in constraint in A and supports in B and C. Since the distributed load q acts only on the span BC, it is possible to reduce the scheme to that of Figure 11.16a, with the elastic compliance of the beam AB represented by the rotational spring. The elementary schemes for resolving the problem are given in Figure 11.33b, where the stiffness k of the spring is equal to $4EI/l$. From Equation 11.34, we obtain $X = ql^2/14$, and hence the moment diagram of Figure 11.33c. The shear emerges as constant in the beam AB and linear in the beam BC (Figure 11.33d). The point of zero shear corresponds to the maximum bending moment (Figure 11.33c), just as the points where the moment becomes zero correspond to the inflections of the elastic line (Figure 11.33e).

Chapter 12

Energy methods for the solution of beam systems

12.1 INTRODUCTION

The principle of virtual work, proposed in Section 8.4 in the context of 3D solids, may be extrapolated, following a similar line of demonstration, to 1D and 2D solids. As regards rectilinear beams, it is sufficient to substitute Equation 8.23 with

$$L_F = -\int_0^l ([\partial]^* \{Q_a\})^T \{\eta_b\} dz \qquad (12.1)$$

whereby, instead of Equation 8.24, we obtain

$$L_F = \int_0^l \{Q_a\}^T [\partial] \{\eta_b\} dz - [\{Q_a\}^T \{\eta_b\}]_0^l \qquad (12.2)$$

0 and l being the coordinates of the beam ends.

We then obtain the equation of the principle of virtual work for a rectilinear beam, subjected to loads distributed over the span and to loads concentrated at the ends,

$$\int_0^l \{Q_a\}^T \{q_b\} dz = \int_0^l \{\mathcal{F}_a\}^T \{\eta_b\} dz + [\{Q_a\}^T \{\eta_b\}]_0^l \qquad (12.3)$$

where, adopting the same nomenclature used in Section 10.3, $\{Q_a\}$ is the vector of the static characteristics, $\{q_b\}$ is the vector of deformation characteristics, $\{\mathcal{F}_a\}$ is the vector of the distributed forces and $\{\eta_b\}$ is the displacement vector.

The fundamental equation (8.26) can then be shown to be valid also in the case of curvilinear beams, by using the mathematical formalism of Section 10.4 and replacing the operator $[\partial]$ with $[\partial][N]$, where $[N]$ is the rotation matrix, and the operator $[\partial]^*$ with $[N]^T[\partial]^*$.

Finally, Equation 8.26 can also be further applied to the case of beam systems, by summing up the contributions of the individual beams. Whereas the integrals of Equation 12.3, already extended to the individual beam of length l, must, in this case, be extended to the entire structure S, the second term of the right-hand side of the equation cancels out in all the internal nodes, for obvious reasons of equilibrium. On the other hand, all the contributions corresponding to the ends that are externally constrained or that are subjected to concentrated loads still remain to be accounted for:

$$\int_S \{Q_a\}^T \{q_b\}ds = \int_S \{\mathscr{F}_a\}^T \{\eta_b\}ds + \sum_i \{Q_{ai}\}^T \{\eta_b\} \tag{12.4}$$

Since the aim of the application of principle of virtual work to statically determinate beam systems is to determine the elastic generalized displacements, it is expedient to consider, for each individual beam or structure, two distinct systems:

1. The **real system** or **system of displacements**
2. The **fictitious system** or **system of forces**

The real system consists simply of the structure under examination, subjected to all the external loads, both mechanical and thermal, including the inelastic settlements, and taking into account the elastic settlements (Figure 12.1a). The fictitious system in turn consists of the same structure, loaded in this case by the single unit force, dual of the elastic displacement sought $\eta^{(r)}$ (Figure 12.1b).

An application of Equation 12.4 yields

$$\int_S N^{(f)}\left(\frac{N^{(r)}}{EA} + \alpha T_0^{(r)}\right)ds + \int_S T^{(f)}\left(\frac{t T^{(r)}}{GA}\right)ds + \int_S M^{(f)}\left(\frac{M^{(r)}}{EI} + 2\frac{\alpha \Delta T^{(r)}}{h}\right)ds$$

$$= 1 \times \eta^{(r)} + \sum_i R_i^{(f)}\left(\eta_{0i}^{(r)} - \frac{R_i^{(r)}}{k_i}\right) \tag{12.5}$$

where
the superscript f denotes the fictitious system or system of forces
the superscript r denotes the real system or system of displacements
$\eta^{(r)}$ designates the real displacement to be determined
R_i indicates the ith constraint reaction
k_i indicates the stiffness
$\eta_{0i}^{(r)}$ denotes the displacement imposed on the corresponding constraint

With the exclusion of the truss structures, it may be stated that the contribution of the third integral on the left-hand side of Equation 12.5, the one corresponding to the bending

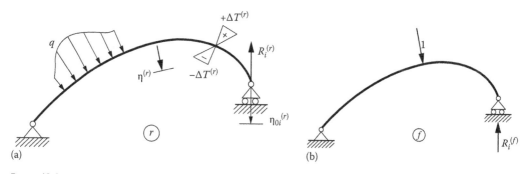

(a)   (b)

Figure 12.1

moment, is usually far greater than that of the two integrals preceding it, which correspond to axial force and shearing force, respectively. Of course, the contributions corresponding to the thermal dilatations and curvatures are absent in the case where there are no thermal loadings $\left(T_0^{(r)} = \Delta T^{(r)} = 0\right)$. In the same way, the summation on the right-hand side of the equation is zero, in the case where there are no elastic or inelastic settlements.

12.2 DETERMINATION OF ELASTIC DISPLACEMENTS IN STATICALLY DETERMINATE STRUCTURES

We shall now show how it is possible to determine the elastic displacements and rotations in the cross sections belonging to statically determinate beam systems by applying Equation 12.5.

As a first elementary example, let us consider the simply supported beam of Figure 12.2a, loaded by a concentrated moment at the right-hand end. We intend to calculate the elastic rotation of this end. To do this, let the fictitious system consist of the same beam loaded at the same end by a unit moment, acting in the same direction as the actual moment (Figure 12.2b). We have, therefore,

$$M^{(r)}(z) = \frac{m}{l} z \tag{12.6a}$$

$$M^{(f)}(z) = \frac{z}{l} \tag{12.6b}$$

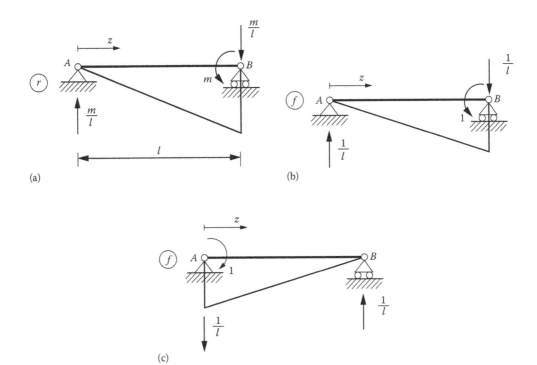

(a)   (b)

(c)

Figure 12.2

whereby, if we apply Equation 12.5, taking into account the absence of axial force, the negligibility of shearing strain and the absence of distortions and settlements, we obtain

$$1 \times \varphi_B = \int_0^l \frac{M^{(f)}M^{(r)}}{EI} \, dz \tag{12.7}$$

Substituting expressions (12.6) into the integral (12.7), we obtain

$$\varphi_B = \frac{ml}{3EI} \tag{12.8}$$

a result already known to us from the treatment of the elastic line (Chapter 10).

We then intend to determine the elastic rotation at the end opposite to the loaded one. In this case, we shall have to consider a different fictitious system, loaded by a unit moment at the end A (Figure 12.2c). Here we have

$$M^{(f)}(z) = \frac{l - z}{l} \tag{12.9}$$

whereby application of the principle of virtual work yields

$$1 \times \varphi_A = \int_0^l \frac{M^{(f)}M^{(r)}}{EI} \, dz \tag{12.10}$$

By substituting the functions (12.9) and (12.6a) into Equation 12.10, we find that

$$\varphi_A = \frac{ml}{6EI} \tag{12.11}$$

In the case of a supported beam subjected to uniform load (Figure 12.3a), the vertical displacement or deflection in the centre can be obtained by means of the fictitious system of Figure 12.3b, consisting of the same beam loaded by a vertical unit force acting in the centre. Given the symmetry of the beam under examination and of the displacement sought, i.e. given the symmetry of the two systems, the real one and the fictitious one, it is possible to evaluate the integrals on half the beam and multiply them by two. The two

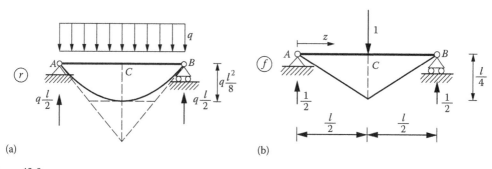

(a)   (b)

Figure 12.3

moments, the real one and the fictitious one, can hence be expressed analytically even just on the left-hand span

$$M^{(r)}(z) = \frac{1}{2}qlz - \frac{1}{2}qz^2 \qquad (12.12a)$$

$$M^{(f)}(z) = \frac{z}{2} \qquad (12.12b)$$

for $0 \leq z \leq l/2$. Applying Equation 12.5, we obtain

$$1 \times \upsilon_c = \frac{2}{EI} \int_0^{l/2} M^{(f)} M^{(r)} dz \qquad (12.13)$$

and, if we substitute Equations 12.12 into Equation 12.13, we can easily obtain

$$\upsilon_c = \frac{5}{384} \frac{ql^4}{EI} \qquad (12.14)$$

Now consider the L-shaped beam of Figure 12.4a, uniformly loaded on the cross member. The determination of the elastic displacement of the roller support on which the upright rests may be obtained by means of the fictitious structure of Figure 12.4b, on which a unit

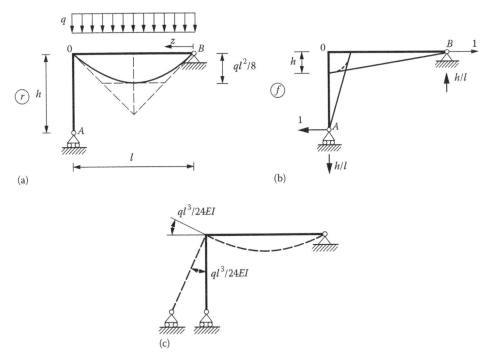

(a)

(b)

(c)

Figure 12.4

horizontal force is applied at the end A. Whereas in the fictitious system the upright is subjected to bending moment, this does not occur in the real system, the upright of which is subjected to axial force alone. Hence, taking into account only the contribution of the cross member, we have

$$M^{(r)}(z) = \frac{1}{2}qlz - \frac{1}{2}qz^2 \qquad (12.15a)$$

$$M^{(f)}(z) = \frac{h}{l}z \qquad (12.15b)$$

whereby we obtain

$$1 \times \delta_A = \frac{1}{EI}\int_0^l \frac{h}{l}z\left(\frac{1}{2}qlz - \frac{1}{2}qz^2\right)dz \qquad (12.16)$$

and thus

$$\delta_A = \frac{ql^3}{24EI}h \qquad (12.17)$$

Note that this displacement is equal to the product of the angle of elastic rotation of the end of a supported beam by the rigid arm of length h provided by the upright (Figure 12.4c).

Let us reconsider the Gerber beam of Figure 12.5a, of which we here intend to evaluate the relative vertical displacement at the double rod (Figures 6.3 and 6.4b). The fictitious structure consists of the same beam, loaded by two unit vertical, equal and opposite forces, acting at the ends of the beams connected together by the double rod (Figure 12.5b). Imposing equilibrium first on the portion CD and then on the portion CA (Figure 12.5c), we obtain the fictitious moment functions:

$$M^{(f)}(z_1) = 3z_1, \quad 0 \le z_1 \le l \qquad (12.18a)$$

$$M^{(f)}(z_2) = z_2, \quad 0 \le z_2 \le 3l \qquad (12.18b)$$

On the other hand, the real moment functions are equal to

$$M^{(r)}(z_1) = \frac{3}{2}qlz_1, \quad 0 \le z_1 \le l \qquad (12.19a)$$

$$M^{(r)}(z_2) = 2qlz_2 - \frac{1}{2}qz_2^2, \quad 0 \le z_2 \le 3l \qquad (12.19b)$$

Applying the principle of virtual work yields the following equation:

$$1 \times \Delta v_c = \frac{1}{EI}\int_0^l \frac{9}{2}qlz_1^2 dz_1 + \frac{1}{EI}\int_0^{3l} z_2\left(2qlz_2 - \frac{1}{2}qz_2^2\right)dz_2 \qquad (12.20)$$

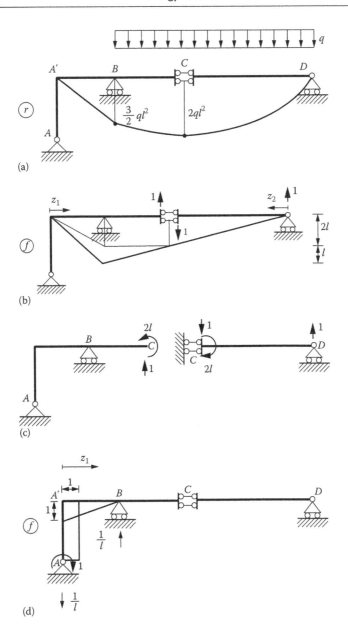

Figure 12.5

Carrying out the calculations, we obtain

$$\Delta v_c = \frac{75}{8} \frac{q l^4}{EI} \tag{12.21}$$

If we wish to know by how much one end is raised and by how much the other end is lowered, it is possible to consider the two unit forces of the scheme of Figure 12.5b separately.

In order to define the rigid rotation of the upright, it is necessary to consider the fictitious system of Figure 12.5d, in which a unit moment is applied to the end A. In this case, the only

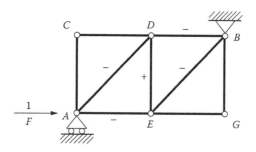

Figure 12.6

contribution to the calculation comes from the portion $A'B$, since only here are both the real moment and the fictitious moment different from zero:

$$1 \times \varphi_A = \frac{1}{EI} \int_0^l \left(\frac{3}{2} q l z_1 \right) \left(1 - \frac{z_1}{l} \right) dz_1 \tag{12.22}$$

Computing, we find the rotation to be positive and hence clockwise as supposed (Figure 6.4b):

$$\varphi_A = \frac{1}{4} \frac{q l^3}{EI} \tag{12.23}$$

Finally, let us consider the truss of Figure 12.6, and let us determine the elastic displacement of the roller support A. We shall therefore have to reconsider the same truss, loaded by a unit force similar to the actual one. In this case, since the axial force is the only static characteristic present, Equation 12.5 reduces as follows:

$$1 \times u_A = \sum_i N_i^{(f)} \frac{N_i^{(r)}}{EA} l_i \tag{12.24}$$

On the basis of the axial forces obtained and listed in Section 6.3 (Figure 6.9), we have

$$u_A EA = 2 \left[\left(-\frac{F}{2} \right) \left(-\frac{1}{2} \right) l \right] + 2 \left[\left(-\frac{F}{2} \sqrt{2} \right) \left(-\frac{1}{2} \sqrt{2} \right) l \sqrt{2} \right] + \left[\left(\frac{F}{2} \right) \left(\frac{1}{2} \right) l \right] \tag{12.25}$$

and thus

$$u_A = \frac{Fl}{EA} \left(\frac{3}{4} + \sqrt{2} \right) \tag{12.26}$$

12.3 RESOLUTION OF STRUCTURES HAVING ONE DEGREE OF STATIC INDETERMINACY

In the case where the structure being examined has one degree of static indeterminacy (Figure 12.7), it is possible to write the equation of congruence using the same calculating procedure adopted in the previous section (Figure 12.1). The procedure will be to equate the elastic displacement produced by the external loads and by the redundant reaction to

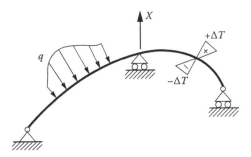

Figure 12.7

zero or to a quantity different from zero (function of the redundant reaction for elastically compliant constraints), according to whether the suppressed constraint is rigid or not.

From the operative point of view, once the equivalent statically determinate structure has been identified, two schemes are resolved:

1. Scheme 0, consisting of the equivalent statically determinate structure, subjected to external loads
2. Scheme 1, consisting of the equivalent statically determinate structure, subjected to the unit redundant reaction

At this point, the system of forces consists of Scheme 1, whilst the system of displacements consists of the superposition of Scheme 0 and Scheme 1, which is multiplied by the redundant unknown X. Equation 12.5 thus yields the displacement η of the constrained point

$$1 \times \eta = \frac{1}{EI} \int_S M^{(1)}(M^{(0)} + XM^{(1)}) \mathrm{d}s \tag{12.27}$$

In the case of a rigid or inelastically compliant constraint, the equation of congruence is obtained by equating the right-hand side of Equation 12.27 to zero or to η_0, where η_0 is the known entity of settlement. In the case of an elastically compliant constraint, the equation of congruence is obtained by equating the right-hand side of Equation 12.27 to $-X/k$, where k is the stiffness of the constraint. In all cases, a linear algebraic equation in the single unknown X is obtained.

In the case of an elastically compliant constraint, we have, for instance,

$$\int_S M^{(0)}M^{(1)}\mathrm{d}s + X \int_S (M^{(1)})^2 \mathrm{d}s = -X\frac{EI}{k} \tag{12.28}$$

whence we obtain

$$X = -\frac{\displaystyle\int_S M^{(0)}M^{(1)}\mathrm{d}s}{\displaystyle\int_S (M^{(1)})^2 \mathrm{d}s + \frac{EI}{k}} \tag{12.29}$$

In the same way, in the case of an inelastically compliant constraint, we obtain

$$X = -\frac{\int_S M^{(0)}M^{(1)}ds - \eta_0 EI}{\int_S (M^{(1)})^2 ds} \tag{12.30}$$

When instead the constraint is rigid ($k \to \infty$, or $\eta_0 \to 0$), both expression (12.29) and expression (12.30) reduce to the following:

$$X = -\frac{\int_S M^{(0)}M^{(1)}ds}{\int_S (M^{(1)})^2 ds} \tag{12.31}$$

In relation to the frame of Figure 12.8a, there are two schemes to be considered to obtain a resolution of the problem using the principle of virtual work. The equivalent statically determinate structure may be obtained, e.g. by eliminating the degree of constraint to the horizontal translation of the hinge A, i.e. transforming the hinge into a roller support (Figure 12.8b and c).

Scheme (Figure 12.8b) thus consists of the equivalent statically determinate structure subjected to the distributed load acting on the overhang CG, while Scheme 1 (Figure 12.8c) consists of the same statically determinate structure, loaded in this case by a unit horizontal force applied at point A. The determination of the constraint reactions of the two schemes is immediate, as is the drawing of the respective moment diagrams (Figure 12.8b and c).

Once a reference system for each beam of the frame has been chosen, it is possible to draw up Table 12.1.

We thus obtain

$$\int_S M^{(0)}M^{(1)}ds = \int_0^l \frac{1}{2}qlz_2(1-2z_2)dz_2 = -\frac{1}{12}ql^4 \tag{12.32a}$$

$$\int_S (M^{(1)})^2 ds = \int_0^l z_1^2 dz_1 + \int_0^l (1-2z_2)^2 dz_2 + \int_0^l z_3^2 dz_3 = l^3 \tag{12.32b}$$

From Equation 12.31, we obtain the redundant unknown

$$X = \frac{1}{12}ql \tag{12.33}$$

The equilibrium scheme of the node C, obtained by superposition of Schemes 0 and $1 \times X$, is shown in Figure 12.8d. Applying once more the principle of superposition, it is possible to find the diagrams of the static characteristics and the deformed configuration, represented in Figure 12.8e through j. The pressure line is represented in Figure 12.8k.

As a second example, consider the asymmetrical portal frame of Figure 12.9a. As an equivalent statically determinate structure, let us choose, from the infinite range of possibilities, the three-hinged arch ACD (Figure 12.9b). In this case, therefore, the structure has been internally disconnected, even though usually, from the point of view of simplicity of calculation, external disconnections are more convenient. Having determined the moment diagrams on Schemes 0 and 1, and having fixed a reference system on each beam, we draw up Table 12.2.

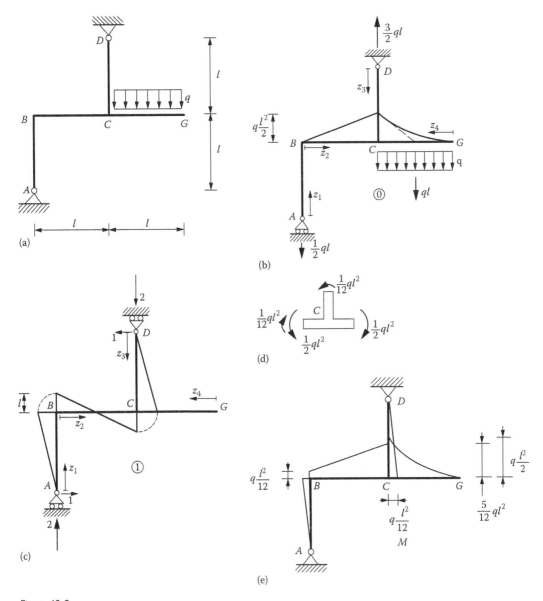

Figure 12.8

(continued)

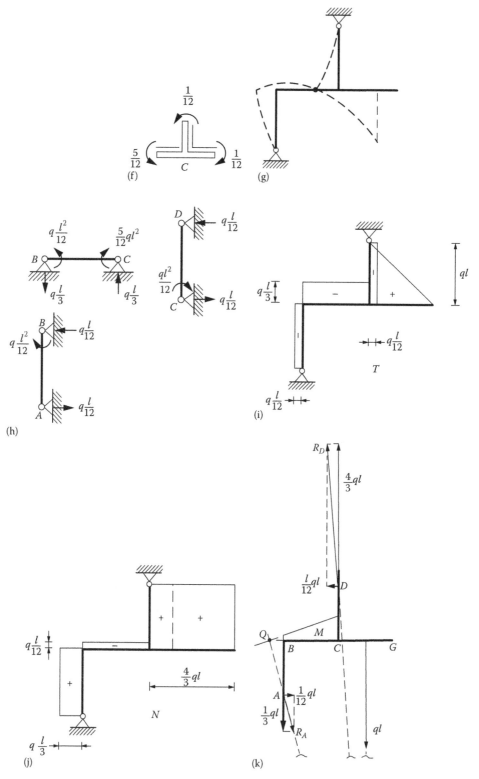

Figure 12.8 (continued)

Table 12.1

Beam	$M^{(0)}$	$M^{(1)}$
AB	0	z_1
BC	$\dfrac{1}{2}qlz_2$	$1-2z_2$
CD	0	z_3
CG	$\dfrac{1}{2}qz_4^2$	0

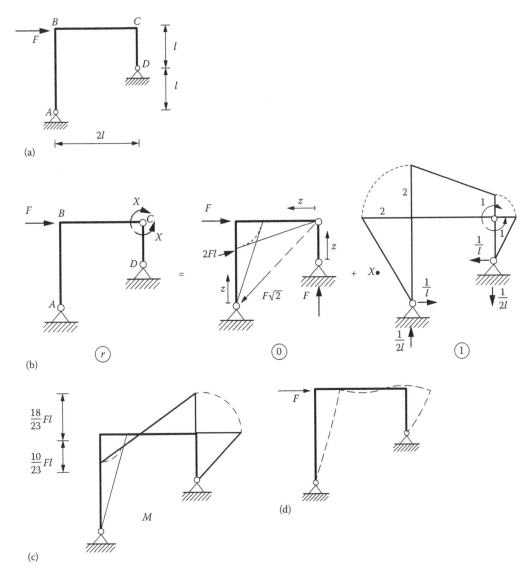

(a)

(b)

(c)

(d)

Figure 12.9

(continued)

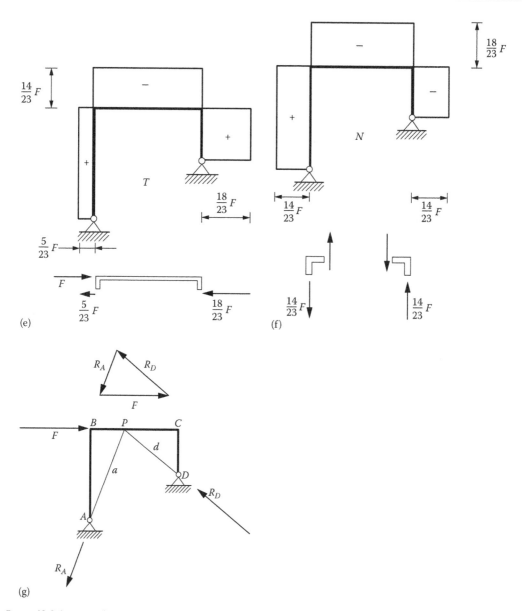

(e)

(f)

(g)

Figure 12.9 (continued)

Table 12.2

Beam	$M^{(0)}$	$M^{(1)}$
AB	Fz	$-\dfrac{z}{l}$
BC	Fz	$-\left(1+\dfrac{z}{2l}\right)$
CD	0	$-\dfrac{z}{l}$

Notice that in the schemes of Figure 12.9b, the moment diagram is shown from the side of the stretched fibres, and that where $M^{(0)}$ and $M^{(1)}$ extend opposite fibres, in the table, the respective functions must give values of opposite sign. The following two integrals are therefore obtained:

$$\int_S M^{(0)}M^{(1)}ds = \int_0^{2l}\left(-\frac{F}{l}z^2\right)dz + \int_0^{2l}(-Fz)\left(1+\frac{z}{2l}\right)dz = -6Fl^2 \tag{12.34a}$$

$$\int_S (M^{(1)})^2 ds = \int_0^{2l}\frac{z^2}{l^2}dz + \int_0^{2l}\left(1+\frac{z^2}{4l^2}+\frac{z}{l}\right)dz + \int_0^l\frac{z^2}{l^2}dz = \frac{23}{3}l \tag{12.34b}$$

From Equation 12.31, we obtain the redundant moment

$$X = \frac{18}{23}Fl \tag{12.35}$$

This calculation is implicitly equivalent to the imposition of angular congruence in the fixed joint-node C:

$$\Delta\varphi_C = 0 \quad \text{or} \tag{12.36a}$$

$$\varphi_{CD} = \varphi_{CA} \tag{12.36b}$$

Applying the principle of superposition, it is possible to find the diagrams of the static characteristics, the deformed configuration and the line of pressure (see Figure 12.9c through g).

As a final example of a mechanically loaded structure having one degree of redundancy, let us refer to the portal frame with oblique stanchion of Figure 12.10a. Let the structure be disconnected externally, so as to transform the hinge D into a horizontally moving roller support (Figure 12.10b). Once the external reactions have been identified on Schemes 0 and 1 and the reference systems have been defined on the individual beams, the determination of the analytical functions $M^{(0)}$ and $M^{(1)}$ is immediate. It is not necessary at this stage to draw the diagram of these functions, which are given in Table 12.3.

Calculation of the integrals

$$\int_S M^{(0)}M^{(1)}ds = \int_0^l l\left(\frac{3}{4}qlz - \frac{1}{2}qz^2\right)dz + \int_0^{l\sqrt{2}}\left(\frac{1}{8}qlz\sqrt{2}\right)\left(z\frac{\sqrt{2}}{2}\right)dz$$

$$= \frac{5+2\sqrt{2}}{24}ql^4 \tag{12.37a}$$

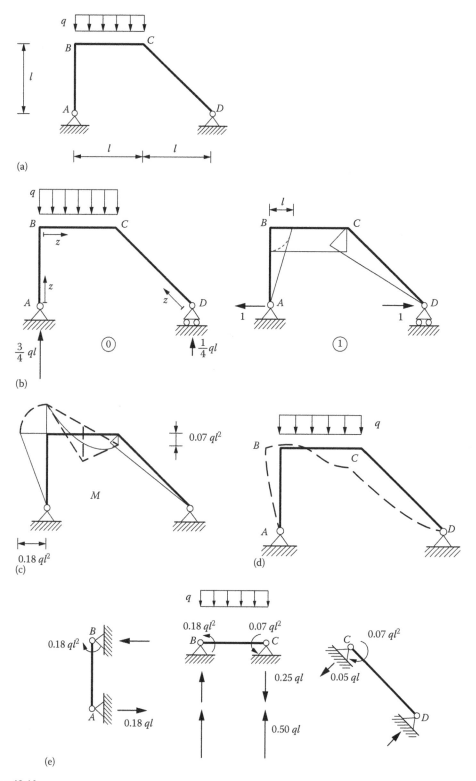

Figure 12.10

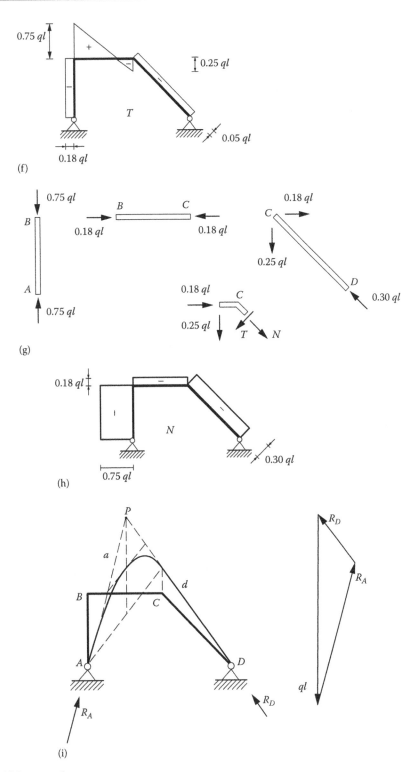

Figure 12.10 (continued)

Table 12.3

Beam	$M^{(0)}$	$M^{(1)}$
AB	0	z
BC	$\dfrac{3}{4}qlz - \dfrac{1}{2}qz^2$	l
CD	$\dfrac{1}{8}qlz\sqrt{2}$	$\dfrac{\sqrt{2}}{2}z$

$$\int_S (M^{(1)})^2 \, ds = \int_0^l z^2 \, dz + \int_0^l l^2 \, dz \int_0^{l\sqrt{2}} \frac{1}{2} z^2 \, dz = \frac{4 + \sqrt{2}}{3} l^3 \tag{12.37b}$$

makes it possible, *via* Equation 12.31, to obtain the redundant reaction

$$X = -\frac{16 + 3\sqrt{2}}{112} ql \tag{12.38}$$

We are thus able to verify that the bending moment in the fixed joint-node C is equal to

$$M_C = Xl + \frac{1}{4}ql^2 = \frac{3(4 - \sqrt{2})}{112} ql^2 \tag{12.39}$$

Applying the principle of superposition, it is possible to find the diagrams of the static characteristics, the deformed configuration and the line of pressure (Figure 12.10c through i).

12.4 RESOLUTION OF STRUCTURES HAVING TWO OR MORE DEGREES OF STATIC INDETERMINACY

In the case of structures having two or more degrees of static indeterminacy, the procedure outlined in the foregoing section can be extended to a pair of fictitious schemes, consisting of the equivalent statically determinate structure, loaded by one redundant unknown at a time. The displacements of the two points in which the disconnection is made can be obtained by applying the principle of virtual work to each fictitious structure.

More precisely, considering the real one (Scheme 0 + X_1 × Scheme 1 + X_2 × Scheme 2) as system of displacements and each of the two fictitious systems as system of forces, we have

$$1 \times \eta_1 = \frac{1}{EI} \int_S M^{(1)}(M^{(0)} + X_1 M^{(1)} + X_2 M^{(2)}) \, ds \tag{12.40a}$$

$$1 \times \eta_2 = \frac{1}{EI} \int_S M^{(2)}(M^{(0)} + X_1 M^{(1)} + X_2 M^{(2)}) \, ds \tag{12.40b}$$

In the case where all the constraints of the structure are rigid, the two relations of congruence, $\eta_1 = \eta_2 = 0$, yield the following two linear algebraic equations:

$$X_1 \int_S (M^{(1)})^2 ds + X_2 \int_S M^{(1)} M^{(2)} ds = -\int_S M^{(1)} M^{(0)} ds \qquad (12.41a)$$

$$X_1 \int_S M^{(2)} M^{(1)} ds + X_2 \int_S (M^{(2)})^2 ds = -\int_S M^{(2)} M^{(0)} ds \qquad (12.41b)$$

If we designate the displacement generated by the redundant reaction $X_2 = 1$ at the point of application and in the direction of the other redundant reaction X_1 as **coefficient of influence** η_{12}, and the displacement generated by $X_1 = 1$ at the point and in the direction of X_2 as η_{21}, from Betti's reciprocal theorem and the principle of virtual work, we can deduce

$$\eta_{12} = \eta_{21} = \frac{1}{EI} \int_S M^{(1)} M^{(2)} ds \qquad (12.42a)$$

while the self-influence coefficients may be expressed as

$$\eta_{11} = \frac{1}{EI} \int_S (M^{(1)})^2 ds \qquad (12.42b)$$

$$\eta_{22} = \frac{1}{EI} \int_S (M^{(2)})^2 ds \qquad (12.42c)$$

Equations 12.41 may thus be cast in the form

$$\eta_{11} X_1 + \eta_{12} X_2 = -\eta_{10} \qquad (12.43a)$$

$$\eta_{21} X_1 + \eta_{22} X_2 = -\eta_{20} \qquad (12.43b)$$

η_{10} and η_{20} being the displacements due to the external load. Resolving the system using Cramer's rule, we obtain

$$X_1 = -\frac{\begin{vmatrix} \eta_{10} & \eta_{12} \\ \eta_{20} & \eta_{22} \end{vmatrix}}{\begin{vmatrix} \eta_{11} & \eta_{12} \\ \eta_{21} & \eta_{22} \end{vmatrix}} = -\frac{\eta_{10}\eta_{22} - \eta_{12}\eta_{20}}{\eta_{11}\eta_{22} - \eta_{12}^2} \qquad (12.44a)$$

$$X_2 = -\frac{\begin{vmatrix} \eta_{11} & \eta_{10} \\ \eta_{21} & \eta_{20} \end{vmatrix}}{\begin{vmatrix} \eta_{11} & \eta_{12} \\ \eta_{21} & \eta_{22} \end{vmatrix}} = -\frac{\eta_{11}\eta_{20} - \eta_{10}\eta_{21}}{\eta_{11}\eta_{22} - \eta_{12}^2} \qquad (12.44b)$$

As an example of the application of the earlier procedure, let us consider the frame with two degrees of redundancy of Figure 12.11a. As an equivalent statically determinate structure, let us consider the same portal frame, with the connecting rod and the constraint to horizontal translation at the feet of the uprights removed (Figure 12.11b). Three schemes are hence to be considered (Figure 12.11b):

1. Scheme 0, with the external load only
2. Scheme 1, with two symmetrical and horizontal unit forces acting at the foot of the uprights
3. Scheme 2, with two symmetrical and horizontal unit forces acting halfway up the uprights

Figure 12.11b also shows the corresponding moment diagram for each of the three schemes. Taking into account symmetry, there are three portions of the structure on which the integrals of Equations 12.42 are to be evaluated. Using a suitable reference system for each portion, we obtain Table 12.4.

There then follows the computation of the coefficients of influence:

$$\int_S (M^{(1)})^2 \, ds = \int_0^{l/2} z^2 \, dz + \int_0^{l/2} \left(\frac{l}{2}+z\right)^2 dz + \int_0^{l/2} l^2 \, dz = \frac{5}{6} l^3 \tag{12.45a}$$

$$\int_S M^{(1)} M^{(2)} \, ds = \int_0^{l/2} z\left(\frac{l}{2}+z\right) dz + \int_0^{l/2} l\left(\frac{l}{2}\right) dz = \frac{17}{48} l^3 \tag{12.45b}$$

$$\int_S M^{(1)} M^{(0)} \, ds = \int_0^{l/2} \left(-\frac{1}{2} q z^2\right)\left(\frac{l}{2}+z\right) dz + \int_0^{l/2} \left(-\frac{1}{8} q l^2\right) l \, dz$$

$$= -\frac{31}{384} q l^4 \tag{12.45c}$$

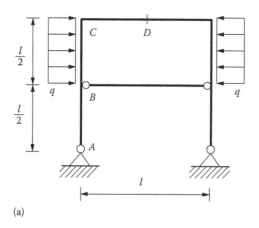

(a)

Figure 12.11

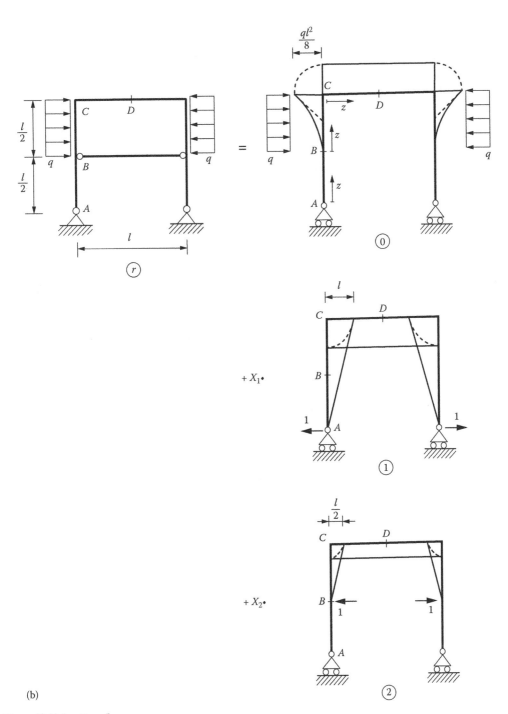

(b)

Figure 12.11 (continued)

(continued)

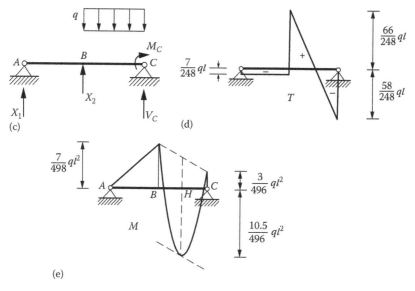

Figure 12.11 (continued)

Table 12.4

Beam	$M^{(0)}$	$M^{(1)}$	$M^{(2)}$
AB	0	z	0
BC	$-\dfrac{1}{2}qz^2$	$\dfrac{l}{2}+z$	z
CD	$-\dfrac{1}{8}ql^2$	l	$\dfrac{l}{2}$

$$\int_S (M^{(2)})^2 \, ds = \int_0^{l/2} z^2 \, dz + \int_0^{l/2} \frac{l^2}{4}\,dz = \frac{l^3}{6} \tag{12.45d}$$

$$\int_S M^{(2)}M^{(0)}\,ds = \int_0^{l/2}\left(-\frac{1}{2}qz^2\right)z\,dz + \int_0^{l/2}\left(-\frac{1}{8}ql^2\right)\frac{l}{2}\,dz$$

$$= -\frac{5}{128}ql^4 \tag{12.45e}$$

Considering Equations 12.42 and 12.43, we finally obtain the following:

$$X_1 = -\frac{7}{248}ql \tag{12.46a}$$

$$X_2 = \frac{73}{248}ql \tag{12.46b}$$

Applying the principle of superposition, it is possible to find the diagrams of the static characteristics (see Figure 12.11c through e).

In the case of a structure having three or more degrees of redundancy, the procedure does not substantially change. Once the equivalent statically determinate structure has been identified, $(n + 1)$ elementary schemes are considered, where n is the degree of redundancy. Applying the principle of virtual work to each scheme, we arrive at a linear algebraic system of n equations in the n unknowns $X_1, X_2, ..., X_n$:

$$\sum_{j=1}^{n} \eta_{ij}X_j = -\eta_{i0}, \quad \text{for } i = 1,2,...,n \tag{12.47}$$

where η_{ij} are the elements of the **influence matrix**

$$\eta_{ij} = \frac{1}{EI}\int_S M^{(i)}M^{(j)}ds, \quad \text{for } i,j = 1,2,...,n \tag{12.48}$$

and η_{i0} are the displacements due to the external load:

$$\eta_{i0} = \frac{1}{EI}\int_S M^{(i)}M^{(0)}ds, \quad \text{for } i = 1,2,...,n \tag{12.49}$$

In the case, for instance, of the frame of Figure 12.12a, it is possible to reduce it to an equivalent statically determinate structure by eliminating the built-in support D at the foot of the upright and by applying the three elementary reactions. There will thus be four schemes to be considered (Figure 12.12b) and Table 12.5, showing the bending moment functions.

The real bending moment may thus be expressed by means of the principle of superposition:

$$M_{AB}^{(r)} = M_{AB}^{(0)} + X_1 M_{AB}^{(1)} + X_2 M_{AB}^{(2)} + X_3 M_{AB}^{(3)}$$

$$= -\frac{m}{2l}z + \frac{1}{2}X_1 z + \frac{1}{2}X_2 z + \frac{1}{2l}X_3 z \tag{12.50a}$$

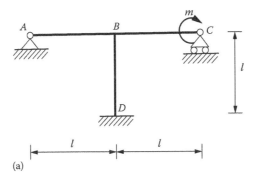

(a)

Figure 12.12

(continued)

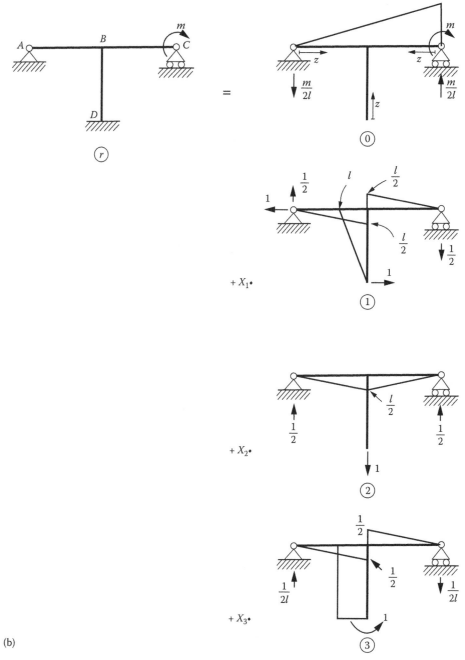

(b)

Figure 12.12 (continued)

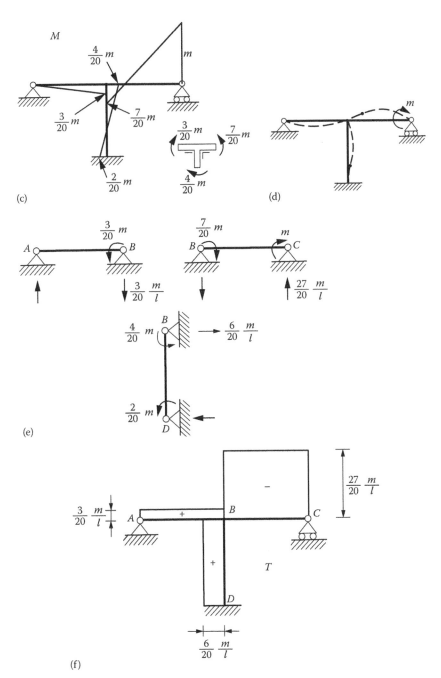

Figure 12.12 (continued)

(continued)

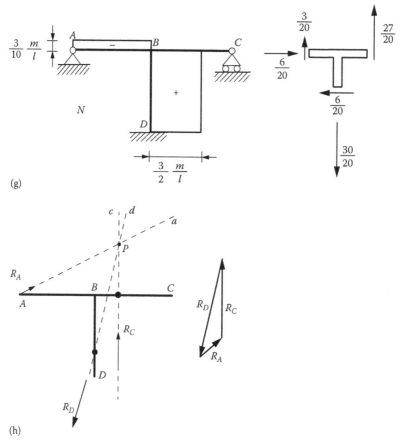

Figure 12.12 (continued)

Table 12.5

Beam	$M^{(0)}$	$M^{(1)}$	$M^{(2)}$	$M^{(3)}$
AB	$-\dfrac{m}{2l}z$	$\dfrac{z}{2}$	$\dfrac{z}{2}$	$\dfrac{z}{2l}$
BC	$-m+\dfrac{m}{2l}z$	$-\dfrac{z}{2}$	$\dfrac{z}{2}$	$-\dfrac{z}{2l}$
CD	0	z	0	1

$$M_{BC}^{(r)} = M_{BC}^{(0)} + X_1 M_{BC}^{(1)} + X_2 M_{BC}^{(2)} + X_3 M_{BC}^{(3)}$$

$$= -m + \frac{m}{2l}z - \frac{1}{2}X_1 z + \frac{1}{2}X_2 z - \frac{1}{2l}X_3 z \tag{12.50b}$$

$$M_{BD}^{(r)} = M_{BD}^{(0)} + X_1 M_{BD}^{(1)} + X_2 M_{BD}^{(2)} + X_3 M_{BD}^{(3)}$$

$$= X_1 z + X_3 \tag{12.50c}$$

The first equation of congruence will therefore be

$$\int_S M^{(1)}M^{(r)}ds = \int_0^l \left(\frac{z}{2}\right)\left(-\frac{m}{2l}z + \frac{1}{2}X_1 z + \frac{1}{2}X_2 z + \frac{1}{2l}X_3 z\right)dz$$

$$+\int_0^l \left(-\frac{z}{2}\right)\left(-m + \frac{m}{2l}z - \frac{1}{2}X_1 z + \frac{1}{2}X_2 z - \frac{1}{2l}X_3 z\right)dz$$

$$+\int_0^l (z)(X_1 z + X_3)dz = 0 \qquad (12.51)$$

which, when the calculations are performed, becomes

$$\frac{1}{4}\left(-\frac{m}{l} + X_1 + X_2 + \frac{X_3}{l}\right)\frac{l^3}{3} - \frac{1}{4}\left(\frac{m}{l} - X_1 + X_2 - \frac{X_3}{l}\right)\frac{l^3}{3} + \frac{m}{2}\frac{l^2}{2} + X_1\frac{l^3}{3} + X_3\frac{l^2}{2} = 0$$

$$(12.52)$$

and hence

$$6X_1 l + 8X_3 + m = 0 \qquad (12.53)$$

Equation 12.53 is satisfied by the values

$$X_1 = -\frac{3}{10}\frac{m}{l} \qquad (12.54a)$$

$$X_3 = \frac{m}{10} \qquad (12.54b)$$

In like manner, it is, on the other hand, possible to obtain the remaining two equations of congruence. Applying the principle of superposition, it is possible to find the diagrams of the static characteristics, the deformed configuration and the line of pressure (Figure 12.12c through i).

12.5 THERMAL DISTORTIONS AND CONSTRAINT SETTLEMENTS

In the case where the redundant structure undergoes thermal distortions, whether spread uniformly over the entire thickness or butterfly-shaped, Equation 12.5 is to be applied, considering the real system as the system of displacements and a statically determinate fictitious system as the system of forces.

Let us take, for instance, the case of the asymmetrical portal frame of Figure 12.13a, subjected to a uniform increase in temperature over the cross member. We shall choose the one of Figure 12.13b as equivalent statically determinate structure, obtained by replacing the hinge D with a roller support. Table 12.6 gives the bending moment and axial force functions, $M^{(1)}$ and $N^{(1)}$, as $M^{(0)} = N^{(0)} = 0$, on account of the absence of external loads of a mechanical nature.

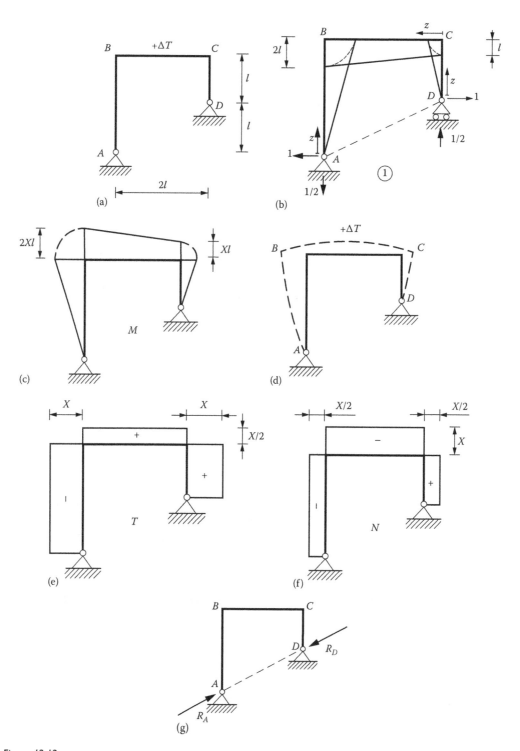

Figure 12.13

Table 12.6

Beam	$M^{(1)}$	$N^{(1)}$
AB	z	$\dfrac{1}{2}$
BC	$1+\dfrac{z}{2}$	1
CD	z	$-\dfrac{1}{2}$

Equation 12.5 yields

$$\int_{BC} N^{(1)}\alpha\Delta T \mathrm{d}s + \int_{S} M^{(1)}\frac{XM^{(1)}}{EI}\mathrm{d}s = 1\times 0 \tag{12.55}$$

and on the basis of Table 12.6

$$2\alpha\Delta Tl + \frac{X}{EI}\left[\int_{0}^{2l} z^2\mathrm{d}z + \int_{0}^{2l}\left(1+\frac{z}{2}\right)^2\mathrm{d}z + \int_{0}^{l} z^2\mathrm{d}z\right] = 0 \tag{12.56}$$

Evaluating the integrals, we obtain

$$X = -\frac{6}{23}\alpha\Delta T\frac{EI}{l^2} \tag{12.57}$$

which corresponds exactly to the shear value on the upright *CD*. Applying the principle of superposition, it is possible to find diagrams of the static characteristics, the deformed configuration and the line of pressure (see Figure 12.13c through g).

As regards statically indeterminate structures having an inelastic constraint settlement, it is necessary to take into account the work performed by the fictitious constraint reaction corresponding to the settlement itself. For example, the portal frame of Figure 12.14a can be rendered statically determinate by replacing the hinge *D* with a horizontally moving roller support (Figure 12.14b). In this case, Equation 12.5 yields

$$1\times 0 + 0\times\eta_0 = \frac{X}{EI}\int_{S}(M^{(1)})^2\mathrm{d}s \tag{12.58}$$

whence it emerges that $X = 0$; i.e. the structure is not subject to internal reactions, since the displacement η_0 of the point *D* can be produced by a simple rigid rotation of the portal frame about the hinge *A* (Figure 12.14c and d).

As regards the statically indeterminate structures with an elastic constraint settlement, it was shown in the introduction to this chapter that it is necessary to take into account the work done by the fictitious constraint reaction acting through the settlement caused by the real constraint reaction.

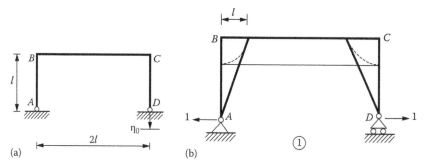

Figure 12.14

Table 12.7

Beam	$M^{(0)}$	$M^{(1)}$
AB	$-\dfrac{m}{2l}z$	$\dfrac{z}{2}$
BC	$-m+\dfrac{m}{2l}z$	$\dfrac{z}{2}$

Consider again the continuous beam on an elastic support of Figure 11.15a. Scheme 0 consists of the beam AC, loaded by the moment m at the end C (Figure 12.15a), whilst Scheme 1 consists of the same beam with a unit load acting in the centre (Figure 12.15b). The moment functions are given in Table 12.7.

Using formula (12.29), we have

$$X = -\frac{\displaystyle\int_S M^{(0)} M^{(1)} \mathrm{d}s}{\displaystyle\int_S (M^{(1)})^2 \mathrm{d}s + \frac{EI}{k}} \tag{12.59}$$

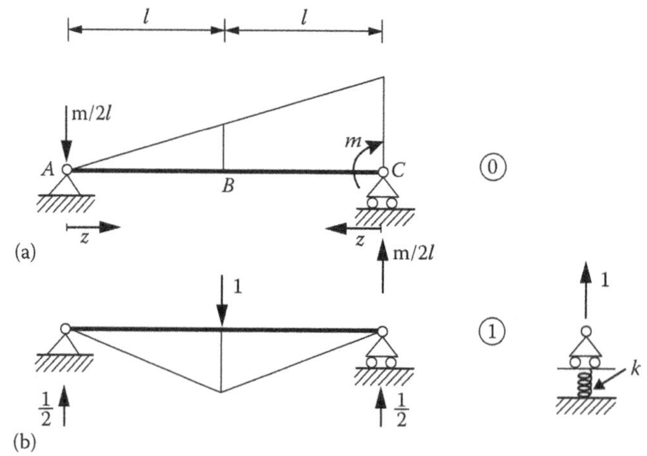

Figure 12.15

where $k = EA/h$ and

$$\int_S M^{(0)}M^{(1)}\mathrm{d}s = \int_0^l \left(-\frac{m}{2l}z\right)\left(\frac{z}{2}\right)\mathrm{d}z + \int_0^l \left(-m + \frac{m}{2l}z\right)\frac{z}{2}\,\mathrm{d}z = -\frac{1}{4}ml^2 \tag{12.60a}$$

$$\int_S (M^{(1)})^2\mathrm{d}s = \int_0^l \frac{z^2}{4}\mathrm{d}z + \int_0^l \frac{z^2}{4}\mathrm{d}z = \frac{l^3}{6} \tag{12.60b}$$

Performing the calculation, we obtain

$$X = \frac{(1/4)ml^2}{(l^3/6) + (EI/k)} \tag{12.61}$$

The two limit cases of an infinitely compliant support and a perfectly rigid support present, respectively, the following vertical reactions V_B:

$$\lim_{k \to 0} X = 0 \tag{12.62a}$$

$$\lim_{k \to \infty} X = \frac{3}{2}\frac{m}{l} \tag{12.62b}$$

Once again, we find the reaction of the central support, already determined following another procedure (Figure 11.14).

12.6 STATICALLY INDETERMINATE TRUSS STRUCTURES

In the case of statically indeterminate truss structures, the application of the principle of virtual work constitutes a highly valid and often rapid method of resolution. The equivalent statically determinate structure is obtained by subtracting a number of bars equal to the degree of redundancy of the structure. These bars, once isolated, must be considered axially compliant under the action of the corresponding redundant reaction.

Let us take as an example the truss structure of Figure 12.16a, subjected to a temperature increase ΔT on the bar CE. When the bar has been isolated and both the bar and the equivalent statically determinate structure have been subjected to the unit fictitious reaction (Figure 12.16b), Equation 12.5 becomes

$$1 \times \Delta\eta^{(r)} = \sum_i \frac{N_i^{(f)}N_i^{(r)}}{EA}l_i \tag{12.63}$$

From the scheme of the bar CE we have, on the other hand,

$$\Delta\eta^{(r)} = \left(\alpha\Delta T - \frac{X}{EA}\right)l\sqrt{2} \tag{12.64}$$

whereby the equation that provides the solution becomes

$$X\left(\sum_i \frac{(N_i^{(t)})^2}{EA}l_i\right) = \left(\alpha\Delta T - \frac{X}{EA}\right)l\sqrt{2} \tag{12.65}$$

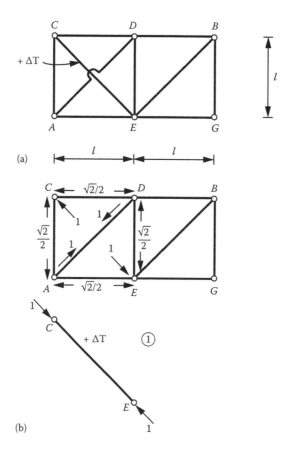

Figure 12.16

Since

$$N_{AC}^{(1)} = N_{CD}^{(1)} = N_{DE}^{(1)} = N_{AE}^{(1)} = \frac{\sqrt{2}}{2} \tag{12.66a}$$

$$N_{AD}^{(1)} = -1 \tag{12.66b}$$

we obtain finally

$$\frac{X}{EA}\left[4 \times \left(\frac{\sqrt{2}}{2}\right)^2 \times l + 1 \times l\sqrt{2}\right] = \left(\alpha\Delta T - \frac{X}{EA}\right)l\sqrt{2} \tag{12.67}$$

and hence

$$X = \frac{2 - \sqrt{2}}{2}\alpha\Delta T EA \tag{12.68}$$

As our second example, let us examine the truss structure of Figure 12.17a, where the upper chord is subjected to a uniform temperature increase ΔT. The bars ED and DC are unloaded by virtue of the equilibrium of the node D, so that the scheme providing the solution reduces to that of Figure 12.17b, where the bar AC has been isolated with respect to the rest of the structure. From the equilibrium of the nodes C and B (Figure 12.17c), the following condition results:

$$N_{AB}^{(1)} = N_{BC}^{(1)} = N_{CE}^{(1)} = N_{EA}^{(1)} = -\frac{\sqrt{3}}{3} \tag{12.69a}$$

$$N_{BE}^{(1)} = \frac{\sqrt{3}}{3} \tag{12.69b}$$

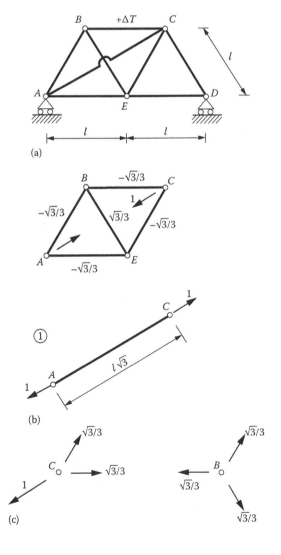

Figure 12.17

Since, by virtue of the fact that the load is of thermal origin, we have $N^{(0)} = 0$, applying the principle of virtual work gives

$$-1 \times \frac{Xl\sqrt{3}}{EA} = \sum_i N_i^{(1)} \left(\frac{XN_i^{(1)}l}{EA} + \alpha\Delta T_i l \right)$$

(12.70)

and, hence, rendering the terms of the summation explicit

$$-\frac{Xl\sqrt{3}}{EA} = \frac{5}{3}\frac{Xl}{EA} - \frac{\sqrt{3}}{3}\alpha\Delta Tl$$

(12.71)

whence we obtain the unknown reaction N_{AC}

$$X = \frac{9 - 5\sqrt{3}}{2}\alpha\Delta TEA$$

(12.72)

The static indeterminacy of the truss structure of Figure 12.18a is due to the continuity of the lower chord. The equivalent statically determinate structure is obtained by inserting

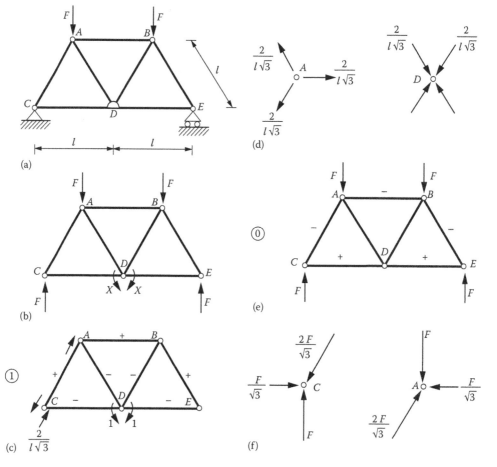

Figure 12.18

a complete hinge at D (Figure 12.18b) and applying the redundant moment X. The fictitious structure of Scheme 1 proves to be loaded in a symmetrical manner by two unit moments (Figure 12.18c). By virtue of the equilibrium of the beams DC and DE, as well as of the nodes A and D (Figure 12.18d), we have

$$N_{AC}^{(1)} = N_{AB}^{(1)} = \frac{2}{l\sqrt{3}} \tag{12.73a}$$

$$N_{AD}^{(1)} = -\frac{2}{l\sqrt{3}} \tag{12.73b}$$

$$N_{CD}^{(1)} = -\frac{1}{l\sqrt{3}} \tag{12.73c}$$

Structure of Scheme 0 (Figure 12.18e) proves, on the other hand, to be loaded only on the external bars:

$$N_{AC}^{(0)} = -\frac{2F}{\sqrt{3}} \tag{12.74a}$$

$$N_{CD}^{(0)} = \frac{F}{\sqrt{3}} \tag{12.74b}$$

$$N_{AB}^{(0)} = -\frac{F}{\sqrt{3}} \tag{12.74c}$$

$$N_{AD}^{(0)} = 0 \tag{12.74d}$$

The schemes of equilibrium of the nodes C and A are represented in Figure 12.18f.
  Application of the principle of virtual work leads to the following equation:

$$
\begin{aligned}
1 \times 0 = &\frac{l}{EA}\left[2N_{AC}^{(0)}N_{AC}^{(1)} + 2N_{CD}^{(0)}N_{CD}^{(1)} + N_{AB}^{(0)}N_{AB}^{(1)}\right] \\
&+ \frac{Xl}{EA}\left[2\left(N_{AC}^{(1)}\right)^2 + 2\left(N_{CD}^{(1)}\right)^2 + \left(N_{AB}^{(1)}\right)^2 + 2\left(N_{AD}^{(1)}\right)^2\right] \\
&+ \frac{X}{EI}\left[2\int_0^l \left(M_{CD}^{(1)}\right)^2 dz\right]
\end{aligned} \tag{12.75}
$$

which allows the determination of the redundant unknown X. We thus have

$$M_{CD}^{(1)} = \frac{z}{l} \tag{12.76}$$

whence

$$\int_0^l \left(M_{CD}^{(1)}\right)^2 dz = \frac{l}{3} \tag{12.77}$$

and, hence, substituting Equations 12.73, 12.74 and 12.77 into Equation 12.75

$$\frac{l}{EA}\left[-\frac{8F}{3l}-\frac{2F}{3l}-\frac{2F}{3l}\right]+\frac{Xl}{EA}\left[\frac{8}{3l^2}+\frac{2}{3l^2}+\frac{4}{3l^2}+\frac{8}{3l^2}\right]+\frac{X}{EI}\left[\frac{2}{3}l\right]=0 \tag{12.78}$$

Finally, we obtain

$$X=\frac{6Fl}{11+\left(\dfrac{l}{\rho}\right)^2} \tag{12.79}$$

where ρ denotes the radius of gyration of the cross section of the bars.

Finally, consider the closed structure of Figure 12.19a, stiffened by a diagonal cross. In order to obtain the equivalent statically determinate structure, let the cross be isolated and

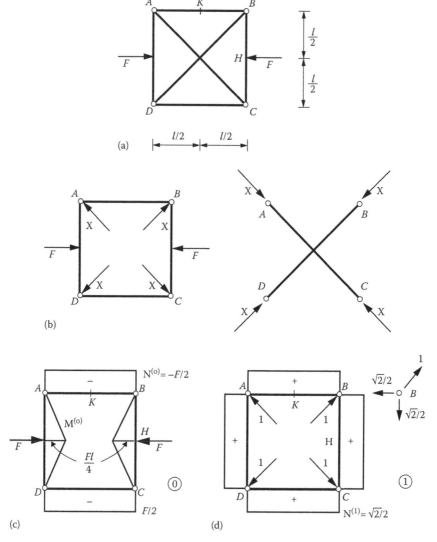

Figure 12.19

subjected to the action of the unknown axial force X, just as the square framework is subjected to equal and opposite loads (Figure 12.19b). The system of Scheme 0 is subjected to a compressive axial force on the beams AB and CD, and to a bending moment on the beams AD and BC (Figure 12.19c). The system of Scheme 1 is subjected to a tensile axial force on all the beams (Figure 12.19d).

The principle of virtual work yields the following condition:

$$-1 \times \frac{Xl\sqrt{2}/2}{EA} = \int_{S/4} \frac{M^{(f)}M^{(r)}}{EI} ds + \int_{S/4} \frac{N^{(f)}N^{(r)}}{EA} ds \qquad (12.80)$$

with

$$M^{(f)} = 0 \qquad (12.81a)$$

$$N^{(f)} = N^{(1)} = \frac{\sqrt{2}}{2} \qquad (12.81b)$$

$$M_{BH}^{(r)} = \frac{F}{2}z \qquad (12.81c)$$

$$M_{BK}^{(r)} = 0 \qquad (12.81d)$$

$$N_{BH}^{(r)} = X\frac{\sqrt{2}}{2} \qquad (12.81e)$$

$$N_{BK}^{(r)} = -\frac{F}{2} + X\frac{\sqrt{2}}{2} \qquad (12.81f)$$

Equation 12.80 thus takes the form

$$-\frac{Xl\sqrt{2}}{2EA} = \frac{l}{2EA}\left(\frac{X}{2} + \frac{X}{2} - \frac{F\sqrt{2}}{4}\right) \qquad (12.82)$$

whence it follows that

$$X = \frac{F(2 - \sqrt{2})}{4} \qquad (12.83)$$

The axial forces in the individual bars are therefore

$$N_{AB} = N_{CD} = -\frac{F}{2} + \frac{F(2 - \sqrt{2})}{4}\frac{\sqrt{2}}{2} = F\frac{\sqrt{2} - 3}{4} \qquad (12.84a)$$

$$N_{AD} = N_{BC} = \frac{F}{4}(\sqrt{2} - 1) \qquad (12.84b)$$

$$N_{BD} = N_{AC} = -\frac{F}{4}(2 - \sqrt{2}) \qquad (12.84c)$$

12.7 ARCHES AND RINGS

In the cases of **arches** and **rings,** and in general of **curvilinear beams,** just as in the previously considered case of truss structures, application of the principle of virtual work proves to be a highly convenient method of solution.

   Consider, for instance, the circular cantilever of Figure 12.20a, subjected to a uniform temperature rise ΔT. Introduce three fictitious schemes, where the cantilever is loaded at the unconstrained end by

   Scheme 1: a unit horizontal force (Figure 12.20b)
   Scheme 2: a unit vertical force (Figure 12.20c)
   Scheme 3: a unit couple (Figure 12.20d)

It is then possible to calculate the generalized displacements of that end. We have in fact

$$N^{(1)} = \cos\varphi \tag{12.85a}$$

$$N^{(2)} = \sin\varphi \tag{12.85b}$$

$$N^{(3)} = 0 \tag{12.85c}$$

and hence

$$1 \times u_B = \int_S N^{(1)}\varepsilon_T \mathrm{d}s \tag{12.86a}$$

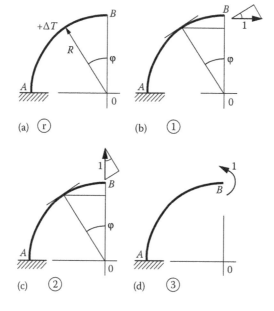

Figure 12.20

$$1 \times \upsilon_B = \int_S N^{(2)} \varepsilon_T \, ds \tag{12.86b}$$

$$1 \times \varphi_B = \int_S N^{(3)} \varepsilon_T \, ds \tag{12.86c}$$

Substituting Equations 12.85 into Equations 12.86, we obtain

$$u_B = \upsilon_B = R\alpha\Delta T, \quad \varphi_B = 0 \tag{12.87}$$

If the same cantilever beam is subjected to a butterfly-shaped thermal variation (Figure 12.21), the previously considered fictitious schemes yield the fictitious moments

$$M^{(1)} = R(\cos\varphi - 1) \tag{12.88a}$$

$$M^{(2)} = R\sin\varphi \tag{12.88b}$$

$$M^{(3)} = 1 \tag{12.88c}$$

and hence the application of the principle of virtual work to each of the three schemes leads to the determination of the displacements of the end B:

$$1 \times u_B = \int_S M^{(1)} \chi_T \, ds \tag{12.89a}$$

$$1 \times \upsilon_B = \int_S M^{(2)} \chi_T \, ds \tag{12.89b}$$

$$1 \times \varphi_B = \int_S M^{(3)} \chi_T \, ds \tag{12.89c}$$

From Equations 12.88 and 12.89, we obtain

$$u_B = 2\alpha \frac{\Delta T}{h} \int_0^{\pi/2} R^2 (\cos\varphi - 1) \, d\varphi \tag{12.90a}$$

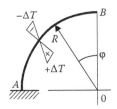

Figure 12.21

$$v_B = 2\alpha \frac{\Delta T}{h} \int\limits_0^{\pi/2} R^2 \sin\varphi \, d\varphi \qquad (12.90b)$$

$$\varphi_B = 2\alpha \frac{\Delta T}{h} \int\limits_0^{\pi/2} R \, d\varphi \qquad (12.90c)$$

and hence

$$u_B = \alpha\Delta T \frac{R^2}{h} (2 - \pi) \qquad (12.91a)$$

$$v_B = 2\alpha\Delta T \frac{R^2}{h} \qquad (12.91b)$$

$$\varphi_B = \pi\alpha\Delta T \frac{R}{h} \qquad (12.91c)$$

To determine the relative horizontal displacement of the ends of the disconnected ring of Figure 12.22a, it is sufficient to consider the fictitious scheme of Figure 12.22b, so that

$$M^{(f)} = M^{(1)} = R(1 - \cos\varphi) \qquad (12.92a)$$

$$M^{(r)} = FM^{(1)} = FR(1 - \cos\varphi) \qquad (12.92b)$$

Application of the principle of virtual work,

$$1 \times \frac{\Delta u}{2} = \int\limits_{S/2} \frac{M^{(f)}M^{(r)}}{EI} \, ds \qquad (12.93)$$

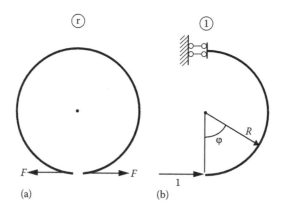

Figure 12.22

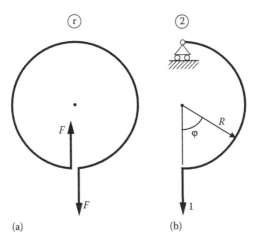

Figure 12.23

on the basis of Equations 12.92, yields the following condition:

$$\frac{\Delta u}{2} = \frac{FR^3}{EI} \int_0^\pi (1 - \cos\varphi)^2 \, d\varphi \tag{12.94}$$

from which we find

$$\Delta u = 3\pi \frac{FR^3}{EI} \tag{12.95}$$

   To determine the relative displacement of the ends of the disconnected ring of Figure 12.23a, two fictitious schemes must be used, one with the unit force horizontal (Figure 12.22b) and the other with the unit force vertical (Figure 12.23b). For the latter scheme, we have

$$M^{(f)} = M^{(2)} = R\sin\varphi \tag{12.96a}$$

$$M^{(r)} = FM^{(2)} = FR\sin\varphi \tag{12.96b}$$

and hence the discontinuity of vertical displacement may be deduced from the following equation:

$$1 \times \frac{\Delta v}{2} = \int_{S/2} \frac{M^{(f)}M^{(r)}}{EI} \, ds \tag{12.97}$$

which, on the basis of Equations 12.96, gives

$$\frac{\Delta v}{2} = \frac{FR^3}{EI} \int_0^\pi \sin^2\varphi \, d\varphi \tag{12.98}$$

and hence

$$\Delta\upsilon = \pi\frac{FR^3}{EI} \tag{12.99}$$

The discontinuity of horizontal displacement is zero owing to skew-symmetry. The absolute displacement may be deduced from the application of the principle of virtual work, considering expression (12.96b) as the real moment and expression (12.92a) as the fictitious moment,

$$1\times u = \int_{S/2}\frac{FM^{(1)}M^{(2)}}{EI}\,\mathrm{d}s \tag{12.100}$$

from which we obtain

$$u = \frac{FR^3}{EI}\int_0^\pi \sin\varphi(1-\cos\varphi)\mathrm{d}\varphi \tag{12.101}$$

and hence

$$u = 2\frac{FR^3}{EI} \tag{12.102}$$

Now consider the statically indeterminate ring of Figure 12.24a, in which the internal connecting rod undergoes a temperature rise ΔT. By virtue of double symmetry, the ring reduces to the quarter of circumference of Figure 12.24b, in which the rod has been isolated and subjected to the redundant reaction $2X_1$. The equivalent statically determinate structure can hence appear as in Figure 12.24c, where the quarter of circumference is subjected also to the second redundant unknown, the moment X_2. The two fictitious structures are represented in Figure 12.24d and e. The real moment is equal to

$$M^{(r)} = X_1 M^{(1)} + X_2 M^{(2)} \tag{12.103a}$$

where

$$M^{(1)} = R\sin\varphi \tag{12.103b}$$

$$M^{(2)} = 1 \tag{12.103c}$$

The first equation of congruence for the connecting rod is written as

$$1\times\left(-\frac{2X_1 R}{EA} + \alpha\Delta TR\right) = \int_0^{\pi/2} M^{(1)}\frac{M^{(r)}}{EI}R\mathrm{d}\varphi \tag{12.104}$$

Performing the calculations, we obtain

$$X_1\left(2\rho^2 + \frac{\pi R^2}{4}\right) + X_2 R = EI\alpha\Delta T \tag{12.105}$$

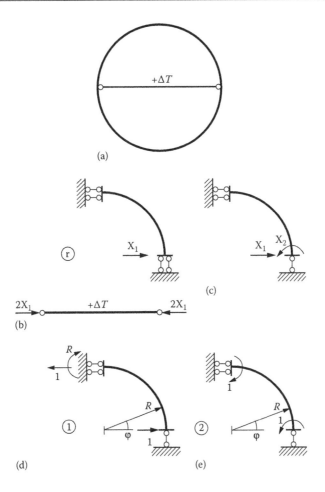

Figure 12.24

Since the radius of gyration of the cross section is much smaller than the dimension R, Equation 12.105 simplifies as follows:

$$\pi R^2 X_1 + 4RX_2 = 4EI\alpha\Delta T \tag{12.106}$$

The second equation of congruence, corresponding to the rotation, is written as

$$1 \times 0 = \int_0^{\pi/2} M^{(2)} \frac{M^{(r)}}{EI} R d\varphi \tag{12.107}$$

From Equations 12.103, we have

$$\int_0^{\pi/2} (X_1 R \sin\varphi + X_2) d\varphi = 0 \tag{12.108}$$

and hence

$$2X_1R + \pi X_2 = 0 \tag{12.109}$$

From Equations 12.106 and 12.109, we obtain the axial force of the rod

$$X_1 = \frac{4\pi EI\alpha\Delta T}{R^2(\pi^2 - 8)} \tag{12.110a}$$

as well as the redundant moment:

$$X_2 = -\frac{2R}{\pi}X_1 \tag{12.110b}$$

The statically indeterminate ring of Figure 12.25a is studied in the same way as for the preceding one. If we imagine turning the scheme by 90°, it is possible to use the fictitious systems of Schemes 1 and 2 of the previous example (Figure 12.24d and e), whilst Scheme 0 is represented in Figure 12.25b,

$$M^{(0)} = \frac{FR}{2}(1 - \cos\varphi) \tag{12.111}$$

The equations of congruence can be written like Equations 12.41, with

$$\int_S (M^{(1)})^2 \, ds = R^3 \int_0^{\pi/2} \sin^2 \varphi \, d\varphi = \frac{\pi}{4}R^3 \tag{12.112a}$$

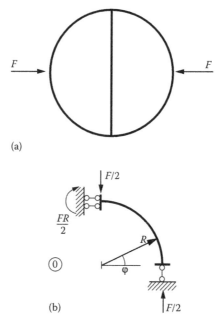

(a)

(b)

Figure 12.25

$$\int_S M^{(1)}M^{(2)}ds = R^2 \int_0^{\pi/2} \sin\varphi d\varphi = R^2 \tag{12.112b}$$

$$\int_S (M^{(2)})^2 ds = R \int_0^{\pi/2} d\varphi = \frac{\pi}{2} R \tag{12.112c}$$

$$\int_S M^{(1)}M^{(0)}ds = \frac{FR^3}{2} \int_0^{\pi/2} \sin\varphi(1-\cos\varphi)d\varphi = \frac{FR^3}{4} \tag{12.112d}$$

$$\int_S M^{(2)}M^{(0)}ds = \frac{FR^2}{2} \int_0^{\pi/2} (1-\cos\varphi)d\varphi = \frac{FR^2}{4}(\pi-2) \tag{12.112e}$$

We have, therefore,

$$\frac{\pi}{4}R^3 X_1 + R^2 X_2 = -\frac{FR^3}{4} \tag{12.113a}$$

$$R^2 X_1 + \frac{\pi}{2}RX_2 = -\frac{FR^2}{4}(\pi-2) \tag{12.113b}$$

From Equations 12.44, we have finally

$$X_1 = F\frac{4-\pi}{8-\pi^2} \tag{12.114a}$$

$$X_2 = FR\frac{\pi^2 - 2\pi - 4}{2(8-\pi^2)} \tag{12.114b}$$

Also the statically indeterminate ring of Figure 12.26a, the orthogonal diaphragms of which are respectively heated and cooled, can be analysed using the method illustrated previously. In this case (Figure 12.26b),

$$M^{(0)} = 0 \tag{12.115a}$$

$$M^{(1)} = R\sin\varphi \tag{12.115b}$$

$$M^{(2)} = 1 \tag{12.115c}$$

$$M^{(3)} = R(1-\cos\varphi) \tag{12.115d}$$

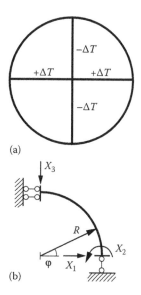

(a)

(b)

Figure 12.26

The ring of Figure 12.27a is subjected to a butterfly-shaped thermal variation. The equivalent statically determinate structure is represented in Figure 12.27b, with

$$M^{(0)} = 0, \quad M^{(1)} = 1 \tag{12.116}$$

The real moment is therefore

$$M^{(r)} = M^{(0)} + X M^{(1)} = X \tag{12.117a}$$

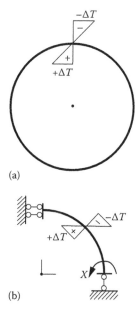

(a)

(b)

Figure 12.27

so that the real thermoelastic curvature is

$$\chi^{(r)} = \frac{X}{EI} + 2\alpha \frac{\Delta T}{h} \qquad (12.117b)$$

The principle of virtual work yields

$$1 \times 0 = \int_{S/4} M^{(1)} \chi^{(r)} ds \qquad (12.118)$$

from which we obtain

$$\left(\frac{X}{EI} + 2\alpha \frac{\Delta T}{h} \right) \frac{\pi R}{2} = 0 \qquad (12.119)$$

and hence

$$X = -2\alpha \Delta T \frac{EI}{h} \qquad (12.120)$$

Consider again the ring of radius R, loaded by three angularly equidistant radial forces (Figure 12.28a). The equivalent statically determinate structure is represented in Figure 12.28b, so that the moments corresponding to Schemes 0 and 1 are the following:

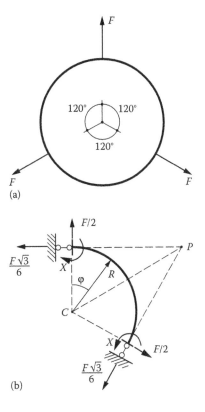

Figure 12.28

$$M^{(0)} = \frac{FR}{2}\sin\varphi - \frac{F\sqrt{3}}{6}R(1 - \cos\varphi) \qquad (12.121a)$$

$$M^{(1)} = 1 \qquad (12.121b)$$

Rotational congruence imposes

$$1 \times 0 = \int_{S/3} M^{(1)} \frac{M^{(0)} + XM^{(1)}}{EI} ds \qquad (12.122)$$

and hence

$$\int_0^{2\pi/3} \left[\frac{FR}{2}\sin\varphi - \frac{F\sqrt{3}}{6}R(1 - \cos\varphi) + X \right] R d\varphi = 0 \qquad (12.123)$$

whence we obtain

$$X = -FR\frac{9 - \pi\sqrt{3}}{6\pi} \qquad (12.124)$$

Let the ring of Figure 12.29a be subjected to the stresses produced by the thermal dilatation of the two orthogonal diaphragms. The equivalent statically determinate structure is represented in Figure 12.29b. From symmetry, it can be reduced to the fictitious system of Scheme 1 of Figure 12.29c. We have therefore

$$M^{(0)} = 0 \qquad (12.125a)$$

$$M^{(1)} = \frac{R}{2}(\sin\varphi + \cos\varphi - 1) \qquad (12.125b)$$

Application of the principle of virtual work yields

$$2 \times \frac{1}{2} \times \left(\alpha\Delta TR - \frac{XR}{EA} \right) = \int_{S/4} X\frac{\left(M^{(1)}\right)^2}{EI} ds \qquad (12.126)$$

from which we obtain

$$\alpha\Delta TR - \frac{XR}{EA} = \frac{XR^3}{4EI} \int_0^{\pi/2} (\sin\varphi + \cos\varphi - 1)^2 d\varphi \qquad (12.127)$$

and hence

$$X = \frac{4\alpha\Delta TEI}{R^2(\pi + 1) + 4\rho^2} \qquad (12.128)$$

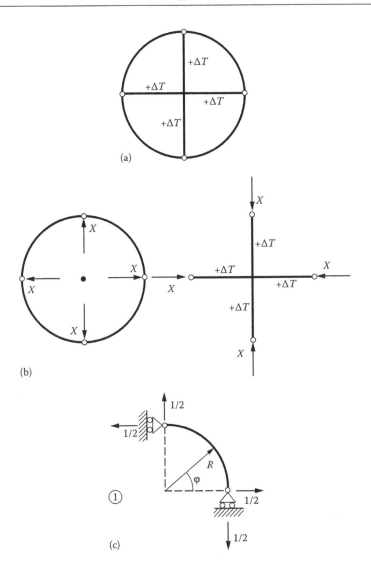

Figure 12.29

Finally let us consider the arch of Figure 12.30a. The equivalent statically determinate structure of Figure 12.30b presents at its lower end an elastically compliant roller support, which simulates the lateral cantilever on which the arch rests. Schemes 0 and 1 give the moments (Figure 12.30c)

$$M^{(0)} = \frac{FR}{2}(1 - \cos\varphi) \tag{12.129a}$$

$$M^{(1)} = -R\sin\varphi \tag{12.129b}$$

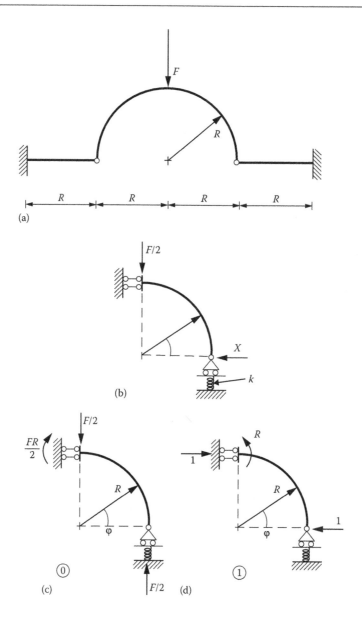

Figure 12.30

and thus we have

$$1 \times 0 = \int_{S/2} M^{(1)} \frac{M^{(0)} + XM^{(1)}}{EI} ds \qquad (12.130)$$

Since

$$\int_{S/2} M^{(0)} M^{(1)} ds = -\frac{FR^3}{4} \qquad (12.131a)$$

$$\int\limits_{S/2} (M^{(1)})^2 \, ds = R^3 \frac{\pi}{4}$$

(12.131b)

we obtain finally $X = F/\pi$. Note how the stiffness k of the spring does not appear in the solution.

12.8 CASTIGLIANO'S THEOREM

In the case where thermal distortions and constraint settlements are absent, the determination of elastic displacements in statically determinate structures can be made using **Castigliano's theorem,** as an alternative to the application of the principle of virtual work proposed in Section 12.2.

Consider a statically determinate structure subjected to n different loads F_1, F_2, ..., F_n (Figure 12.31). The principle of superposition makes it possible to express the n generalized displacements dual of the forces, η_1, η_2, ..., η_n, *via* the coefficients of influence η_{ij}:

$$\eta_i = \sum_{j=1}^{n} \eta_{ij} F_j, \quad \text{for } i = 1,2,...,n$$

(12.132)

Clapeyron's theorem then gives the strain energy of the structure in the form

$$L_{\text{def}} = \frac{1}{2} \sum_{i=1}^{n} F_i \eta_i$$

(12.133a)

which, taking into account Equations 12.132, becomes

$$L_{\text{def}} = \frac{1}{2} \sum_{i=1}^{n} \sum_{j=1}^{n} F_i F_j \eta_{ij}$$

(12.133b)

Finally, deriving the strain energy with respect to each force F_i, $i = 1, 2, ..., n$, we have

$$\frac{\partial L_{\text{def}}}{\partial F_i} = \sum_{j=1}^{n} \eta_{ij} F_j, \quad \text{for } i = 1,2,...,n$$

(12.134)

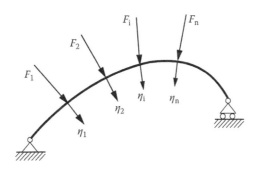

Figure 12.31

and, hence, on the basis of Equation 12.132, we find

$$\frac{\partial L_{\text{def}}}{\partial F_i} = \eta_i, \quad \text{for } i = 1,2,\ldots,n \tag{12.135}$$

Relation (12.135) represents the statement of Castigliano's theorem: **the derivative of the strain energy with respect to the magnitude of an applied force is equal to the global elastic displacement, dual with respect to the same force.**

In the case where one wishes to calculate a generic displacement which does not correspond to an applied force, it is possible to apply a fictitious force, corresponding to the displacement sought, and, once the partial derivative of the work has been obtained, to make the magnitude of the earlier force tend to zero.

It is possible to demonstrate the perfect equivalence of Castigliano's theorem with the principle of virtual work. In fact, in the case where axial force and shearing force give a negligible contribution to the strain energy, we have

$$L_{\text{def}} = \frac{1}{2} \int_S \frac{M^2}{EI} \, ds \tag{12.136}$$

and, hence, on the basis of Equation 12.135,

$$\eta_i = \frac{1}{2EI} \frac{\partial}{\partial F_i} \int_S M^2 \, ds \tag{12.137}$$

Carrying the differential operator under the integral sign, we obtain

$$\eta_i = \frac{1}{EI} \int_S M \frac{\partial M}{\partial F_i} \, ds \tag{12.138}$$

Since then, the real moment M may be interpreted as the sum of n partial moments, each generated by the generic force F_i,

$$M = M^{(r)} = \sum_{i=1}^n F_i M^{(i)} \tag{12.139}$$

Equation 12.138 becomes

$$\eta_i = \frac{1}{EI} \int_S M^{(r)} M^{(i)} \, ds \tag{12.140}$$

in which relation we can recognize the equation of the principle of virtual work, amply discussed and applied hitherto.

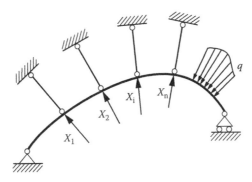

Figure 12.32

12.9 MENABREA'S THEOREM

Menabrea's theorem may be derived directly from Castigliano's theorem, even though originally the two theorems were demonstrated independently.

Menabrea's theorem is also called the **theorem of minimum strain energy** and refers to redundant structures (Figure 12.32). It states in fact that, **given a structure with n degrees of redundancy, the n values of the redundant unknowns make the strain energy of the structure a minimum.**

Considering the equivalent statically determinate structure and the n redundant unknowns $X_1, X_2, ..., X_n$, Castigliano's theorem gives

$$\eta_i = \frac{\partial L_{\text{def}}}{\partial X_i}, \quad \text{for } i = 1, 2, ..., n \tag{12.141a}$$

while the conditions of congruence, in the case where all the constraints are rigid, are

$$\eta_i = 0, \quad \text{for } i = 1, 2, ..., n \tag{12.141b}$$

By the transitive law, we obtain therefore

$$\frac{\partial}{\partial X_i} L_{\text{def}}(X_1, X_2, ..., X_n) = 0, \quad \text{for } i = 1, 2, ..., n \tag{12.142}$$

The foregoing equation confirms the statement of Menabrea's theorem, L_{def} being a positive definite quadratic function in the variables X_i.

Appendix A: Calculation of the internal reactions in a circular arch subjected to a radial hydrostatic load

A.I ANALYTICAL METHOD

The differential equation (5.9) in the case of Figure 5.4, whereby

$$p = m = 0 \tag{A.1a}$$

$$q(\vartheta) = -\gamma R(1 - \cos \vartheta) \tag{A.1b}$$

reduces to the following form:

$$\frac{d^3 M}{d\vartheta^3} + \frac{dM}{d\vartheta} = \gamma R^3 \sin \vartheta \tag{A.2}$$

For the calculation of the complete integral, i.e. of the integral of the associated homogeneous equation, consider the characteristic equation

$$\lambda^3 + \lambda = 0 \tag{A.3}$$

which presents the following roots:

$$\lambda_1 = 0, \quad \lambda_2 = +i, \quad \lambda_3 = -i$$

where i is the imaginary unit. As is known from mathematical analysis, the complete integral therefore takes the form

$$M_g(\vartheta) = C_1 + C_2 \cos \vartheta + C_3 \sin \vartheta \tag{A.4}$$

As regards the particular solution, since we are considering a case in which the known term is of the sort

$$b(\vartheta) = P_m(\vartheta) e^{\alpha \vartheta} \tag{A.5}$$

where $P_m(\vartheta)$ indicates a polynomial of the mth order, with $m = 0$, and $\alpha = i$, this may be sought in the form

$$M_0(\vartheta) = \vartheta^r R_m(\vartheta) e^{\alpha \vartheta} \tag{A.6}$$

where $r = 1$ is the multiplicity with which $\alpha = i$ is the solution of the characteristic equation, and $R_m(\vartheta)$ is a polynomial in ϑ of the order $m = 0$, and thus in this case is a constant. We have therefore

$$M_0(\vartheta) = a\vartheta \sin \vartheta \tag{A.7}$$

Substituting Equation A.7 in Equation A.2, we obtain the coefficient a. Differentiating Equation A.7 sequentially, we have in fact

$$\frac{dM_0}{d\vartheta} = a(\vartheta \cos \vartheta + \sin \vartheta) \tag{A.8a}$$

$$\frac{d^2 M_0}{d\vartheta^2} = a(-\vartheta \sin \vartheta + 2 \cos \vartheta) \tag{A.8b}$$

$$\frac{d^3 M_0}{d\vartheta^3} = a(-\vartheta \cos \vartheta - 3 \sin \vartheta) \tag{A.8c}$$

and thus substituting Equations A.8a and A.8c in Equation A.2

$$-2a \sin \vartheta = \gamma R^3 \sin \vartheta \tag{A.9}$$

or

$$a = -\frac{\gamma R^3}{2} \tag{A.10}$$

Summing the complete integral (A.4) and the particular solution (A.7), with the coefficient a given by Equation A.10, we can write finally

$$M(\vartheta) = C_1 + C_2 \cos \vartheta + C_3 \sin \vartheta - \frac{\gamma R^3}{2} \vartheta \sin \vartheta \tag{A.11}$$

The arbitrary constants C_1, C_2, C_3 are calculated using the following boundary conditions:

$$M_A = 0 \tag{A.12a}$$

$$T_A = 0 \tag{A.12b}$$

$$M_B = 0 \tag{A.12c}$$

since in A there is a roller support and in B a hinge (Figure 5.4). Recalling the relation (5.6), which links shear and bending moment, we have

$$M(0) = C_1 + C_2 = 0 \tag{A.13a}$$

$$T(0) = \frac{C_3}{R} = 0 \tag{A.13b}$$

$$M\left(-\frac{\pi}{2}\right) = C_1 - C_3 - \frac{\pi\gamma R^3}{4} = 0 \tag{A.13c}$$

The linear algebraic system (A.13) admits of the following roots:

$$C_1 = \frac{\gamma R^3}{4}\pi \tag{A.14a}$$

$$C_2 = -\frac{\gamma R^3}{4}\pi \tag{A.14b}$$

$$C_3 = 0 \tag{A.14c}$$

Finally we obtain from Equation A.11

$$M(\vartheta) = \frac{\gamma R^3}{4}(\pi - \pi\cos\vartheta - 2\vartheta\sin\vartheta) \tag{A.15}$$

Applying Equations 5.6 and 5.8, we obtain also the shearing force and the axial force

$$T(\vartheta) = -\frac{\gamma R^2}{4}\left[(2 - \pi)\sin\vartheta + 2\vartheta\cos\vartheta\right] \tag{A.16}$$

$$N(\vartheta) = \frac{\gamma R^2}{4}\left[-4 + \pi\cos\vartheta + 2\vartheta\sin\vartheta\right] \tag{A.17}$$

A.2 DIRECT METHOD

The equations of equilibrium to horizontal translation, vertical translation and rotation about point B, respectively, appear as follows (Figure A.1):

$$H_A + H_B = \int_0^{\pi/2} |q(\vartheta)|\sin\vartheta\, R d\vartheta \tag{A.18a}$$

$$V_B = \int_0^{\pi/2} |q(\vartheta)|\cos\vartheta\, R d\vartheta \tag{A.18b}$$

$$H_A R = \int_0^{\pi/2} |q(\vartheta)|\cos\vartheta\, R^2 d\vartheta \tag{A.18c}$$

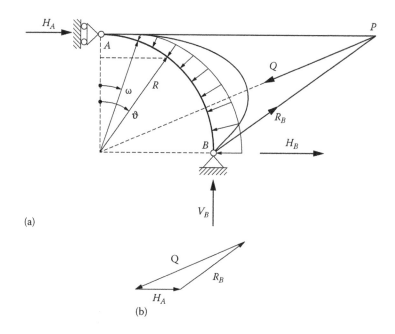

(a)

(b)

Figure A.1

Computing the integrals, we obtain a linear algebraic system in the three unknowns H_A, H_B, V_B,

$$H_A + H_B = \frac{\gamma R^2}{2} \tag{A.19a}$$

$$V_B = \gamma R^2 \left(1 - \frac{\pi}{4}\right) \tag{A.19b}$$

$$H_A = \gamma R^2 \left(1 - \frac{\pi}{4}\right) \tag{A.19c}$$

which gives the solution

$$H_A = \gamma R^2 \left(1 - \frac{\pi}{4}\right) \tag{A.20a}$$

$$H_B = \gamma R^2 \left(\frac{\pi}{4} - \frac{1}{2}\right) \tag{A.20b}$$

$$V_B = \gamma R^2 \left(1 - \frac{\pi}{4}\right) \tag{A.20c}$$

The bending moment acting in a generic cross section of angular coordinate ϑ is obtained by summing the contributions which precede the cross section itself (Figure A.1)

$$M(\vartheta) = -H_A R(1 - \cos \vartheta) + \int_0^\vartheta |q(\omega)| R^2 \sin(\vartheta - \omega) d\omega \qquad (A.21)$$

Substituting the distribution (A.1b) into the integral, we have

$$\int_0^\vartheta |q(\omega)| R^2 \sin(\vartheta - \omega) d\omega$$

$$= \gamma R^3 \int_0^\vartheta \sin(\vartheta - \omega) d\omega - \gamma R^3 \int_0^\vartheta \cos \omega \sin(\vartheta - \omega) d\omega \qquad (A.22)$$

Applying the well-known trigonometric formulas, we obtain

$$\int_0^\vartheta |q(\omega)| R^2 \sin(\vartheta - \omega) d\omega$$

$$= \gamma R^3 \int_0^\vartheta (\sin \vartheta \cos \omega - \cos \vartheta \sin \omega) d\omega$$

$$- \gamma R^3 \int_0^\vartheta \cos \omega (\sin \vartheta \cos \omega - \cos \vartheta \sin \omega) d\omega$$

$$= \gamma R^3 \left\{ \sin \vartheta [\sin \omega]_0^\vartheta + \cos \vartheta [\cos \omega]_0^\vartheta \right.$$

$$\left. - \frac{1}{2} \sin \vartheta \left[\omega + \frac{1}{2} \sin 2\omega \right]_0^\vartheta - \frac{1}{4} \cos \vartheta [\cos 2\omega]_0^\vartheta \right\}$$

$$= \gamma R^3 \left\{ \sin^2 \vartheta + \cos \vartheta (\cos \vartheta - 1) \right.$$

$$\left. - \frac{1}{2} \sin \vartheta \left(\vartheta + \frac{1}{2} \sin 2\vartheta \right) - \frac{1}{4} \cos \vartheta (\cos 2\vartheta - 1) \right\} \qquad (A.23)$$

Finally

$$M(\vartheta) = -\gamma R^3 \left(1 - \frac{\pi}{4}\right)(1 - \cos \vartheta)$$

$$+ \gamma R^3 \left(1 - \cos \vartheta - \frac{1}{2} \vartheta \sin \vartheta - \frac{1}{4} \sin \vartheta \sin 2\vartheta \right.$$

$$\left. - \frac{1}{4} \cos \vartheta \cos 2\vartheta + \frac{1}{4} \cos \vartheta \right) \qquad (A.24)$$

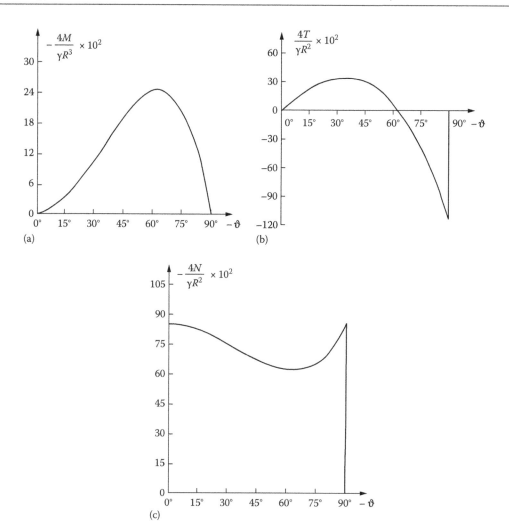

Figure A.2

The expression (A.24) reduces to Equation A.15, just as Equations A.16 and A.17 for shearing force and axial force can be found again using the direct method.

Figure A.2a through c represent the variations of $M(\vartheta)$, $T(\vartheta)$, $N(\vartheta)$, respectively. It may be noted how the bending moment in each case stretches the fibres at the intrados, the configuration of the pressure line being that of Figure A.1, and how it presents a maximum absolute value for $\vartheta \simeq 62°$, where the shear vanishes. It is further to be noted how the axial force is always compressive, with a minimum absolute value for $\vartheta \simeq 60°$ and equal maximum absolute values at the two extremes.

Appendix B: Calculation of the internal reactions in a circular arch subjected to a uniformly distributed vertical load

B.1 ANALYTICAL METHOD

The differential equation (5.9) in the case of Figure 5.8, whereby

$$m = 0 \tag{B.1a}$$

$$p = q_0 \cos \vartheta \sin \vartheta \tag{B.1b}$$

$$q = -q_0 \cos^2 \vartheta \tag{B.1c}$$

reduces to the following form:

$$\frac{d^3 M}{d\vartheta^3} + \frac{dM}{d\vartheta} = -3R^2 q_0 \sin \vartheta \cos \vartheta \tag{B.2}$$

The complete integral is the same as in the case studied in Equation A.4, whilst the particular solution is to be sought in the form

$$M_0(\vartheta) = a \cos 2\vartheta \tag{B.3}$$

The known term being expressible as

$$b(\vartheta) = -\frac{3}{2} R^2 q_0 \sin 2\vartheta \tag{B.4}$$

Differentiating Equation B.3 sequentially we obtain in fact

$$\frac{dM_0}{d\vartheta} = -2a \sin 2\vartheta \tag{B.5a}$$

$$\frac{d^2 M_0}{d\vartheta^2} = -4a \cos 2\vartheta \tag{B.5b}$$

$$\frac{d^3 M_0}{d\vartheta^3} = 8a \sin 2\vartheta \tag{B.5c}$$

and thus substituting Equations B.5a and B.5c in Equation B.2

$$6a\sin 2\vartheta = -\frac{3}{2}R^2 q_0 \sin 2\vartheta \tag{B.6}$$

or

$$a = -\frac{q_0 R^2}{4} \tag{B.7}$$

The solution thus appears as follows:

$$M(\vartheta) = C_1 + C_2 \cos\vartheta + C_3 \sin\vartheta - \frac{q_0 R^2}{4}\cos 2\vartheta \tag{B.8}$$

The arbitrary constants C_1, C_2, C_3 are calculated via the boundary conditions (A.12)

$$M(0) = C_1 + C_2 - \frac{q_0 R^2}{4} = 0 \tag{B.9a}$$

$$T(0) = \frac{C_3}{R} = 0 \tag{B.9b}$$

$$M\left(-\frac{\pi}{2}\right) = C_1 - C_3 + \frac{q_0 R^2}{4} = 0 \tag{B.9c}$$

The linear algebraic system (B.9) possesses the following roots:

$$C_1 = -\frac{q_0 R^2}{4} \tag{B.10a}$$

$$C_2 = \frac{q_0 R^2}{4} \tag{B.10b}$$

$$C_3 = 0 \tag{B.10c}$$

so that the solution becomes

$$M(\vartheta) = -\frac{q_0 R^2}{4} + \frac{q_0 R^2}{2}\cos\vartheta - \frac{q_0 R^2}{4}\cos 2\vartheta \tag{B.11}$$

or

$$M(\vartheta) = -\frac{q_0 R^2}{2}\left(1 - \cos\vartheta - \sin^2\vartheta\right) \tag{B.12}$$

Applying Equations 5.6 and 5.8, we obtain also the shearing force and the axial force

$$T(\vartheta) = -\frac{q_0 R}{2}(\sin\vartheta - 2\sin\vartheta\cos\vartheta) \tag{B.13}$$

$$N(\vartheta) = -\frac{q_0 R}{2}(\cos\vartheta + 2\sin^2\vartheta) \tag{B.14}$$

B.2 DIRECT METHOD

The equations of equilibrium with regard to horizontal translation, vertical translation and rotation about point B, respectively, appear as follows:

$$H_A = H_B \tag{B.15a}$$

$$V_B = q_0 R \tag{B.15b}$$

$$H_A R = q_0 \frac{R^2}{2} \tag{B.15c}$$

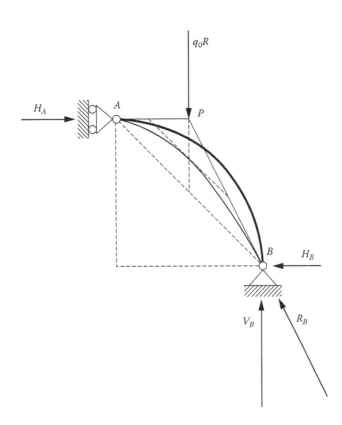

Figure B.1

from which the constraint reactions are obtained as follows:

$$H_A = \frac{1}{2} q_0 R \tag{B.16a}$$

$$H_B = \frac{1}{2} q_0 R \tag{B.16b}$$

$$V_B = q_0 R \tag{B.16c}$$

The bending moment acting in a generic section of angular coordinate ϑ d is obtained by summing the contributions which precede the section itself (Figure 5.8)

$$M(\vartheta) = -H_A R(1 - \cos \vartheta) + \frac{1}{2} q_0 R^2 \sin^2 \vartheta \tag{B.17}$$

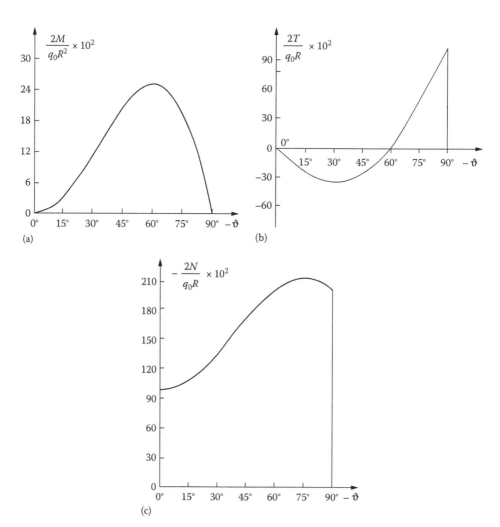

Figure B.2

Substituting Equation B.16a in Equation B.17 we obtain again Equation B.12.

In the same way, the shearing force is

$$T(\vartheta) = -H_A \sin\vartheta + \left(q_0 R \sin\vartheta\right)\cos\vartheta \qquad \text{(B.18)}$$

just as the axial force is

$$N(\vartheta) = -H_A \cos\vartheta - \left(q_0 R \sin\vartheta\right)\sin\vartheta \qquad \text{(B.19)}$$

Equations B.18 and B.19 coincide with Equations B.13 and B.14, respectively.

Figure B.1a through c represent, in order, the variations of $M(\vartheta)$, $T(\vartheta)$, $N(\vartheta)$. It may be noted that the bending moment in each case stretches the fibres at the extrados, the configuration of the pressure line being that of Figure B.2, and that it presents a maximum for $\vartheta \simeq 60°$ where the shear vanishes. It should further be noted how the axial force is always compressive, with a minimum absolute value at the end A and a maximum for $\vartheta \simeq 75°$.

Appendix C: Anisotropic material

C.I ANISOTROPIC ELASTIC CONSTITUTIVE LAW

The matrix expression of elastic potential is given by Equation 8.44a, which represents a quadratic form in the components of strain, just as the elastic constitutive law is given by Equation 8.44b, which links components of stress and components of strain. In both formulas, there appears the Hessian matrix $[H]$, which, in the case of isotropic material, has been rendered explicit in the expression (8.74). The expression of complementary elastic potential, on the other hand, is given by Equation 8.49, just as the inverse constitutive law is formally represented by Equation 3.16 and rendered explicit for isotropic material in Equation 8.73.

Whereas in the case of isotropic material, the inverse matrix $[H]^{-1}$ is a function only of the two parameters E and v, in the case of anisotropic material the independent parameters can even amount to 21, as has been seen in Chapter 8. There are, however, a number of intermediate cases, related to the properties of symmetry of the material. It is possible to show how, if there is a plane of symmetry $z = 0$, the material in the point under consideration is called monoclinic, and the relation (8.46) reduces to the following:

$$\begin{bmatrix} \varepsilon_x \\ \varepsilon_y \\ \varepsilon_z \\ \gamma_{xy} \\ \gamma_{xz} \\ \gamma_{yz} \end{bmatrix} = \begin{bmatrix} C_{11} & C_{12} & C_{13} & C_{14} & 0 & 0 \\ C_{12} & C_{22} & C_{23} & C_{24} & 0 & 0 \\ C_{13} & C_{23} & C_{33} & C_{34} & 0 & 0 \\ C_{14} & C_{24} & C_{34} & C_{44} & 0 & 0 \\ 0 & 0 & 0 & 0 & C_{55} & C_{56} \\ 0 & 0 & 0 & 0 & C_{56} & C_{66} \end{bmatrix} \begin{bmatrix} \sigma_x \\ \sigma_y \\ \sigma_z \\ \tau_{xy} \\ \tau_{xz} \\ \tau_{yz} \end{bmatrix} \tag{C.1}$$

with 13 independent parameters.

If there are two perpendicular planes of symmetry, it is possible to demonstrate that there exists then a third plane of symmetry which is perpendicular to both. The relation (8.46), in the reference system oriented according to the principal directions of the material, takes the following form:

$$\begin{bmatrix} \varepsilon_x \\ \varepsilon_y \\ \varepsilon_z \\ \gamma_{xy} \\ \gamma_{xz} \\ \gamma_{yz} \end{bmatrix} = \begin{bmatrix} C_{11} & C_{12} & C_{13} & 0 & 0 & 0 \\ C_{12} & C_{22} & C_{23} & 0 & 0 & 0 \\ C_{13} & C_{23} & C_{33} & 0 & 0 & 0 \\ 0 & 0 & 0 & C_{44} & 0 & 0 \\ 0 & 0 & 0 & 0 & C_{55} & 0 \\ 0 & 0 & 0 & 0 & 0 & C_{66} \end{bmatrix} \begin{bmatrix} \sigma_x \\ \sigma_y \\ \sigma_z \\ \tau_{xy} \\ \tau_{xz} \\ \tau_{yz} \end{bmatrix} \tag{C.2}$$

with nine independent parameters, and the material at the point under consideration is said to be **orthotropic**. It should be noted that, in this case, there is no interaction between dilations and shearing stresses, just as there is none between normal stresses and shearing strains.

If the material is **transversely isotropic**, i.e. if the properties are the same in all the directions that define one of the three principal planes, for example the plane $z = 0$, the relation (8.46) then presents only five independent parameters:

$$
\begin{bmatrix} \varepsilon_x \\ \varepsilon_y \\ \varepsilon_z \\ \gamma_{xy} \\ \gamma_{xz} \\ \gamma_{yz} \end{bmatrix} = \begin{bmatrix} C_{11} & C_{12} & C_{13} & 0 & 0 & 0 \\ C_{12} & C_{11} & C_{13} & 0 & 0 & 0 \\ C_{13} & C_{13} & C_{33} & 0 & 0 & 0 \\ 0 & 0 & 0 & 2(C_{11}-C_{12}) & 0 & 0 \\ 0 & 0 & 0 & 0 & C_{55} & 0 \\ 0 & 0 & 0 & 0 & 0 & C_{55} \end{bmatrix} \begin{bmatrix} \sigma_x \\ \sigma_y \\ \sigma_z \\ \tau_{xy} \\ \tau_{xz} \\ \tau_{yz} \end{bmatrix}
\tag{C.3}
$$

If, finally, the planes of symmetry are infinite, i.e. the material is transversely isotropic on any plane, we have

$$
\begin{bmatrix} \varepsilon_x \\ \varepsilon_y \\ \varepsilon_z \\ \gamma_{xy} \\ \gamma_{xz} \\ \gamma_{yz} \end{bmatrix} = \begin{bmatrix} C_{11} & C_{12} & C_{12} & 0 & 0 & 0 \\ C_{12} & C_{11} & C_{12} & 0 & 0 & 0 \\ C_{12} & C_{12} & C_{11} & 0 & 0 & 0 \\ 0 & 0 & 0 & 2(C_{11}-C_{12}) & 0 & 0 \\ 0 & 0 & 0 & 0 & 2(C_{11}-C_{12}) & 0 \\ 0 & 0 & 0 & 0 & 0 & 2(C_{11}-C_{12}) \end{bmatrix} \begin{bmatrix} \sigma_x \\ \sigma_y \\ \sigma_z \\ \tau_{xy} \\ \tau_{xz} \\ \tau_{yz} \end{bmatrix}
\tag{C.4}
$$

the material being completely isotropic at the point under consideration, with $C_{11} = 1/E$ and $C_{12} = -v/E$ (Equation 8.73).

C.2 ORTHOTROPIC MATERIAL

Fibre-reinforced materials, which are by now extensively used in all types of engineering sectors, are generally orthotropic, or, at least, transversely isotropic. The principal planes of the material are naturally defined by the directions of the reinforcing fibres. The technical constants of these materials are the normal elastic modulus and the shear modulus, as well as the Poisson ratios, according to the following explicit compliance matrix:

$$
[H]^{-1} = \begin{bmatrix} \dfrac{1}{E_1} & -\dfrac{v_{21}}{E_2} & -\dfrac{v_{31}}{E_3} & 0 & 0 & 0 \\ -\dfrac{v_{12}}{E_1} & \dfrac{1}{E_2} & -\dfrac{v_{32}}{E_3} & 0 & 0 & 0 \\ -\dfrac{v_{13}}{E_1} & -\dfrac{v_{23}}{E_2} & \dfrac{1}{E_3} & 0 & 0 & 0 \\ 0 & 0 & 0 & \dfrac{1}{G_{12}} & 0 & 0 \\ 0 & 0 & 0 & 0 & \dfrac{1}{G_{13}} & 0 \\ 0 & 0 & 0 & 0 & 0 & \dfrac{1}{G_{23}} \end{bmatrix}
\tag{C.5}
$$

The Poisson ratio ν_{ij} represents the transverse dilation in the j direction, when the material is stressed in the i direction

$$\nu_{ij} = -\frac{\varepsilon_j}{\varepsilon_i} \tag{C.6}$$

By virtue of the symmetry of the compliance matrix, we have

$$\frac{\nu_{ij}}{E_i} = \frac{\nu_{ji}}{E_j}, \quad \text{for } i, j = 1, 2, 3 \tag{C.7}$$

so that there are only nine independent parameters, m is known. The symmetry of the matrix $[H]^{-1}$ ensures, on the other hand, that Betti's Reciprocal Theorem is satisfied.

The elements of the stiffness matrix $[H]$ are obtained by inverting the compliance matrix, and are as follows:

$$H_{11} = \frac{1 - \nu_{23}\nu_{32}}{E_2 E_3 \Delta} \tag{C.8a}$$

$$H_{12} = \frac{\nu_{21} + \nu_{31}\nu_{23}}{E_2 E_3 \Delta} = \frac{\nu_{12} + \nu_{32}\nu_{13}}{E_1 E_3 \Delta} \tag{C.8b}$$

$$H_{13} = \frac{\nu_{31} + \nu_{21}\nu_{32}}{E_2 E_3 \Delta} = \frac{\nu_{13} + \nu_{12}\nu_{23}}{E_1 E_2 \Delta} \tag{C.8c}$$

$$H_{22} = \frac{1 - \nu_{13}\nu_{31}}{E_1 E_3 \Delta} \tag{C.8d}$$

$$H_{23} = \frac{\nu_{32} + \nu_{12}\nu_{31}}{E_1 E_3 \Delta} = \frac{\nu_{23} + \nu_{21}\nu_{13}}{E_1 E_2 \Delta} \tag{C.8e}$$

$$H_{33} = \frac{1 - \nu_{12}\nu_{21}}{E_1 E_2 \Delta} \tag{C.8f}$$

$$H_{44} = G_{12} \tag{C.8g}$$

$$H_{55} = G_{13} \tag{C.8h}$$

$$H_{66} = G_{23} \tag{C.8i}$$

with

$$\Delta = \frac{1 - \nu_{12}\nu_{21} - \nu_{23}\nu_{32} - \nu_{31}\nu_{13} - 2\nu_{21}\nu_{32}\nu_{13}}{E_1 E_2 E_3} \tag{C.9}$$

As in the case of isotropic material, also in the more general case of orthotropic material the elastic constants must respect certain conditions, so as to render both the stiffness matrix and the compliance matrix positive definite. From Equations C.8 there follows

$$(1 - v_{23}v_{32}), \quad (1 - v_{13}v_{31}), \quad (1 - v_{12}v_{21}) > 0 \tag{C.10a}$$

$$\bar{\Delta} = 1 - v_{12}v_{21} - v_{23}v_{32} - v_{31}v_{13} - 2v_{21}v_{32}v_{13} > 0 \tag{C.10b}$$

via the relations of symmetry (C.7), the inequalities (C.10a) may be reproposed in the following forms:

$$|v_{21}| < \left(\frac{E_2}{E_1}\right)^{1/2}, \quad |v_{12}| < \left(\frac{E_1}{E_2}\right)^{1/2} \tag{C.11a}$$

$$|v_{32}| < \left(\frac{E_3}{E_2}\right)^{1/2}, \quad |v_{23}| < \left(\frac{E_2}{E_3}\right)^{1/2} \tag{C.11b}$$

$$|v_{13}| < \left(\frac{E_1}{E_3}\right)^{1/2}, \quad |v_{31}| < \left(\frac{E_3}{E_1}\right)^{1/2} \tag{C.11c}$$

just as Equation C.10b can alternatively be expressed as follows:

$$v_{21}v_{32}v_{13} < \frac{1 - v_{21}^2(E_1/E_2) - v_{32}^2(E_2/E_3) - v_{13}^2(E_3/E_1)}{2} < \frac{1}{2} \tag{C.12}$$

As an example, it is possible to mention that of a composite material made up of boron fibres embedded in an epoxy polymer matrix, for which $v_{12} \simeq 2, E_1/E_2 \simeq 10$, so that the second of inequalities (C.11a) is satisfied. The coefficient of transverse contraction, though appearing surprisingly high, is consistent with the conditions obtained previously. On the other hand, the mutual coefficient $v_{21} \simeq 0.22$ satisfies the first of inequalities (C.11a).

C.3 STRESS–STRAIN RELATIONS FOR PLANE STRESS CONDITIONS

In the case of a plane stress condition (in the plane $z = 0$) the relation (C.2) reduces to the following:

$$\begin{bmatrix} \varepsilon_1 \\ \varepsilon_2 \\ \gamma_{12} \end{bmatrix} = \begin{bmatrix} \dfrac{1}{E_1} & -\dfrac{v_{21}}{E_2} & 0 \\ -\dfrac{v_{12}}{E_1} & \dfrac{1}{E_2} & 0 \\ 0 & 0 & \dfrac{1}{G_{12}} \end{bmatrix} \begin{bmatrix} \sigma_1 \\ \sigma_2 \\ \tau_{12} \end{bmatrix} \tag{C.13a}$$

whilst its inverse is

$$\begin{bmatrix} \sigma_1 \\ \sigma_2 \\ \tau_{12} \end{bmatrix} = \begin{bmatrix} H_{11} & H_{12} & 0 \\ H_{12} & H_{22} & 0 \\ 0 & 0 & H_{44} \end{bmatrix} \begin{bmatrix} \varepsilon_1 \\ \varepsilon_2 \\ \gamma_{12} \end{bmatrix} \tag{C.13b}$$

with

$$H_{11} = \frac{E_1}{1 - \nu_{12}\nu_{21}}$$ (C.14a)

$$H_{12} = \frac{\nu_{12}E_2}{1 - \nu_{12}\nu_{21}} = \frac{\nu_{21}E_1}{1 - \nu_{12}\nu_{21}}$$ (C.14b)

$$H_{22} = \frac{E_2}{1 - \nu_{12}\nu_{21}}$$ (C.14c)

$$H_{44} = G_{12}$$ (C.14d)

Frequently the principal directions of orthotropy 12 do not coincide with the directions of the XY coordinate axes, which are the geometrically natural directions for solving the problem (Figure C.1). For this reason, it is necessary to determine a relation that can connect stresses and strains in the principal system of the material with stresses and strains in the coordinate system of the body.

Recalling relations (2.33) introduced in the framework of the geometry of areas, the transformation equation that expresses the stresses in the system 12 as functions of the stresses in the system XY is as follows:

$$\begin{bmatrix} \sigma_1 \\ \sigma_2 \\ \tau_{12} \end{bmatrix} = \begin{bmatrix} \cos^2\vartheta & \sin^2\vartheta & -2\sin\vartheta\cos\vartheta \\ \sin^2\vartheta & \cos^2\vartheta & 2\sin\vartheta\cos\vartheta \\ \sin\vartheta\cos\vartheta & -\sin\vartheta\cos\vartheta & \cos2\vartheta \end{bmatrix} \begin{bmatrix} \sigma_x \\ \sigma_y \\ \tau_{xy} \end{bmatrix}$$ (C.15)

where ϑ is the angle between the axes X and 1 (Figure C.1). The transformation corresponding to strain is analogous

$$\begin{bmatrix} \varepsilon_1 \\ \varepsilon_2 \\ \frac{1}{2}\gamma_{12} \end{bmatrix} = \begin{bmatrix} \cos^2\vartheta & \sin^2\vartheta & -2\sin\vartheta\cos\vartheta \\ \sin^2\vartheta & \cos^2\vartheta & 2\sin\vartheta\cos\vartheta \\ \sin\vartheta\cos\vartheta & -\sin\vartheta\cos\vartheta & \cos2\vartheta \end{bmatrix} \begin{bmatrix} \varepsilon_x \\ \varepsilon_y \\ \frac{1}{2}\gamma_{xy} \end{bmatrix}$$ (C.16)

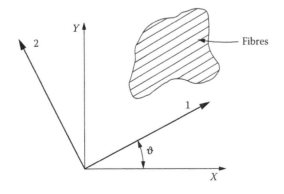

Figure C.1

The two foregoing transformations can be put in a compact form

$$\{\sigma\}_{12} = [T]\{\sigma\}_{XY} \tag{C.17a}$$

$$\{\varepsilon\}_{12} = [R][T][R]^{-1}\{\varepsilon\}_{XY} \tag{C.17b}$$

where

$$\{\varepsilon\}_{XY}^{\mathrm{T}} = [\varepsilon_x, \varepsilon_y, \gamma_{xy}] \tag{C.18a}$$

$$\{\varepsilon\}_{12}^{\mathrm{T}} = [\varepsilon_1, \varepsilon_2, \gamma_{12}] \tag{C.18b}$$

and $[R]$ is the so-called **Reuter matrix**

$$[R] = \begin{bmatrix} 1 & 0 & 0 \\ 0 & 1 & 0 \\ 0 & 0 & 2 \end{bmatrix} \tag{C.19}$$

which serves to consider the shearing strain and not its half, which appears as an off-diagonal term in the strain tensor.

Casting equation (C.13a) in the form

$$\{\varepsilon\}_{12} = [C]\{\sigma\}_{12} \tag{C.20}$$

and substituting Equations C.17 in Equation C.20, we obtain

$$[R][T][R]^{-1}\{\varepsilon\}_{XY} = [C][T]\{\sigma\}_{XY} \tag{C.21}$$

Premultiplying both sides of Equation C.21 by $[R][T]^{-1}[R]^{-1}$, we have

$$\{\varepsilon\}_{XY} = [R][T]^{-1}[R]^{-1}[C][T]\{\sigma\}_{XY} \tag{C.22}$$

Since it is possible to show that

$$[R][T]^{-1}[R]^{-1} = [T]^{\mathrm{T}} \tag{C.23}$$

Finally we can write

$$\{\varepsilon\}_{XY} = [T]^{\mathrm{T}}[C][T]\{\sigma\}_{XY} \tag{C.24}$$

or

$$\{\varepsilon\}_{XY} = [C^*]\{\sigma\}_{XY} \tag{C.25}$$

The compliance matrix, rotated, may be cast in the following form:

$$[C^*] = \begin{bmatrix} \dfrac{1}{E_x} & -\dfrac{v_{yx}}{E_y} & \dfrac{\eta_{x.xy}}{G_{xy}} \\[2ex] -\dfrac{v_{xy}}{E_x} & \dfrac{1}{E_y} & \dfrac{\eta_{y.xy}}{G_{xy}} \\[2ex] \dfrac{\eta_{xy.x}}{E_x} & \dfrac{\eta_{xy.y}}{E_y} & \dfrac{1}{G_{xy}} \end{bmatrix}$$

(C.26)

which is symmetrical by virtue of Betti's Reciprocal Theorem. Herein there appear **Lekhnitski's coefficients,** with the following physical meaning: $\eta_{ij,i}$, dilation in the i direction caused by the shearing stress τ_{ij},

$$\eta_{ij,i} = \frac{\varepsilon_i}{\gamma_{ij}}$$

(C.27)

$\eta_{i,ij}$, shearing strain of the axes ij caused by the normal stress in the i direction,

$$\eta_{i,ij} = \frac{\gamma_{ij}}{\varepsilon_i}$$

(C.28)

Note that the rotated compliance matrix $[C^*]$ is a fully populated matrix, unlike the principal compliance matrix $[C]$. Notwithstanding this, the independent parameters remain tour $(E_1, E_2, v_{12}, G_{12})$

$$\frac{1}{E_x} = \frac{1}{E_1}\cos^4\vartheta + \left(\frac{1}{G_{12}} - \frac{2v_{12}}{E_1}\right)\sin^2\vartheta\cos^2\vartheta + \frac{1}{E_2}\sin^4\vartheta$$

(C.29a)

$$v_{xy} = E_x\left[\frac{v_{12}}{E_1}(\sin^4\vartheta + \cos^4\vartheta) - \left(\frac{1}{E_1} + \frac{1}{E_2} - \frac{1}{G_{12}}\right)\sin^2\vartheta\cos^2\vartheta\right]$$

(C.29b)

$$\frac{1}{E_y} = \frac{1}{E_1}\sin^4\vartheta + \left(\frac{1}{G_{12}} - \frac{2v_{12}}{E_1}\right)\sin^2\vartheta\cos^2\vartheta + \frac{1}{E_2}\cos^4\vartheta$$

(C29c)

$$\frac{1}{G_{xy}} = 2\left(\frac{2}{E_1} + \frac{2}{E_2} + \frac{4v_{12}}{E_1} - \frac{1}{G_{12}}\right)\sin^2\vartheta\cos^2\vartheta + \frac{1}{G_{12}}(\sin^4\vartheta + \cos^4\vartheta)$$

(C.29d)

$$\eta_{xy,x} = E_x\left[\left(\frac{2}{E_1} + \frac{2v_{12}}{E_1} - \frac{1}{G_{12}}\right)\sin\vartheta\cos^3\vartheta - \left(\frac{2}{E_2} + \frac{2v_{12}}{E_1} - \frac{1}{G_{12}}\right)\sin^3\vartheta\cos\vartheta\right]$$

(C.29e)

$$\eta_{xy,y} = E_y\left[\left(\frac{2}{E_1} + \frac{2v_{12}}{E_1} - \frac{1}{G_{12}}\right)\sin^3\vartheta\cos\vartheta - \left(\frac{2}{E_2} + \frac{2v_{12}}{E_1} - \frac{1}{G_{12}}\right)\sin\vartheta\cos^3\vartheta\right]$$

(C.29f)

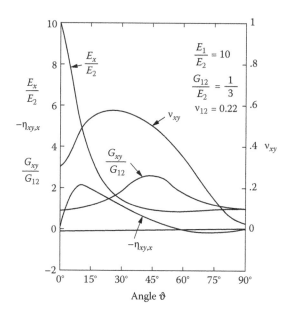

Figure C.2

The apparent parameters (C.29) are plotted as functions of the angle ϑ, in Figure C.2, in the case of the **epoxy–boron** composite material, already considered. It may be noted that

1. The shear modulus G_{xy} is maximum for $\vartheta = 45°$.
2. The coefficient $\eta_{xy,x}$ vanishes, obviously, for $\vartheta = 0°$ and $\vartheta = 90°$.
3. The normal elastic modulus E_y varies identically as E_x, exchanging ϑ with $90° - \vartheta$ (the same applies for ν_{yx} and $\eta_{xy,y}$).
4. E_x may be less than both E_1 and E_2, just as it may be greater than both (cf. what occurs in the case where $\vartheta \simeq 60°$). In other words, the extreme values of the parameters are not found necessarily in the principal directions of the material.

C.4 STRENGTH CRITERIA FOR ORTHOTROPIC MATERIALS

Since in orthotropic materials strength varies with the variation of direction, the direction of maximum stress may not be the most dangerous one.

Let different properties be assumed in tension and in compression (Figure C.3):

X_t is the tensile strength in the direction 1
X_c is the compressive strength in the direction 1
Y_t is the tensile strength in the direction 2
Y_c is the compressive strength in the direction 2
S is the shear strength

Note that the foregoing strengths have been defined in the principal directions of the material, and that of course they vary as the coordinate axes vary.

The **criterion of maximum stress** requires that all the following inequalities should be satisfied at the same time:

$$-X_c < \sigma_1 < X_t \tag{C.30a}$$

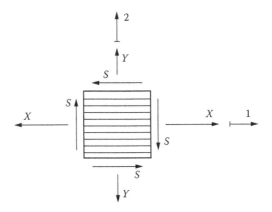

Figure C.3

$$-Y_c < \sigma_2 < Y_t \qquad\qquad\qquad\qquad\qquad\qquad\text{(C.30b)}$$

$$|\tau_{12}| < S \qquad\qquad\qquad\qquad\qquad\qquad\text{(C.30c)}$$

If only one of Equations C.30 is not satisfied, the critical condition of the material is assumed according to the mechanism of rupture associated with X_c, X_t, Y_c, Y_t or S. Hence interaction is not assumed to exist between the various modes of rupture.

Let us consider a material that is fibre-reinforced in one direction, submitted to a condition of uniaxial stress inclined at an angle ϑ with respect to the fibres (Figure C.4). The stresses in the principal reference system of the material are obtained from Equations C.15:

$$\sigma_1 = \sigma_x \cos^2 \vartheta \qquad\qquad\qquad\qquad\qquad\qquad\text{(C.31a)}$$

$$\sigma_2 = \sigma_x \sin^2 \vartheta \qquad\qquad\qquad\qquad\qquad\qquad\text{(C.31b)}$$

$$\tau_{12} = \sigma_x \sin\vartheta \cos\vartheta \qquad\qquad\qquad\qquad\qquad\qquad\text{(C.31c)}$$

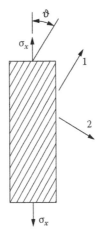

Figure C.4

Substituting Equations C.31 in inequalities (C.30), we obtain three mutually competing criteria:

$$-\frac{X_c}{\cos^2\vartheta} < \sigma_x < \frac{X_t}{\cos^2\vartheta} \tag{C.32a}$$

$$-\frac{Y_c}{\sin^2\vartheta} < \sigma_x < \frac{Y_t}{\sin^2\vartheta} \tag{C.32b}$$

$$-\frac{S}{\sin\vartheta\cos\vartheta} < \sigma_x < \frac{S}{\sin\vartheta\cos\vartheta} \tag{C.32c}$$

These criteria are plotted in Figure C.5 for a **glass–epoxy** composite having the following properties:

$X_c = 150\,\text{ksi}$

$X_t = 150\,\text{ksi}$

$Y_c = 20\,\text{ksi}$

$Y_t = 4\,\text{ksi}$

$S = 6\,\text{ksi}$

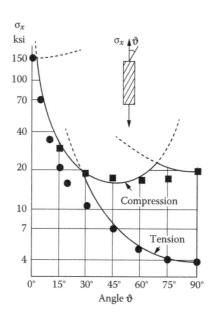

Figure C.5

Hence the uniaxial strength is represented as a function of the angle ϑ in Figure C.5 (some experimental results are also given). The criterion of maximum stress, both tensile and compressive, consists in fact of three curves, the bottom one of which in each case governs the rupture phenomenon.

The **Tsai-Hill criterion** instead consists of a single curve devoid of cusps which presents the following general form:

$$(G+H)\sigma_1^2 + (F+H)\sigma_2^2 + (F+G)\sigma_3^2 - 2H\sigma_1\sigma_2 - 2G\sigma_1\sigma_3 - 2F\sigma_2\sigma_3 + 2L\tau_{12}^2 + 2M\tau_{13}^2 + 2N\tau_{23}^2 = 1$$

$$(C.33)$$

The parameters F, G, H, L, M, N are correlated with the strengths X, Y, S introduced previously. If only τ_{12} acts, we have in fact

$$2L = \frac{1}{S^2} \tag{C.34}$$

just as, if only σ_1 acts, or, respectively σ_2 or σ_3

$$G + H = \frac{1}{X^2} \tag{C.35a}$$

$$F + H = \frac{1}{Y^2} \tag{C.35b}$$

$$F + G = \frac{1}{Z^2} \tag{C.35c}$$

where Z indicates the strength in the direction 3, normal to the stress plane. From Equations C.35 it follows that

$$2H = \frac{1}{X^2} + \frac{1}{Y^2} - \frac{1}{Z^2} \tag{C.36a}$$

$$2G = \frac{1}{X^2} + \frac{1}{Z^2} - \frac{1}{Y^2} \tag{C.36b}$$

$$2F = \frac{1}{Y^2} + \frac{1}{Z^2} - \frac{1}{X^2} \tag{C.36c}$$

If the body is assumed to be transversely isotropic in the plane 23, then we have $Y = Z$, and thus Equation C.33 reduces to the following form:

$$\frac{\sigma_1^2}{X^2} + \frac{\sigma_2^2}{Y^2} - \frac{\sigma_1\sigma_2}{X^2} + \frac{\tau_{12}^2}{S^2} = 1 \tag{C.37}$$

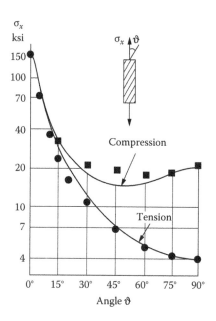

Figure C.6

Finally, substituting Equations C.31 in Equation C.37, we obtain

$$\frac{\cos^4 \vartheta}{X^2} + \frac{\sin^4 \vartheta}{Y^2} + \left(\frac{1}{S^2} - \frac{1}{X^2}\right) \sin^2 \vartheta \cos^2 \vartheta = \frac{1}{\sigma_x^2} \tag{C.38}$$

The criterion (C.38) is represented in Figure C.6 for the **glass–epoxy** composite, both in the case of tension and in that of compression. It is possible to note an excellent agreement between theory and experimentation. In particular, a notable improvement has been obtained in comparison with the criterion of maximum stress for $\vartheta \simeq 30°$ where the latter fails by approximately 100% (Figure C.5).

Appendix D: Heterogeneous beam

D.I MULTILAYER BEAM IN FLEXURE

Let us consider a beam having rectangular cross section, consisting of n layers of different materials (Figure D.1a). If, as in the case of homogeneous material, we assume the conservation of the plane sections and thus the linear variation of axial dilation

$$\varepsilon_z = \frac{1}{E_1}(by + c) \tag{D.1}$$

and consider the conditions of equivalence (9.16a and 9.16d), we have

$$\sum_{i=1}^{n} \int_{A_i} (by + c)\frac{E_i}{E_1} dA = b\sum_{i=1}^{n} \frac{E_i}{E_1} \int_{A_i} y dA + c\sum_{i=1}^{n} \frac{E_i}{E_1} \int_{A_i} dA = 0 \tag{D.2a}$$

$$\sum_{i=1}^{n} \int_{A_i} (by + c)y\frac{E_i}{E_1} dA = b\sum_{i=1}^{n} \frac{E_i}{E_1} \int_{A_i} y^2 dA + c\sum_{i=1}^{n} \frac{E_i}{E_1} \int_{A_i} y dA = M_x \tag{D.2b}$$

Equation D.2 can take the form

$$bS_x + cA = 0 \tag{D.3a}$$

$$bI_x + cS_x = M_x \tag{D.3b}$$

where the usual symbols introduced in Chapter 2, corresponding to the geometry of areas, must be translated into the ones corresponding to the geometry of masses, each elementary area dA being weighed via the ratio of the moduli of elasticity E_i/E_1. If we define the **centroid of the cross section** as the point of the Y axis for which $S_x = 0$, results are obtained that are formally analogous to those obtained in Section 9.4

$$\varepsilon_z^i = \frac{1}{E_1}\frac{M_x}{I_x}y \tag{D.4a}$$

$$\sigma_z^i = \frac{E_i}{E_1}\frac{M_x}{I_x}y \tag{D.4b}$$

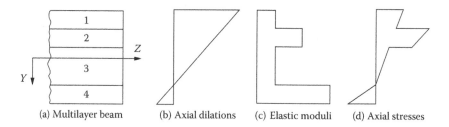

(a) Multilayer beam   (b) Axial dilations   (c) Elastic moduli   (d) Axial stresses

Figure D.1

Whilst Equation D.4a represents the supposed linear function (Figure D.1b), Equation D.4b represents a step function, the stresses being greater in more rigid materials (Figure D.1c), as well as in the points farther from the neutral axis (Figure D.1d).

As regards the strain condition of the beam, it may be stated, in an analogous way to the case of homogeneous material, that the curvature of the centroidal axis is

$$\chi_x = \frac{M_x}{E_1 I_x} \tag{D.5}$$

where of course the moment of inertia is the one corresponding to the elementary areas weighed via the moduli of elasticity

$$I_x = \sum_{i=1}^{n} \frac{E_i}{E_1} \int_{A_i} y^2 \mathrm{d}A \tag{D.6}$$

D.2 REINFORCED CONCRETE

Reinforced-concrete beams may be considered as multilayer beams. Steel bars have the function of withstanding tensile forces and are thus usually embedded in the concrete on the side of the fibres in tension. Concrete and steel present excellent adherence and the same coefficient of thermal expansion, so that their dilation tends to occur without differential displacements or discontinuities.

The basic hypotheses for the statics of reinforced concrete are the following:

1. Concrete behaves like a linear elastic material in compression, whilst it is **non-traction-bearing**. In other words, concrete presents a zero elastic modulus in tension.
2. Steel behaves as a linear elastic material both in compression and in tension.
3. The steel bars cannot slip inside the concrete.
4. The cross section of the beam remains plane.

In bending, the neutral axis divides the section into two parts: one part is in compression with elastic modulus E_c, whilst the other is in tension with zero elastic modulus (Figure D.2). The steel bars are below the neutral axis, in the part in tension, and present an elastic modulus E_s. On the other hand, the position of the neutral axis and thus the area of material with elastic modulus E_c is *a priori* unknown. Hence this area has to be identified first, and then the formulas seen in the previous section are applied.

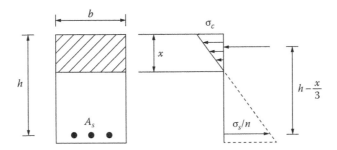

Figure D.2

The condition of axial equivalence for the rectangular cross section in Figure D.2 is

$$\sigma_s A_s = \sigma_c \frac{bx}{2} \tag{D.7}$$

where
  σ_s is the tensile stress in the steel
  σ_c is the maximum compressive stress in the upper edge of the concrete
  A_s is the area of steel
  b is the width of the beam
  x is the unknown distance of the neutral axis from the upper edge of the beam

The condition of equivalence corresponding to the bending moment is written as

$$M = \sigma_s A_s \left(h - \frac{x}{3} \right) \tag{D.8}$$

where
  M is the moment applied to the section
  $(h - x/3)$ is the arm contained between the two resultants of the tensile and compressive
  forces, respectively

Note that the thickness of the concrete cover does not enter into this calculation.
  The condition of linear variation of the axial dilations gives

$$\frac{\varepsilon_s}{\varepsilon_c} = \frac{h - x}{x} \tag{D.9}$$

where
  ε_s is the dilation of the bars
  whilst ε_c is the dilation of the concrete at the upper edge of the section (Figure D.2)

Introducing the stresses we have

$$\sigma_s = n\sigma_c \frac{h - x}{x} \tag{D.10}$$

where n is the ratio between the elastic modulus of the steel and the elastic modulus of the compressed concrete,

$$n = \frac{E_s}{E_c} \simeq 10 \tag{D.11}$$

Combining Equations D.7 and D.10, we obtain

$$n\sigma_c A_s \frac{h-x}{x} = \frac{\sigma_c bx}{2} \tag{D.12}$$

or

$$\frac{1}{2}bx^2 - nA_s(h-x) = 0 \tag{D.13}$$

The positive root x of the quadratic equation (D.13) gives the position of the neutral axis.

From Equations D.7 and D.8 it is possible to obtain the stresses in the concrete and in the steel as functions of the distance x

$$\sigma_s = \frac{M}{A_s\left(h - x/3\right)} \tag{D.14a}$$

$$\sigma_c = \frac{2M}{bx\left(h - x/3\right)} \tag{D.14b}$$

Equations D.14 resolve the problem of verifying the strength of the reinforced section. If, instead, we are faced with the design problem, the unknowns to be determined are h, x, A_s. Equation D.10 then transforms as follows:

$$x = \frac{n\sigma_c h}{\sigma_s + n\sigma_c} \tag{D.15}$$

Equations D.7 and D.15 yield on the other hand

$$A_s = \frac{n\sigma_c^2}{2\sigma_s(\sigma_s + n\sigma_c)} bh \tag{D.16}$$

Equation D.8, via Equations D. 15 and D.16, becomes

$$M = \left(1 - \frac{\alpha}{3}\right)\beta\sigma_s bh^2 \tag{D.17}$$

with

$$\alpha = \frac{n\sigma_c}{\sigma_s + n\sigma_c} \tag{D.18a}$$

$$\beta = \frac{n\sigma_c^2}{2\sigma_s(\sigma_s + n\sigma_c)} \tag{D.18b}$$

From Equation D.17 we obtain finally

$$b = \left(\frac{M}{(1 - (\alpha/3))\beta\sigma_s b} \right)^{1/2} \tag{D.19}$$

The admissible values of σ_c and σ_s having been assigned, Equation D.19 gives the depth of the beam, whilst Equations D.15 and D.16 give the position of the neutral axis and the area of steel, respectively.

Appendix E: Heterogeneous plate

Let us consider a **laminate,** i.e. a multilayer plate, in which each layer (or lamina) is orthotropic in a particular principal orientation (Figure E.1). The stress–strain relation in the principal coordinates of each layer is of the type represented by Equations C.13b and C.14. In an external reference system the inverse relation of Equation C.25 is presented as follows:

$$\{\sigma\}_{XY} = [H^*]\{\varepsilon\}_{XY} \tag{E.1}$$

with

$$H_{11}^* = H_{11}\cos^4\vartheta + 2(H_{12} + 2H_{44})\sin^2\vartheta\cos^2\vartheta + H_{22}\sin^4\vartheta \tag{E.2a}$$

$$H_{12}^* = (H_{11} + H_{22} - 4H_{44})\sin^2\vartheta\cos^2\vartheta + H_{12}(\sin^4\vartheta + \cos^4\vartheta) \tag{E.2b}$$

$$H_{22}^* = H_{11}\sin^4\vartheta + 2(H_{12} + 2H_{44})\sin^2\vartheta\cos^2\vartheta + H_{22}\cos^4\vartheta \tag{E.2c}$$

$$H_{14}^* = (H_{11} - H_{12} - 2H_{44})\sin\vartheta\cos^3\vartheta + (H_{12} - H_{22} + 2H_{44})\sin^3\vartheta\cos\vartheta \tag{E.2d}$$

$$H_{24}^* = (H_{11} - H_{12} - 2H_{44})\sin^3\vartheta\cos\vartheta + (H_{12} - H_{22} + 2H_{44})\sin\vartheta\cos^3\vartheta \tag{E.2e}$$

$$H_{44}^* = (H_{11} + H_{22} - 2H_{12} - 2H_{44})\sin^2\vartheta\cos^2\vartheta + H_{44}(\sin^4\vartheta + \cos^4\vartheta) \tag{E.2f}$$

Also in the case of the multilayer plate it is possible to formulate Kirchhoff's hypothesis, already described in Section 10.10. In the case where the plate presents a membrane regime, in addition to a flexural regime, Equations 10.158 are completed as follows:

$$\begin{bmatrix} \varepsilon_x \\ \varepsilon_y \\ \gamma_{xy} \end{bmatrix} = \begin{bmatrix} \varepsilon_x^0 \\ \varepsilon_y^0 \\ \gamma_{xy}^0 \end{bmatrix} + z \begin{bmatrix} \chi_x \\ \chi_y \\ \chi_{xy} \end{bmatrix} \tag{E.3}$$

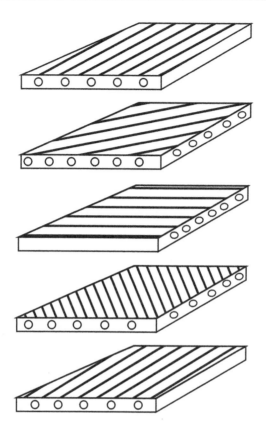

Figure E.1

where ε_x^0, ε_y^0, γ_{xy}^0 are the strains of the middle plane. Substituting Equation E.3 in Equation E.1, the stresses in the kth layer can then be expressed as functions of the strains and curvatures of the middle plane

$$\{\sigma\}_{XY}^k = [H^*]^k(\{\varepsilon^0\}_{XY} + z\{\chi\}_{XY}) \tag{E.4}$$

Since the stiffness matrix $[H^*]^k$ can vary from layer to layer, the variation of the stresses through the thickness of the laminate is not necessarily linear, although the variation of the strains is (Figure D.1).

The internal forces and moments acting in the laminate are obtained by integration of the stresses that develop in each lamina

$$\begin{bmatrix} N_x \\ N_y \\ N_{xy} \end{bmatrix} = \int_{-h/2}^{+h/2} \begin{bmatrix} \sigma_x \\ \sigma_y \\ \tau_{xy} \end{bmatrix} dz = \sum_{k=1}^{n} \int_{z_{k-1}}^{z_k} \begin{bmatrix} \sigma_x \\ \sigma_y \\ \tau_{xy} \end{bmatrix}^k dz \tag{E.5a}$$

$$\begin{bmatrix} M_x \\ M_y \\ M_{xy} \end{bmatrix} = \int_{-h/2}^{+h/2} \begin{bmatrix} \sigma_x \\ \sigma_y \\ \tau_{xy} \end{bmatrix} z\,dz = \sum_{k=1}^{n} \int_{z_{k-1}}^{z_k} \begin{bmatrix} \sigma_x \\ \sigma_y \\ \tau_{xy} \end{bmatrix}^k z\,dz \tag{E.5b}$$

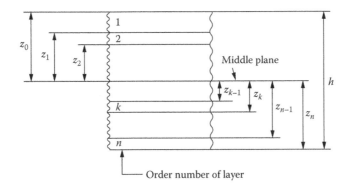

Figure E.2

where z_k and z_{k-1} are defined in Figure E.2, with $z_0 = -h/2$. Substituting Equation E.4 in Equations E.5, we obtain

$$\{N\} = \sum_{k=1}^{n} [H^*]^k \left(\{\varepsilon^0\}_{XY} \int_{z_{k-1}}^{z_k} dz + \{\chi\}_{XY} \int_{z_{k-1}}^{z_k} z\,dz \right) \tag{E.6a}$$

$$\{M\} = \sum_{k=1}^{n} [H^*]^k \left(\{\varepsilon^0\}_{XY} \int_{z_{k-1}}^{z_k} z\,dz + \{\chi\}_{XY} \int_{z_{k-1}}^{z_k} z^2\,dz \right) \tag{E.6b}$$

Finally we can write

$$\begin{bmatrix} N_x \\ N_y \\ N_{xy} \\ \cdots \\ M_x \\ M_y \\ M_{xy} \end{bmatrix} = \begin{bmatrix} A_{11} & A_{12} & A_{14} & \vdots & B_{11} & B_{12} & B_{14} \\ A_{12} & A_{22} & A_{24} & \vdots & B_{12} & B_{22} & B_{24} \\ A_{14} & A_{24} & A_{44} & \vdots & B_{14} & B_{24} & B_{44} \\ \cdots & \cdots & \cdots & & \cdots & \cdots & \cdots \\ B_{11} & B_{12} & B_{14} & \vdots & D_{11} & D_{12} & D_{14} \\ B_{12} & B_{22} & B_{24} & \vdots & D_{12} & D_{22} & D_{24} \\ B_{14} & B_{24} & B_{44} & \vdots & D_{14} & D_{24} & D_{44} \end{bmatrix} \begin{bmatrix} \varepsilon^0_x \\ \varepsilon^0_y \\ \gamma^0_{xy} \\ \cdots \\ \chi_x \\ \chi_y \\ \chi_{xy} \end{bmatrix} \tag{E.7}$$

with

$$A_{ij} = \sum_{k=1}^{n} H_{ij}^{*k}(z_k - z_{k-1}) \tag{E.8a}$$

$$B_{ij} = \frac{1}{2} \sum_{k=1}^{n} H_{ij}^{*k} \left(z_k^2 - z_{k-1}^2 \right) \tag{E.8b}$$

$$D_{ij} = \frac{1}{3} \sum_{k=1}^{n} H_{ij}^{*k} \left(z_k^3 - z_{k-1}^3 \right) \tag{E.8c}$$

Expressions (E.8a) represent membrane stiffness, expressions (E.8c) represent flexural stiffness, whilst expressions (E.8b) represent the mutual stiffness or coupling stiffness, which produces flexure and/or torsion of the laminate subjected to tension, as well as stretching of the middle plane when the laminate is bent.

Appendix F: Finite difference method

F.1 TORSION OF BEAMS OF GENERIC CROSS SECTION ($\nabla^2\omega = 0$)

As we have seen in Section 9.7, the warping function, introduced in treating the problem of torsion of beams of generic cross section, is **harmonic** and thus satisfies Laplace's equation (9.98) with the boundary condition (9.106). Here we shall mention the solution based on the **Finite Difference Method**, which is a numerical method of discretization that is useful for dealing with problems for which closed-form solutions are not possible.

If a regular function $y(x)$ presents, in a series of equidistant points, the values $y_0, y_1, y_2, ...,$ for $x = 0$, $x = \delta$, $x = 2\delta$, ..., the first differential at these points can be approximated as follows:

$$(\Delta y)_{x=0} = y_1 - y_0 \tag{F.1a}$$

$$(\Delta y)_{x=\delta} = y_2 - y_1 \tag{F.1b}$$

$$(\Delta y)_{x=2\delta} = y_3 - y_2 \tag{F.1c}$$

$$\vdots$$

Dividing the differentials (F.1) by the length δ of the intervals, we obtain an approximate value of the first derivative at the corresponding points, in the form of an incremental ratio:

$$\left(\frac{dy}{dx}\right)_{x=0} \simeq \frac{y_1 - y_0}{\delta} \tag{F.2a}$$

$$\left(\frac{dy}{dx}\right)_{x=\delta} \simeq \frac{y_2 - y_1}{\delta} \tag{F.2b}$$

$$\left(\frac{dy}{dx}\right)_{x=2\delta} \simeq \frac{y_3 - y_2}{\delta} \tag{F.2c}$$

$$\vdots$$

It is then possible to approximate the second differentials using the first differentials

$$(\Delta^2 y)_{x=\delta} = (\Delta y)_{x=\delta} - (\Delta y)_{x=0} = y_2 - 2y_1 + y_0 \tag{F.3a}$$

$$(\Delta^2 y)_{x=2\delta} = y_3 - 2y_2 + y_1 \tag{F.3b}$$

$$(\Delta^2 y)_{x=3\delta} = y_4 - 2y_3 + y_2 \tag{F.3c}$$

$$\vdots$$

so that the second derivatives, calculated again as incremental ratios, appear as follows:

$$\left(\frac{d^2 y}{dx^2}\right)_{x=\delta} \simeq \frac{(\Delta^2 y)_{x=\delta}}{\delta^2} = \frac{y_2 - 2y_1 + y_0}{\delta^2} \tag{F.4a}$$

$$\left(\frac{d^2 y}{dx^2}\right)_{x=2\delta} \simeq \frac{y_3 - 2y_2 + y_1}{\delta^2} \tag{F.4b}$$

$$\left(\frac{d^2 y}{dx^2}\right)_{x=3\delta} \simeq \frac{y_4 - 2y_3 + y_2}{\delta^2} \tag{F.4c}$$

$$\vdots$$

If we have a function of two variables $\omega(x, y)$, it is possible to calculate the partial derivatives using expressions similar to Equations F.2 and F.4. Let us consider, for instance, a generic compact cross section (Figure F.1) and the superposition of a regular square-mesh grid of nodal points. We can then approximate the values of the partial derivatives of the function ω in the generic point 0

$$\frac{\partial \omega}{\partial x} \simeq \frac{\omega_1 - \omega_0}{\delta}, \quad \frac{\partial \omega}{\partial y} \simeq \frac{\omega_2 - \omega_0}{\delta} \tag{F.5a}$$

$$\frac{\partial^2 \omega}{\partial x^2} \simeq \frac{\omega_1 - 2\omega_0 + \omega_3}{\delta^2}, \quad \frac{\partial^2 \omega}{\partial y^2} \simeq \frac{\omega_2 - 2\omega_0 + \omega_4}{\delta^2} \tag{F.5b}$$

Using the foregoing expressions and similar ones, the differential equation (9.98) is transformed into a system of algebraic finite difference equations of the type

$$\frac{1}{\delta^2}(\omega_1 + \omega_2 + \omega_3 + \omega_4 - 4\omega_0) = 0 \tag{F.6}$$

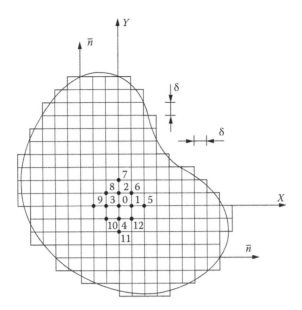

Figure F.1

As many equations as there are unknowns ω_i will thus be obtained. For each node of the grid, in fact, we have one unknown and one equation. This equation is the field one (F.6) if the node is internal, whilst it represents the discretized form of the integrodifferential boundary condition (9.106) if the node belongs to the boundary.

F.2 PLATES IN FLEXURE ($\nabla^4 w = q/D$)

The Finite Difference Method is applied to advantage also for the approximate numerical solution of the plate equation (10.207), which is a fourth-order differential equation, like the equation governing the Airy stress function Φ for plane problems.

Consider once more the grid of Figure F.1. The second partial derivatives in the points 0, 1, 3 may be approximated, respectively, as follows:

$$\left(\frac{\partial^2 w}{\partial x^2}\right)_0 \simeq \frac{1}{\delta^2}(w_1 - 2w_0 + w_3) \tag{F.7a}$$

$$\left(\frac{\partial^2 w}{\partial x^2}\right)_1 \simeq \frac{1}{\delta^2}(w_5 - 2w_1 + w_0) \tag{F.7b}$$

$$\left(\frac{\partial^2 w}{\partial x^2}\right)_3 \simeq \frac{1}{\delta^2}(w_0 - 2w_3 + w_9) \tag{F.7c}$$

so that the fourth partial derivative becomes

$$\left(\frac{\partial^4 w}{\partial x^4}\right)_0 = \frac{\partial^2}{\partial x^2}\left(\frac{\partial^2 w}{\partial x^2}\right)_0$$

$$\simeq \frac{1}{\delta^2}\left[\left(\frac{\partial^2 w}{\partial x^2}\right)_1 - 2\left(\frac{\partial^2 w}{\partial x^2}\right)_0 + \left(\frac{\partial^2 w}{\partial x^2}\right)_3\right]$$

$$\simeq \frac{1}{\delta^4}\left(6w_0 - 4w_1 - 4w_3 + w_5 + w_9\right) \tag{F.8a}$$

Likewise we have

$$\left(\frac{\partial^4 w}{\partial y^4}\right)_0 \simeq \frac{1}{\delta^4}\left(6w_0 - 4w_2 - 4w_4 + w_7 + w_{11}\right) \tag{F.8b}$$

$$\left(\frac{\partial^4 w}{\partial x^2 \partial y^2}\right)_0 \simeq \frac{1}{\delta^4}\left[4w_0 - 2\left(w_1 + w_2 + w_3 + w_4\right) + w_6 + w_8 + w_{10} + w_{12}\right] \tag{F.8c}$$

Substituting Equations F.8 in Equation 10.204, we obtain the finite difference equation corresponding to the node 0

$$20w_0 - 8\left(w_1 + w_2 + w_3 + w_4\right) + 2\left(w_6 + w_8 + w_{10} + w_{12}\right) + w_5 + w_7 + w_9 + w_{11} = \frac{q_0}{D} \tag{F.9}$$

Also in this case, as many algebraic equations will be obtained as there are unknowns. For each node within the grid we have in fact a field equation (F.9), whilst for the boundary nodes it is possible to write two kinematic conditions

$$w_i = 0, \quad \frac{\partial w_i}{\partial n} = 0 \tag{F.10}$$

which represent a built-in edge of the plate, or three static conditions corresponding to the free-edge loadings, which are the shearing force, given by Equations 10.190, as well as the bending and twisting moments, given by Equations 10.203. These loadings can be expressed as linear combinations of the second and third partial derivatives of the function w, and thus involve a further two fictitious nodes, which are outside the domain of interest. Likewise, the second of the two kinematic conditions (F.10) involves a fictitious supplementary node, outside the domain.

Appendix G: Torsion of multiply connected thin-walled cross sections

The problem of torsion of **doubly connected** closed thin-walled sections has been dealt with in Section 9.9 and solved by applying Equations 9.136 and 9.140.

Consider, instead, the case of a **triply connected** cross section consisting of a tubular element with a diaphragm (Figure G.1). Let τ_i, and b_i be the shearing stress and the thickness in each of the portions that make up the cross section. From the **hydrodynamic analogy**, the products $\tau_1 b_1$, $\tau_2 b_2$, $\tau_3 b_3$, are constant in each point of each portion, and

$$\tau_1 b_1 = \tau_2 b_2 + \tau_3 b_3 \tag{G.1}$$

If $h(s)$ denotes the distance of the centroid from the generic tangent to the mid-line, by equivalence we have

$$\tau_1 b_1 \int_1 h(s)\,ds + \tau_2 b_2 \int_2 h(s)\,ds + \tau_3 b_3 \int_3 h(s)\,ds = M_z \tag{G.2}$$

Using Equation G.1 and indicating with Ω_1 and Ω_2 the areas enclosed in the circuits 1–3 and 2–3, we obtain

$$2\left(\tau_1 b_1 \Omega_1 + \tau_2 b_2 \Omega_2\right) = M_z \tag{G.3}$$

The unit angle of torsion is in general expressed by Equation 9.109. Introducing the factor of torsional rigidity (9.140) and expressing the thickness $b(s)$ on the basis of Equation 9.136, we have

$$\Theta = \frac{M_z}{4 G \Omega^2} \oint_{\mathscr{C}} \frac{ds}{\left(M_z / 2\Omega \tau_{zs}\right)} = \frac{1}{2 G \Omega} \oint_{\mathscr{C}} \tau_{zs}\,ds \tag{G.4}$$

or

$$\oint_{\mathscr{C}} \tau_{zs}\,ds = 2 G \Theta \Omega \tag{G.5}$$

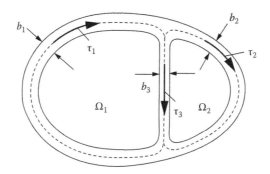

Figure G.1

This relation applies to any closed line, also in the case of multiply connected cross sections. If it is applied to circuits 1–3 and 2–3, in the case where the thicknesses b_1, b_2, b_3, and therefore also the stresses τ_1, τ_2, τ_3, are constant in each of the three portions 1, 2, 3, we have

$$\tau_1 s_1 + \tau_3 s_3 = 2G\Theta\Omega_1 \tag{G.6}$$

$$\tau_2 s_2 - \tau_3 s_3 = 2G\Theta\Omega_2 \tag{G.7}$$

The linear algebraic system consisting of the four Equations G.1, G.3, G.6, G.7, in the four unknowns τ_1, τ_2 τ_3, Θ provides tits solution to the problem.

Index

Printed and bound by CPI Group (UK) Ltd, Croydon, CR0 4YY

01/11/2024

01782605-0012